Ecological impact assessment

Jo Treweek

Institute of Terrestrial Ecology, Monks Wood
Komex Clarke Bond, Bristol

Blackwell
Science

© 1999 by
Blackwell Science Ltd
Editorial Offices:
Osney Mead, Oxford OX2 OEL
25 John Street, London WC1N 2BL
23 Ainslie Place, Edinburgh EH3 6AJ
350 Main Street, Malden
  MA 02148 5018, USA
54 University Street, Carlton
  Victoria 3053, Australia
10, rue Casimir Delavigne
  75006 Paris, France

Other Editorial Offices:
Blackwell Wissenschafts-Verlag GmbH
Kurfürstendamm 57
10707 Berlin, Germany

Blackwell Science KK
MG Kodenmacho Building
7–10 Kodenmacho Nihombashi
Chuo-ku, Tokyo 104, Japan

First published 1999

Set by Setrite Typesetters, Hong Kong
Printed and bound in Great Britain
by MPG Books Ltd, Bodmin, Cornwall

The Blackwell Science logo is a
trade mark of Blackwell Science Ltd,
registered at the United Kingdom
Trade Marks Registry

DISTRIBUTORS

Marston Book Services Ltd
PO Box 269
Abingdon, Oxon OX14 4YN
(*Orders*: Tel: 01235 465500
          Fax: 01235 465555)

USA
Blackwell Science, Inc.
Commerce Place
350 Main Street
Malden, MA 02148 5018
(*Orders*: Tel: 800 759 6102
          781 388 8250
          Fax: 781 388 8255)

Canada
Login Brothers Book Company
324 Saulteaux Crescent
Winnipeg, Manitoba R3J 3T2
(*Orders*: Tel: 204 837 2987)

Australia
Blackwell Science Pty Ltd
54 University Street
Carlton, Victoria 3053
(*Orders*: Tel: 3 9347 0300
          Fax: 3 9347 5001)

A catalogue record for this title
is available from the British Library

ISBN 0-632-03738-5

Library of Congress
Cataloging-in-publication Data

Treweek, Joanna.
    Ecological impact assessment/
    Joanna Treweek.
        p.   cm.
    ISBN 0-632-03738-5
    1. Ecological assessment (Biology)
    I. Title.
    QH541.15.E22 T74 1999
    333.95'14–dc21
                          98-30307
                              CIP

For further information on
Blackwell Science, visit our website:
www.blackwell-science.com

Contents

Preface

Ecological impact assessment (EcIA) is the process of identifying, quantifying and evaluating the potential impacts of defined actions on ecosystems or their components.

Reconciling economic growth and development with the conservation of world biodiversity demands resource-based approaches to environmental analysis that are not constrained by artificial boundaries. EcIA has a vital part to play in implementing the principles of sustainability and 'wise use' which form the foundations of the Rio Declaration on Environment and Development. One of the main applications of EcIA has been in support of environmental impact assessment (EIA), which is now enshrined in the legislation of many countries. EIA is intended to cater for the integrated assessment of a range of social, biophysical and economic issues, and by providing analytical procedures for studying relationships between organisms and their environment, ecological science has an obvious role. As well as being an integral part of EIA, however, EcIA techniques have much to offer in land use and natural resource planning generally as well as integrated pollution control (IPC). There is also a close relationship between EcIA and some aspects of ecological risk assessment.

Official guidance on appropriate techniques for EcIA has been limited, and in many countries the lack of any formalized procedures has resulted in neglect of some key issues. The ability of ecologists to make sound predictions about environmental impacts has been hampered by the inherent complexities of ecological systems, the lack of access to relevant information and by legislative and procedural constraints. Practitioners in the field must become well versed in a wide range of ecological subjects, as well as techniques for policy- and decision-making and sometimes economic analysis too.

This book provides a background to the theory and practice of EcIA for ecologists, environmental managers, decision-makers and students. Legislative contexts and frameworks for EcIA are summarized and the basic elements of the EcIA process are outlined. The scientific basis for techniques of survey, monitoring, prediction and ecological evaluation is explored. The book concludes with chapters on geographical information systems (GIS) and the principles of survey design and analysis.

Hopefully, the book will help to ensure that ecological issues are given full and careful consideration in all areas of environmental planning and management.

Acknowledgements

There are many people I would like to thank for their help and support.

On the domestic front, Stuart, his parents, my parents, John, Yen and our children all supported each other and me throughout and I could not have produced this book without them.

It was a pleasure to work with Ian Sherman, Susan Sternberg and Katrina McCallum at Blackwell Science and I am grateful for their patience, commitment, advice and practical assistance.

Many colleagues and friends at ITE Monks Wood had input into the book and provided examples of relevant research or experience. I would especially like to thank John Sheail, Mike Roberts, Brian Eversham, Owen Mountford, Richard Pywell, Chris Preston, Henry Arnold, Jane Croft, Peter Hankard, Sarah Manchester and Richard Wadsworth.

I am also very grateful to the following people at Komex International for their commitment to science in a commercial world and for giving me their time even when it was so short: especially Paul Hardisty, Tom Parker, Rob Bracken, Joseph Wells, John Brownlee, John Watson, Catherine Watson and Miles Tindall.

I would like to thank David Hill, John Rotifer, Ian Sherman and John Sheail for their constructive criticism of earlier drafts and their positive encouragement.

Members of the Ecology Section of the International Association for Impact Assessment (IAIA), Stewart Thompson, Joshua Schachter, Penny Angold, Andrea Bagri, Olivia Bina, Ann Dom, Daryl Buck and Trudy Watt all helped with illustrative material and advice. Thanks to Neil Veitch for sorting out the cover illustration.

The following publishers gave permission to reproduce figures: Academic Press Ltd, Cambridge University Press, McGraw-Hill, Elsevier Science Ltd, The University of Chicago Press, IBN-DLO Institute for Forestry and Nature Research, the Netherlands, the RSPB and Kluwer Academic Publishers.

1 Introduction

Ecological impact assessment (EcIA) is the process of identifying, quantifying and evaluating the potential impacts of defined actions on ecosystems or their components. If properly implemented it provides a scientifically defensible approach to ecosystem management.

It is often used in conjunction with environmental impact assessment (EIA) studies with a broader remit, which also considers the social and economic consequences of development activities. This book focuses largely on the theory and practice of EcIA carried out for purposes of EIA, as it is in this context most experience of EcIA has been gained. However, EcIA has potential applications in all areas of natural resource management and planning. A number of other forms of environmental legislation have also generated a demand for objective and consistent methods of EcIA, particularly those relating to regulation of industrial activity. Legislation on integrated pollution control (IPC), for example, increasingly demands ecological analysis and/or risk assessment.

Demand for EIA and EcIA as tools for environmental management has grown rapidly. Much of the impetus for their development has come from concern about increasing pressures on natural resources and native wildlife throughout the world. Study of the fossil record suggests that current rates of extinction are unprecedented. Every year at least one vertebrate species becomes extinct (Chiras 1988) and the global extinction rate for all species may exceed 17 000 per year (Wilson 1990; Lubchenko *et al.* 1991). Not only are individual habitats and species declining to the point where their survival is threatened, but the general diversity and abundance of wildlife is decreasing as a consequence of intensifying human exploitation. Loss of biological diversity is cause for concern in its own right, but also because it may be symptomatic of environmental degradation which could have serious consequences for human wellbeing and survival. Human habitats are also threatened!

Concern about progressive loss of biological diversity has resulted in initiatives to conserve 'important' elements of the world's 'natural' heritage. The main focus of conservation effort has been the establishment and maintenance of protected areas or nature reserves, intended to sustain representative examples of species and habitats. The overall degree of protection afforded by this approach to nature protection varies between countries, but the area conserved is invariably low relative to that actively managed and used by people for economic reasons. Protected areas cover only approximately 6% of the world's land mass (Mason & Sadoff 1994) and formal protection does not necessarily imply absence of ongoing economic activity. National parks, for example, are not immune to new development and many are subject to considerable pressure from recreational activities. The effectiveness of protective measures invariably depends on the ability to regulate external pressures, but this is becoming increasingly difficult. The viability of nature reserves in most countries depends on the maintenance of external, independent wildlife populations and gene pools, but these are increasingly threatened by habitat loss and fragmentation, which restrict species distributions and reduce the viability of remaining populations. Furthermore, the widespread and trans-boundary effects of pollution make it difficult to achieve protection simply by designating sites as nature reserves. In the Netherlands, many seminatural grasslands have been so altered by the effects of atmospheric nitrogen deposition that the maintenance of characteristic communities has proved almost impossible. It has been recognized for some years that the 'setting aside of land for nature reserves is an inadequate means of conserving even a representative sample' of the flora and the fauna of such a small and populous country as the UK (Tubbs & Blackwood 1971). The same applies to most countries where human population pressure is reducing the availability of habitat and resources for wildlife conservation.

Designation of protected areas alone is therefore failing to safeguard biological diversity, and measures of more general applicability are needed to regulate threats. The most enduring protection for wildlife habitat may come, not from 'preservation' in the strict sense, but from 'multiple use management', which is designed in a participatory manner and accounts for the needs of those whose livelihoods are affected. Integrating the goals of conservation into mainstream economic development can be one way of ensuring the long-term survival of a protected area's ecosystems. Mason and Sadoff (1994) cite the example of the Sagamartha (Mount Everest) National Park in Nepal, where indigenous Sherpas live and work in the park as guides and porters, combining both conservation and economic development goals.

Many threats to the longer-term stability of ecosystems are individually insignificant but collectively serious. By their very definition, cumulative ecological effects are not attributable to any one source or action and cannot be regulated in isolation. This makes cross-sectoral approaches to natural resource management essential. Organizations such as the World Bank and the World Commission on Environment and Development (the Brundtland Commission) are not alone in concluding that 'environmental issues must be addressed as part of overall economic policy' (Wood & Djeddour 1992).

In its broadest definition, 'ecology' can be considered to include 'just about everything involving man and his environment' (Peters 1991). EcIA techniques are used primarily to predict the consequences of human activities for organisms other than people, but it is becoming increasingly difficult to ignore the huge importance of people either as agents of ecological change or as receptors of environmental impact. While the United Nations Environment Programme's (UNEP's) definition of 'Environmental Risk Assessment', for example, draws a distinction between 'ecological' and 'human health' risk assessment, it also acknowledges the obvious links between them, principally via resources such as air, water, soil and food (UNEP-IETC 1996). There are vast numbers of people who depend, for their livelihoods, on the viability of specific ecosystems, whether these are rivers, wetlands, rainforests or oceans. Figure 1.1 shows people pulling in their fishing nets off the coast of Togo. On the other hand, many ecosystems have evolved away from a wholly 'natural' state to a seminatural condition in which they are actually dependent on long traditions of human intervention or management for maintenance of their characteristic features. The majority of wildlife habitat in the UK, for example, is seminatural and will tend to alter, degrade or disappear unless managed in some way. Figure 1.2 shows traditionally managed hay meadows, which develop characteristic floras and faunas.

The relationships between economic status and environmental preservation are far from straightforward, but are certainly intimate. Rogers

Fig. 1.1 Pulling in fishing nets off the Togolese Coast.

Fig. 1.2 Hand-built hayricks in the UK. (Courtesy of Owen Mountford.)

(1996b) points out that many of the world's poorest people live in some of the 'least resilient and most threatened environments', like tropical forests and semiarid zones. Figure 1.3 shows deforestation for agriculture in the Highlands of Central Madagascar. At the same time, many of the modern technologies that accompany economic growth are agents

Fig. 1.3 Deforested mountainside, Highlands of Central Madagascar. (©1997 Joshua Schachter.)

Fig. 1.4 Oil pipelines, Ecuadorian Amazon Basin, Ecuador. (©1997 Joshua Schachter.)

of environmental destruction (Fig. 1.4). Both poverty and wealth can be intimately associated with environmental damage. It is therefore difficult to see how the safeguard of environmental resources can be achieved without careful consideration of their social and economic context. One of the main reasons why EIA (or environmental assessment, EA) has become so prevalent, is because it addresses the relationship between biophysical, social and economic factors explicitly. Before it is possible to address ecological, economic and social issues in an integrated fashion, however, specific issues in these three areas must be identified, measured and evaluated with clarity and consistency.

EcIA is used to identify, predict and evaluate the ecological consequences of defined development actions, whether these are localized, individual projects or regional development plans. It can be applied at a range of scales, to assess effects on individual species or on whole ecosystems. There are a number of legal frameworks for EcIA. It may be practised as an integral part of EIA, in support of industrial regulatory procedures or as part of land use planning. Whatever its scale of application, EcIA is based on identification of key ecosystem components and understanding of the factors or processes that determine their interactions with each other, so that impacts of specific activities can be superimposed on baseline conditions and their potential effects forecast (Treweek 1995). EcIA must invariably look beyond artificial boundaries like factory fences or national frontiers to interpret impacts. Essentially, it demands a resource-based approach to analysis.

EcIA is firmly rooted in ecological science, drawing on traditional techniques of survey, monitoring, functional analysis and predictive modelling. In addition, however, EcIA requires evaluation of the implications of any predicted outcomes. It is this aspect of evaluation which distinguishes EcIA from the pure science of ecology and which has created demand for new approaches to the ways in which ecological information is handled. It is common, but not axiomatic, for 'ecological impacts' to be interpreted in terms of 'nature conservation' or 'wildlife value', but scales of value determined on the basis of other social or economic considerations can also be used (Brown & Moran 1993), for example impacts on wetland areas might be measured in terms of reduced capacity to control pollution. Ultimately, it is necessary to take the findings of EcIA and to consider them in conjunction with the findings of other categories of assessment for decisions to be made. Ecological outcomes must therefore be translated into a common language or scale for comparison with other findings, whether these are of a social, economic or political nature. In short, EcIA should provide a scientifically defensible rationale for decision making and for environmental management.

1.1 The EcIA process

In essence, EcIA involves finding answers to the following questions.
- What is where?
- Why is it there?
- What happens if ...?
- So what if it does?

The idealized EcIA process is flexible, iterative, proactive and based on accurate, consistent, transparent and defensible methods. This is only achievable if EcIA forms a part of integrated systems of environmental regulation, based on strategic planning and regulation of environmental quality and operating standards in relation to quality objectives or standards (Box 1.1) These might include targets for endangered species recovery programmes or nationally agreed objectives for the preservation of critical natural capital. Most systems of environmental assessment and regulation fall a long way short of this model, particularly with respect to the adjustment of control measures to tackle unforeseen departures from acceptable standards of environmental quality.

The EcIA process therefore incorporates monitoring and feedback at all stages to characterize the 'state of the environment' (Treweek 1995). Ideally, all studies carried out for purposes of EcIA would form part of a coherent monitoring programme with a planned duration. Otherwise, it is unlikely to be possible to characterize baseline conditions or to distinguish between the consequences of natural variation and proposed

---

Box 1.1 Idealized system of environmental regulation as a context for EcIA.

Strategic planning
- Issues are identified
- Desired environmental outcomes are determined

Objective setting
- Desired environmental outcomes are converted to clearly defined objectives
- Factors contributing to the achievement of these objectives are identified

Performance standards
- Standards are set to monitor performance relative to achievement of objectives

Control measures
- Specific conditions are placed on consents to use or affect resources, for example through production of conservation management plans

Monitoring
- Performance is monitored in relation to standards and achievement of objectives

Review/regulation
- Control measures are adjusted if performance falls below standard

---

Box 1.2 Principles for EcIA implementation.

- EcIA should begin as early as possible in the design or planning of proposed activities to ensure that key issues are identified and data requirements can be met
- The spatial and temporal scope of EcIA study should be consistent with both proposed actions and ecosystem limits
- EcIA should be based on reliable and relevant data and experience
- EcIA should ensure that all potentially significant impacts of proposed actions are identified and evaluated
- Assumptions and confidence limits should be specified for predictions
- EcIA should ensure that significant adverse impacts are avoided, minimized or compensated for using effective and well-tested techniques
- EcIA should address all relevant (and related) factors, for example cumulative, long-term and large-scale effects
- EcIA should result in environmental management that takes account of the significance of residual effects and acknowledged uncertainties in prediction and mitigation
- EcIA should be based on assessment and evaluation informed by results of reliable monitoring
- EcIA should take account of 'limits of acceptable change' (LAC) where possible

actions. However, the need for straightforward legislative and administrative procedures often constrains EcIA within boundaries (in time and/or space) that are not determined objectively or scientifically and do not match ecosystem limits. In practical terms, the effectiveness of EcIA therefore depends on the extent to which legislative and procedural frameworks support the adoption of certain principles as outlined in Box 1.2. Some of the current legislative contexts for EcIA are explored in Chapter 2.

Because EIA and related legislation has been the main driving force behind the development of EcIA as a formal discipline, it has been customary to address the key questions listed above in sequential procedural stages commensurate with those in EIA. These are summarized in Box 1.3 and are considered in more detail in following chapters.

Box 1.3 Common procedural steps in EcIA.

Scoping
- Interpretation of the proposal and its associated sources of ecological stress or disturbance
- Description, characterization, inventory of potential ecological receptors in receiving environment
- Preliminary impact screening to establish the range of possible interactions between potential stressors and ecosystem components or receptors
- Defining study limits, deciding how stressor/receptor interactions should be studied, identifying valued ecosystem components, establishing the range and focus of further study

Focusing
- Identifying valued ecosystem components (these might be species, habitats or important ecological processes or functions)
- Refining scope of study

Impact assessment
- Baseline studies to characterize affected ecosystems (their condition or state in the absence of any proposed action)
- Impact prediction and assessment (to identify and predict impacts on selected ecosystem components by comparing against baseline)

Impact mitigation
- Mitigation (to redress significant adverse effects)

Impact evaluation
- Impact evaluation to determine the significance and importance of predicted, residual, ecological impacts, measuring against both internal and external standards or criteria

Monitoring and feedback
- Ecological monitoring to strengthen the knowledge base and provide opportunities for corrective action in the light of unforeseen outcomes
- Feedback to assess proposal implementation and compliance

1.1.1 Scoping

Scoping is used to define appropriate EcIA study limits, given the nature of potential interactions between a proposal and the receiving environment. Establishing the distributions of potentially affected ecosystems and their components (determining 'what is where') is an essential step in defining the scope of an EcIA. Ecosystems are not static, either spatially or temporally. In the absence of historic information or monitoring data, it is not always easy to quantify temporal trends in distribution. Comprehensive, up-to-date national monitoring data makes it possible to determine what can be expected to occur in any given area of search, thereby improving the chances of appropriate survey design and reducing the risk that important ecological components may be missed during biodiversity inventories.

In addition to the collation of existing, background information, scoping for EcIA relies on standard ecological techniques of survey and inventory. Surveys provide information on the arrangement and organization of ecosystem components in an area, while inventories generally entail the collection, sorting, cataloguing and mapping of specific entities such as 'genes, individuals, populations, species, habitats, landscapes and ecosystems' (Stork & Samways 1995). To date there has been a tendency for inventories to focus on more 'macro' elements of ecosystems (habitats, vegetation and vertebrate species). The micro flora and fauna may actually have much more to offer in terms of environmental diagnostics, but are assessed much less often than 'charismatic mega-vertebrates'. EcIA inventories focusing primarily on genes are practically unheard of, although some have addressed the particular problems of genetically isolated populations of animals. In EcIA, 'top-down' approaches to survey and inventory are the norm, which place species in the context of habitats and larger landscape units (e.g. 'ecoregions'). New technologies have revolutionized our ability to survey effectively over large areas and have opened up access to new sources of data. Together, remotely sensed data and geographical information systems (GIS) have made it possible to extend ecological survey and monitoring from local to more global scales at relatively low cost, so that local distributions can be placed in a wider spatial context for assessment and evaluation.

It is important to consider the distributions of ecological components with respect to 'zones of impact' when deriving suitable study limits for EcIA. This requires analysis of the spatial and temporal relationships between proposal activities and ecological components. EcIA scoping therefore requires, not only characterization of the receiving environment, but an understanding of the proposal and the impacts it may generate. Some form of ecological impact screening and exposure assessment therefore usually forms part of EcIA scoping.

1.1.2 Focusing

Focusing procedures are necessary to refine the scope of the EcIA, as it is never possible to study all components and functions of the receiving environment in detail. It is important to get 'focusing' right, as it has a major influence on the way in which impact prediction or assessment is carried out. 'Focusing' is the subject of Chapter 4. Various approaches can be used to select the 'valued' components and functions (VECs), which will form the main focus of subsequent studies. These involve a form of 'evaluation' that relates, not to the significance of impacts, as discussed in Chapter 6, but to the inherent perceived value of receptors or their usefulness as indicators.

The need to explain observed distributions and to understand ecosystem 'behaviour' means that ecological functions and processes may be selected to form the focus of EcIA studies, as well as valued species and habitats.

1.1.3 Impact prediction

Prediction lies at the heart of EcIA, but is also generally acknowledged to be one of its weakest areas. The complexity of ecosystems, sampling limitations and lack of opportunities for follow-up monitoring mean that predictions for EcIA must often be made with considerable uncertainty. There has been a tendency for EcIA practitioners to identify the possibility that defined impacts might occur, but not to estimate the probability that they will. Considerable progress in this respect has been made through the development of techniques for ecological risk assessment, applied largely to pollution control and remediation. Impact prediction is considered in Chapter 5.

Before attempting to predict the likely outcome of a defined impact or ecosystem change, it is necessary to understand ecosystem function, in order to explain and account for any distributions that have been observed. The ecological processes that will drive any change must be defined and any existing variability accounted for so that changes can be attributed to defined actions or stressors. This is the basis for defining the baseline conditions that would pertain in the absence of any imposed change. For most ecosystems, our ability to explain and quantify ecosystem function is relatively limited. On the other hand, some correlations between species distributions and physical factors are well understood: clear relationships between climate, soil and plant distributions, for example, can be identified. Impact assessment should model or predict ecological outcomes relative to baseline, taking account of their likely frequency, range, magnitude and severity.

EcIA relies on prediction, but also provides numerous 'case studies', which enable applied ecologists to build on their understanding of ecosystem response to defined stresses. Again, the ability to capitalize on this experience is currently hampered by a general lack of monitoring or follow-up.

1.1.4 Evaluation

Predicted ecological impacts can be interpreted statistically, or evaluated against ecological thresholds or limits. Some of the criteria used in EcIA evaluation are discussed in Chapter 6. However, the values placed on ecosystem components, processes and functions are determined ultimately by social, political and economic factors. Ecological information must therefore be provided in such a form that comparisons can be made with other categories of impact, for example to decide whether ecological considerations should override engineering constraints or social factors in selecting the preferred route option for a road. Before it is possible to make this kind of judgement, some sort of ecological evaluation is required so that the ecological significance of predicted impacts can be estimated using consistent criteria and methods before comparisons are made with impacts in other sectors. It might be necessary to compare the adverse effects of different siting options on the availability of breeding habitat for two species of bird, for example, to decide which of two habitat types is more valuable or to establish whether a predicted habitat loss will have significant impacts on the viability of a species. The results of ecological evaluation feed into a decision-making process which must also take account of social, economic and ethical factors in deciding, for example, which of the two bird species or habitat types has most value.

1.1.5 Mitigation

Some adverse ecological impacts can be reduced through mitigation. Mitigation includes any deliberate action undertaken to ameliorate adverse effects, whether by controlling impacts by source or the exposure of ecological receptors. The effectiveness of mitigation determines the magnitude and significance of residual ecological impacts. If mitigation is 100% successful, then there will be no adverse consequences associated with the action. In reality, this is highly unlikely. Chapter 7 outlines the main approaches to ecological mitigation and also summarizes factors that should be taken into account when evaluating the effectiveness of mitigation measures.

1.1.6 Monitoring

Monitoring (see Chapter 8) enables us to distinguish between natural and anthropogenic changes. There may be considerable year-to-year variations in ecosystems or their components, which conceal long-term underlying trends. Without monitoring to provide reliable information about the status of wildlife habitats and species, how can we ever be sure whether an observed change is significant? National ecological monitoring schemes are becoming more prevalent as pressure grows to inventory the world's biological resources. These should generate some of the information needed to develop reliable ecological baselines. At a more local level, follow-up of post-development ecological impacts is required to strengthen our knowledge of impacts, the reliability of EcIA prediction and the effectiveness of mitigation measures. Some of the many good reasons why ecologists continue to demand greater investment in monitoring programmes are outlined in Chapter 8.

1.1.7 Geographical information systems

Chapter 9 provides information on the application of GIS in EcIA.

GIS have much to offer as an operational platform for impact assessment and monitoring. Use of GIS has also opened up access to new sources of data, which have made regional ecological studies a realistic prospect. GIS therefore have particular relevance to assessment of cumulative effects and strategic ecological assessment.

1.1.8 Survey design and analysis

Chapter 10 considers EcIA techniques. In particular, it discusses some of the scientific, statistical and practical problems associated with survey design and analysis for EcIA. No attempt is made to provide an exhaustive account of practical survey methods, as these are often geographically or subject specific and are covered in other texts. EcIAs are associated with particular problems of statistical design and analysis. Most of the techniques that have been developed specifically for EcIA depend on a coherent monitoring framework, which is often lacking. Methods for building ecological criteria into decision-making tools also require further development.

1.2 Recommended reading

Causes of environmental degradation

Swanson, T. (ed.) (1996) *The Economics of Environmental Degradation. Tragedy for the Commons?* Edward Elgar, Cheltenham.

Global perspectives on biodiversity conservation

Krattiger, A.F., McNeely, J.A., Lesser, W.H., Miller, K.R., St Hill, Y. & Senanayake,
   R. (eds) (1994) *Widening Perspectives on Biodiversity*. IUCN/IAE, Gland.

Introduction to impact assessment

Vanclay, F. & Bronstein, D.A. (eds) (1995) *Environmental and Social Impact
   Assessment*. John Wiley and Sons, Chichester.

2 Legislative contexts for ecological impact assessment

2.1 Introduction

This chapter considers how various forms of environmental legislation have generated an application and a need for ecological impact assessment (EcIA) techniques. The main focus is on environmental impact assessment (EIA) and related legislation as this has provided the main context for formal application of EcIA. Detailed accounts of EIA legislation in different countries can be found elsewhere (e.g. Wood 1995). Differences in legislative requirements for EIA between countries are considered only insofar as they influence the demand for, or approaches to, EcIA. Other legislative contexts for EcIA have been generated by new global initiatives to regulate trans-boundary impacts, and to conserve the world's biological diversity (see section 2.4). Integrated pollution control (IPC) is another potential application for EcIA techniques and this is discussed in section 2.5.

2.2 Environmental impact assessment

A considerable body of environmental law has now accumulated which addresses the complex relationship between human economic development and the state of the environment. In particular, legislation for EIA has helped to bring environmental considerations to the top of the political agenda in many countries. It was the US National Environmental Policy Act (NEPA) of 1969 that first established a legislative requirement for potential environmental impacts to be assessed by proponents of development actions. This Act set a precedent for adoption of similar forms of legislation on EIA in other countries and the use of some form of EIA is now widely accepted to be prerequisite for effective environmental planning and management. By the early 1990s, over 40 countries had legislated for it (Robinson 1992).

EIA is defined as the process of identifying, estimating and evaluating the environmental consequences of current or proposed actions (Vanclay & Bronstein 1995). The term environmental assessment (EA) is also in common use. The two terms are largely interchangeable: preferences vary, but EIA is perhaps more prevalent. EIA is one of the few formal procedures that permits explicit analysis of the relationship between economic activity and the conservation of natural resources, providing a common platform for the integrated assessment of social, biophysical and economic issues. In theory, it can act as a bridge that 'integrates the science of environmental analysis with the politics of resource management' (Smith 1993). However, early interpretations of EIA considered the 'environment' in largely biophysical terms and tended to neglect the consequences of proposals for human health and wellbeing, resulting in neglect of key social issues. The extent to which EIA is intended to address social issues still varies considerably between countries. While the assessment of impacts on people is often addressed through EIA, some countries have elected to legislate separately for social impact assessment (SIA). A range of categories and disciplines have emerged which extend the principles of EIA to forecast the likely impacts, for example development aid ('development impact assessment'), the introduction of new technologies ('technology assessment' or TA) and the introduction of new policies ('policy assessment'). In generic terms, these are generally referred to as forms of 'impact assessment' (IA) (Vanclay & Bronstein 1995).

A number of categories of impact are also used to distinguish impact assessment disciplines. For example, impacts on human health may be addressed through 'environmental health assessment' or 'health impact assessment', on economic circumstances through 'economic assessment', on noise levels through 'noise assessment' and so on. The various forms of EIA or IA likely to be encountered are summarized in Table 2.1. They can be distinguished in terms of their application, the sources of impact they address, their main impact receptors and the main types of impact that occur.

Ecological impacts are often regarded as just one other category of impact, EcIA being used as a subdiscipline of EIA to identify, predict and evaluate the ecological impacts of proposed development actions. Applying such a distinction, 'noise' would be regarded as 'ecological' in impact if it were disrupting the breeding success of birds, but 'environmental' in impact if it were disturbing people. Such partitioning of impact categories in EIA can sometimes result in the neglect of important linkages and interrelationships. As mentioned earlier, most ecosystems are affected to some extent by anthropogenic influences (not always negatively). Theoretically, EIA provides scope for integrated assessment across categories and it is important not to neglect this role.

Internationally, increasing emphasis is being given to the role of EIA in promoting sustainable development. In this context, EIA has a potential role in the following (after Sadler 1996):
• safeguarding valued ecological processes and heritage areas, safeguarding traditional ecological knowledge;

Table 2.1 Aspects, applications and definitions of environmental impact assessment (EIA) categories.

Aspect of process	Application	Terminology used for EIA category
Application	General	Impact assessment (IA), environmental assessment (EA) or environmental impact assessment (EIA)
	Global, strategic (to policies, plans, programmes, regions, countries)	Strategic environmental assessment (SEA)
	Local, specific (to individual projects/sites)	Project-EA or -EIA
Sources of impact (actions)	Specific projects and activities	Project-EA or -EIA
	Policies, plans or programmes of activity	SEA
	Combined or associated projects or activities	Sectoral EIA (EA), cumulative effects assessment (CEA)
	Development aid	Development impact assessment
	Introduction of new technology	Technology assessment (TA)
	Introduction of new policy or changes in policy	Policy assessment
Impact receptors	The 'whole environment' (the combined biophysical and human environment)	EIA or EA
	Environmental 'media', e.g. water, air	Addressed as subcategories of EIA or ecological impact assessment (EcIA)
	Ecosystems and their components ('habitats', species, populations/ metapopulations, communities, etc.)	EcIA or ecological assessment (EcA)
	People (individuals)	Addressed through EIA or social impact assessment (SIA)
	People (populations, communities)	Demographic impact assessment
	People (health)	Health impact assessment (HIA)
Impact categories (general)	Economic/fiscal	Economic impact assessment
	Ecological	EcIA
	Cumulative	Cumulative effects assessment
	Social and cultural	SIA
	Health	HIA
	Climate	Climate impact assessment
	Noise	Noise impact assessment

Impacts may be direct, indirect, synergistic, trans-boundary, cumulative, short- or long-term, immediate or delayed.

- avoiding irreversible and unacceptable loss or deterioration of natural capital;
- ensuring economic development is adjusted to the potentials and capacities of the resource base;
- optimizing natural resource use, conservation and management opportunities;
- protecting human health and community wellbeing;
- addressing distributional concerns related to the disruption of people and traditional lifestyles.

Many of these goals reflect a desire to preserve natural resources and ecosystem integrity, something that depends on information and understanding about ecosystem function and the ability to make sound predictions about ecosystem change. For this reason, EIA frequently demands some form of ecological assessment or evaluation.

2.2.1 The EIA process

An international study of the effectiveness of environmental assessment (Sadler 1996) investigated EIA principles and procedures as developed, applied and legislated for throughout the world. The study identified the basic components of an idealized EIA process, as summarized in Table 2.2. These components will not always be required explicitly by the legislation or may be only partially implemented in different countries.

While EIA legislation differs considerably in detail between countries, the effectiveness of EIA in promoting environmentally sensitive development invariably depends on certain key factors. The following guiding principles apply to the EIA process wherever it is practised (after Sadler 1996).

EIA should be applied:
• to all development projects or activities likely to cause significant adverse impacts;
• as a primary instrument for environmental management to ensure that impacts of development are minimized, avoided or compensated for;
• so that the scope of study is consistent with the nature of activities being assessed and commensurate with the likely issues and impacts;
• on the basis of well-defined roles, rules and responsibilities for key factors.

EIA should be undertaken:
• throughout the project cycle, beginning as early as possible in the design phase;
• using 'best practice' science and mitigation technology;
• in accordance with established procedures and project-specific terms of reference;
• to provide appropriate opportunities for public involvement of communities, groups and parties directly affected by or with an interest in the project and/or its environmental impacts.

EIA should address:
• all relevant (and related) factors, for example including cumulative, long-term and large-scale effects;
• alternative designs, locations and technological approaches to that proposed;

Table 2.2 The environmental impact assessment (EIA) process.

Stage	Purpose	Additional considerations
Inception: framing of proposal	Intention to undertake action is stated Intended actions are outlined	Was environmental advice taken in early design phase?
Screening to determine need for EIA (in this context, distinct from 'impact screening')	Proposed action(s) are assessed to determine if full EIA is required Some legislation makes explicit demands for screening and gives guidance on 'indicative thresholds' Decision is taken concerning need for formal EIA Preliminary EIA, full EIA or no EIA may be required	Clear definition/description of proposed actions is needed If existing knowledge is scanty, 'mini' EIA may be required to establish need for more detailed assessment Be aware of potential cumulative effects
Scoping	Identify key issues and impacts to establish the study limits	Do not be constrained by artificial boundaries Consider all aspects and review current state of knowledge
Assessment	Survey, measure and predict likely impacts on environmental components selected at the scoping phase	Draw on all relevant sources of information Use best practice survey methodology Be aware of possible unexpected 'finds' Define confidence limits for methodologies and results
Evaluation	Determine the significance and social importance of measured impacts	Use appropriate scale of analysis Quantify significance if possible, before imposing subjective judgements to estimate importance
Mitigation	Specify measures to prevent, minimize and offset or otherwise compensate for environmental loss and damage	Indicate likely effectiveness based on experience elsewhere Identify any redistributional effects (temporal or spatial)
Report results	Produce environmental statement (ES) to document findings of EIA	Include all relevant data Interpret results for non-specialist readers
Decision-making	Determine whether proposal should proceed Proposal approved, rejected or modifications required	Aspects of EIA should be repeated if necessary to provide information needed to inform decision-making
ES review	Ensure report meets terms of reference and complies with legislation and standards of good practice	Review criteria should be consistent and objective
Establish terms and conditions for consent	Re-define proposal to take account of EIA findings, e.g. to avoid serious adverse impacts or to incorporate suitable mitigation	Time should be allowed for any re-definition
Monitoring	Check actions comply with terms and conditions Monitor environmental effects	Monitoring is essential to check accuracy of predictions and improve the knowledge base

Continued

Table 2.2 (*Continued*).

Stage	Purpose	Additional considerations
Follow-up management	Put contingency plans in place to deal with breakdowns in mitigation Address unanticipated impacts	Things do not always go as planned or expected!
Audit	Document longer-term outcomes Check performance against environmental quality standards	Improve EIA and project planning by learning from experience

- sustainability considerations including, assimilative capacity, carrying capacity and biological diversity.

EIA should result in:

- accurate and relevant information about the magnitude, significance and likelihood of occurrence of potential risks and consequences of a proposed undertaking and its alternatives;
- the preparation of an impact statement or report that presents this information in a clear, understandable and relevant form for decision making;
- identification of confidence limits for predictions;
- clarification of areas of agreement/disagreement among the parties involved in the process.

EIA should provide the basis for:

- environmentally sound decision making in which terms and conditions are clearly specified and enforced;
- the design, planning and construction of development projects that meet environmental standards and management objectives;
- an appropriate follow-up process with requirements for monitoring, management and audit;
- follow-up requirements that are based on the significance of potential effects and on the uncertainties associated with prediction and mitigation;
- learning from experience with a view to making future improvements to design of projects or the application of the EIA process.

Many of these principles depend, in the first instance, on a sound legislative base with clear purpose and requirements, and on procedural controls that ensure that the levels of EIA undertaken, the scope of any studies undertaken and the timetables for their completion are appropriate to the circumstances.

2.2.2 Project-EIA

EIA has been applied most commonly to individual project proposals

or actions, to ensure that their environmental implications are given full and careful consideration before development consent is determined.

In the European Union (EU), for example, Council Directive 85/337/EEC of 27 June 1985 ('The Directive') on 'the assessment of the effects of certain public and private projects on the environment' (OJ No. L 175, 5.7.1985) was intended to ensure that the decision-making authorities would be provided with the information needed to make a well-informed decision about the acceptability of a specific project in full knowledge of the facts regarding potential impacts on the environment.

The Directive divides projects into categories always having significant effects on the environment, and for which EIA will be required in all cases (Annex I projects) and those which will only require EIA in some circumstances (Annex II projects). Under the terms of the Directive, the eligibility or qualification of Annex II projects for EIA is determined by the individual Member States, which are required to consider the 'size, nature and location' of proposed projects in establishing appropriate criteria or thresholds (sometimes called indicative thresholds). A recent review of the Directive brought more projects under Annex I, increasing the number of development types for which EIA is mandatory. Indicative thresholds for Schedule II projects have also been revised, and criteria have been established for exclusive thresholds to identify projects that will be automatically excluded from EIA (see section 2.3.2).

For projects where EIA has been required by the decision-making authorities, the Directive then specifies the minimum information that should be provided concerning the project and its likely effects. The UK's version of the Directive requires impact assessments to identify, describe and assess 'in an appropriate manner', the direct and indirect effects of a project on the factors listed in Box 2.1.

With respect to ecological effects, the legislation does provide scope for a thorough and holistic approach to assessment, the Directive indicating that the 'description of the likely significant effects' of a proposed project on the environment should cover its 'direct effects and any indirect, secondary, cumulative, short-, medium- and long-term, permanent and temporary, positive and negative effects'. In common with EIA legislation in many countries, however, the wording of the legislation is open to interpretation, and it is common for an absolute minimum of information to be supplied. The general failure to consider cumulative and indirect effects, in particular, has resulted in a number of initiatives to implement a separate discipline of 'cumulative effects assessment' or to ensure a more holistic approach through strategic environmental assessment (SEA). Specific guidance on the implementation of the Directive with respect to EcIA might help to ensure a more rigorous interpretation.

Box 2.1 Factors to be addressed in UK project EIAs.

- Human beings, fauna and flora
- Soil, water, air, climate and the landscape
- The interaction between the above factors
- Material assets and the cultural heritage

Project proponents are required to summarize relevant information in an environmental statement (ES), providing at least:

- A description of the project comprising information on the site, design and size of the project
- A description of the measures envisaged to avoid, reduce and, if possible, remedy significant adverse effects
- The data necessary to identify and assess the main effects that the project is likely to have on the environment
- A non-technical summary of the above information

While there are considerable differences between countries in the scope of project-EIA (i.e. the range of actions and impact types covered by legislation) and the methods used to assess and evaluate impacts, remarkably similar problems with the assessment of ecological impacts have been encountered in most of them. EIA confined to the project level is arguably less effective for ecological considerations than it is for any other impact category. Commonly encountered problems are listed in Box 2.2.

Pritchard (1993) emphasized that, while project-level EIA might have a very important part to play in the regulation of ecological effects, it

Box 2.2 Problems with EcIA for EIA at the project level.

- Failure to go beyond the 'site boundary' or project-derived limits
- Inability to assess impacts across national boundaries
- Problems with extending study limits due to restricted site ownership
- Lack of long-term datasets and baseline information to provide an evaluation context for localized, short-term results
- Failure to identify or measure cumulative ecological effects
- Inability to consider alternative sites or methods in detail
- Failure to evaluate ecological impacts in a wider context (only local implications considered)
- Failure to address the additive ecological effects of projects that are individually exempt from EIA
- Failure to recognize 'saturation thresholds' above which environmental resilience breaks down, or carrying capacities are exceeded
- Failure to address 'time- or space-crowded' impacts

will always be flawed if the strategic decisions that generate individual projects are intrinsically environmentally harmful, or are progressed without regard to the overall harm they may cause. This is especially likely where there are issues considered to be of overriding importance such as generation of jobs or economic gain.

2.2.3 Strategic environmental assessment

The effectiveness of EIA in reducing environmental degradation at source is greater the earlier it is used in the decision-making process. The term strategic environmental assessment (SEA) has been coined to refer to the application of EIA in strategic planning and policy making (Ortolano & Shepherd 1995). Theoretically, it is possible to assert that project-EIA and SEA are distinct only in terms of their scope and scale of application and that SEA simply extends the principles of EIA to cover development policies, plans and programmes (Sadler 1993). In practice, however, SEA has been legislated for separately and has evolved its own, distinct approaches and techniques.

Provision for SEA exists in a number of countries, including the US (Webb & Sigal 1992), Canada (Ortolano & Shepherd 1995), Australia and New Zealand (Wood 1992) and the Netherlands (Verheem 1992). Within the EU, the desirability of extending the principles of EIA from individual projects to policies, plans and programmes has been discussed for some years, and a draft Directive on SEA has been under consultation since 1991. In addition to the Netherlands, provisions for assessing the environmental impacts of some policies, plans and programmes do already exist in Belgium, Denmark, Finland, France, Germany, Spain, Sweden and the UK (EIA Centre 1995). However, practical implementation of SEA has tended to lag behind formal provision for it in the legislation. In Canada, for example, despite the stated commitment of the federal government to assessing the environmental impacts of policy initiatives, the federal government has yet to make it a legislated, enforceable requirement (Gibson 1993). In general, governments throughout the world have been very reluctant to demand EIA for new policies and legislation, despite its probable benefits in improving the quality of decisions about alternative options and proposals. The flagship for 'fully tiered' EIA is California, where several hundred SEAs have been undertaken to date (Wood 1995).

Whereas project-level EIA is essentially reactive, SEA permits forward planning to avoid impacts on sensitive environmental components and to plan development that is compatible with conservation of natural resources. Theoretical models of forward planning are based on a hierarchical or tiered approach, starting with the formulation of a policy, plans and then programmes within which individual projects may be implemented (Wood 1995). Lee and Wood (1978), Lee and Walsh (1992) and Wood (1995) have illustrated this with reference to national, regional and local planning of transport, where a national transport policy would provide the context for a 'long-term national roads plan' including 'coordinated and timed objectives' for implementing that part of the policy relating to road transport. Such a plan might spawn a 'road building programme' and, finally, the project-level construction of a road or road section (Wood 1995). Clearly, EIA of separate sections of individual roads cannot be used to estimate the overall consequences of a national road-building programme for a country's biological diversity. It certainly cannot be used to evaluate the relative costs and benefits of travel by road and rail.

SEA can therefore provide a framework for the assessment and regulation of ecological impacts that tend to fall through the project-level EIA-net. Ecological impacts are often subtle, complex, indirect, delayed or expressed at a considerable distance from their source. Ecological impacts that are individually insignificant and may remain undetected in case-by-case appraisal, for example, might be addressed more effectively through SEA (Eberhardt 1976). Much of the pressure to legislate for SEA has come from ecologists and conservationists concerned about progressive and chronic damage to the 'natural heritage' due to cumulative effects. Cumulative impacts expressed at a regional, national or international scale can only be controlled by planning or managing development at that scale (Sonntag et al. 1987). Ecologists need to be able to design impact studies with adequate coverage both to capture wider-scale impacts and to place site-specific impacts in their wider context. In many cases, effective control of ecological impacts requires trans-boundary planning and cooperation. Ecologists also need to take a long-term perspective to place short-term, or immediate effects in the context of long-term trends or cycles. Artificial boundaries (in time and space) are rarely consistent with natural variation, making a more strategic and synoptic approach to EcIA fundamental to its maturation as an effective tool for environmental management.

Strategic ecological assessment (SEcA) might therefore be needed to address impacts on:
• single resources within single regions;
• single resources among multiple regions;
• multiple resources in a single region;
• multiple resources in multiple regions;
and to answer assessment questions like:
• What proportion of a country's wetlands constitute critical natural capital?
• What would the overall implications of a change in transport policy be for protected species in a region?

- Will an alternative policy approach result in less biodiversity loss?
- Is the proposed policy, plan or programme consistent with local, regional, national or international agreements on biodiversity conservation?

Some form of strategic assessment would improve the ability to avoid valued sites, protected habitats and species and excessive exploitation of limited natural resources. International obligations with respect to responsible or 'wise use' of natural resources would appear to make SEA imperative.

The main advantages of SEA for addressing ecological impacts are listed in Box 2.3.

SEcA demands comprehensive, up-to-date, national data on the state of natural resources and on the status and distributions of wildlife habitats and species. This is currently lacking in many countries. However, legislation for SEcA could also act as a catalyst for strengthening the knowledge base. International action to conserve biological diversity has also helped in this regard, as countries have been obliged to undertake inventories of their biological diversity, to assess the status of endemic species and to establish defensible thresholds for species and habitats.

SEA and associated SEcA are most important for those activities which exploit natural resources directly (primary industries such as forestry,

---

Box 2.3 Advantages of SEA as a context for EcIA.

- A resource-based approach
- Avoidance of adverse impacts on valued, sensitive or threatened areas/ecosystem components
- Effective consideration of alternative sites
- Earlier recognition of potential ecological constraints in siting and design of infrastructure
- Provision of contextual information needed to focus ecological survey and analysis on key components
- Ability to characterize baseline conditions
- Ability to institute suitable data collection and monitoring programmes
- Consideration of the effects of associated projects
- Assessment of cumulative effects
- Assessment of synergistic effects
- Assessment of trans-boundary effects
- Assessment of ecological impacts at appropriate scale
- Ability to manage regions according to principles of landscape ecology
- Scope for setting limits of acceptable change and deriving viability thresholds

fishing, mining and mineral extraction) and which have significant consequences for land use over large areas (such as transport). Increasingly,
however, SEA is also necessary to take account of potentially adverse
cumulative effects, especially those which might result in chronic damage
(slow releases of radiation, chronic pollution of groundwater resources,
'nibbling' of habitats due to progressive infrastructure development,
etc.).

Therivel and Thompson (1996) give a comprehensive review of the
role of SEA in nature conservation, using case studies and examples from
a number of countries. Examples where ecological considerations have
been addressed through SEA include:

• SEA of petroleum activities in the Barents Sea, Norway—predicted
impacts of oil spills on seabirds and fish eggs and larvae (UNECE
1992);
• SEA of German windfarms—ecological constraint mapping using
endangered bird species and valuable biotopes as indicators
(Kleinschmidt 1994);
• SEA of Firth of Forth transport strategy (Scotland)—planning to
minimize impacts of new road works on natural features of significance
for biodiversity, especially loss, disturbance or fragmentation of areas
important for nature conservation (Scottish Office 1994).

In addition, many papers have been published reiterating the need
for more strategic approaches to the assessment and regulation of
development to tackle ecological problems.

For example, Thompson et al. (1995) explored the potential application of SEA to the farming of Atlantic salmon Salmo salar in
mainland Scotland, primarily as a means of dealing with serious
cumulative impacts on benthic organisms, seabirds and native, wild
populations of fish. Marine fish farms have proliferated in ecologically
sensitive waters in Scotland. Indicative thresholds were set so as to
exclude the majority of farms, which can now be found in almost all
mainland sea loch systems. Project-EIA is currently failing to tackle
cumulative effects due to uncontrolled proliferation in important coastal
habitats. It is also failing to address the disturbance to native wildlife
(seabirds and seals) caused by operational activities, the knock-on effects
of associated infrastructure development (particularly for access and
transport), the excessive use of wild fish stocks to feed captive fish,
the possible cumulative and synergistic effects of chemical releases,
trans-boundary pollution effects, effects on genetic constitutions of wild
salmon populations and on population dynamics, including disruptions
to predator–prey relationships. Such effects cannot be tackled effectively
through assessment of individual projects: consideration of long-term,
cumulative, interactive and knock-on effects is only possible if a more
strategic approach is taken.

The increase in the licensing for oil and gas exploration and production in UK waters in recent years, coupled with a move into inshore waters (some of which are ecologically sensitive), has also been a cause of considerable concern to nature conservationists. Again, licensing has continued without any overall assessment of cumulative impact. In early 1997 (in response to pressure from non-government organizations and pressure groups), the EU ruled that the UK Government must implement the 1985 EIA Directive (85/337/EEC) with regard to offshore oil and gas development (Green 1995). A consortium of wildlife conservation groups presented the Government with a detailed proposal to reform the licensing process to take account of both the EIA Directive and the Government's international commitments with regard to conservation of biological diversity. The consortium called for a full SEA of the continental shelf to assess its biological resources and their sensitivity. The SEA would identify areas too important to develop as well as areas where information is inadequate to estimate vulnerability with confidence or where technology is inadequate for safe and sensitive development. In such areas, the precautionary principle should apply.

There has also been much interest in strategic assessment of the ecological impacts of transport networks. Bina *et al.* (1997) carried out a pilot study to try and quantify the ecological impacts of trans-European transport networks. They used a geographical information systems (GIS)-based approach to identify areas designated for conservation of birds ('important bird areas' or IBAs), which would be within 2 or 10 km of proposed routes and found that a significant number of sites would be potentially affected. Other studies have reinforced the need for SEcA to ensure that international obligations to conserve biological diversity are compatible with development plans and programmes and the development of reliable methodologies is a research priority.

2.3 Differences in formal EIA procedures: implications for EcIA

Clearly, EIA legislation varies considerably in detail between countries, even in the EU where Member States are bound by the requirements of the same Directive. These differences have implications for the way in which EcIA is practised in different countries.

2.3.1 Responsibility for undertaking EIA

In some countries EIA is the responsibility of the proponents of development actions (whether these are individuals, companies, local or national governments). In others it is initiated independently. The former, more prevalent, approach carries obvious risks of bias, but has the advantage of devolving the cost of EIA away from governments and tax payers. In theory, it might also facilitate dialogue between developers and ecological consultants and provide scope for better mutual understanding of constraints when carrying out EcIA. Most evidence to date, however, suggests that proponents of development actions make poor use of ecological advice and often consult too late. There is also a suspicion that development proponents place ecologists under pressure (however subtle) to downplay the severity of adverse ecological impacts. Another obvious problem is the desire of development proponents to minimize expenditure on EcIA, which often results in serious under-resourcing. Hard evidence is very difficult to come by, but it has been estimated that, on average, about 1% of total development project budgets are allocated to EIA. Amounts allocated to EcIA are relatively very small indeed.

Where EIA is the primary responsibility of bodies with vested interest, it is important that there should be at least some element of independent review. With respect to EcIA, independent review might help to ensure that best-practice and up-to-date methods and data are used. It is particularly valuable for decision-makers who may have no other effective means of checking that they have been provided with all relevant information. Not all decision-making authorities have suffi-cient resources to employ ecological specialists, in which case they may not be aware of important omissions or methodological shortcomings. Furthermore, many planning authorities are under-resourced and have insufficient time or staff to devote to 'vetting' of environmental state-ments (ESs) or monitoring the EIA process.

2.3.2 Eligibility for EIA (indicative thresholds)

There is enormous variation in the extent to which eligibility of develop-ment actions for EIA is specified by legislation. Differing interpretations of the same legislation are common, implying some considerable ambiguity of meaning. The European Community (EC) Directive, for example, initially listed 'Annex I' project types for which EIA would be mandatory and Annex II projects for which EIA would be discretionary. For Annex II projects the criteria used to determine eligibility were not always clear or consistent. In most Member States of the EU, thresholds of eligibility for EIA were legally binding, but the UK only put advisory criteria in place, leaving it up to competent authorities to decide whether different Annex II projects should be subject to EIA (CEC 1993). A recent revision of the Directive has necessitated revising and strengthening indicative thresholds and introducing 'exclusive' thresholds for most project categories, below which the need for EIA can be ruled out in advance unless a proposal is likely to affect a 'sensitive' area. Meanwhile, eligibility or 'indicative' thresholds for the same project types vary considerably between Member States. The threshold for industrial and domestic waste disposal facilities, for example, is 25 000 tonnes per annum in the Netherlands, Belgium and Ireland, but 75 000 tonnes per annum in the UK (Wathern & Russell 1993).

There are a number of examples where ecological criteria have been used to determine indicative thresholds for EIA (Table 2.3), although the use of social and economic criteria is more common. EIA is often required in situations where sites designated for nature conservation are affected. In the UK, EIA is now required in circumstances where developments are likely to affect sites designated for nature conservation. Recent debate suggests that the new exclusive thresholds for EIA will not apply in the 'sensitive areas' listed in Box 2.4 and consideration of the need for EIA will be on a case-by-case basis. This at least ensures that the environmental implications of actions affecting designated areas are addressed, even if it implies that designation may not be acting as an effective deterrent to development! Many local planning authorities in the UK also insist on EIA for developments affecting sites of county importance for nature conservation. Even though there is no strict legal requirement, many developers will in any case opt to undertake EIA in such circumstances so as to avoid jeopardizing their chances of gaining development consent.

Likewise, in the Ukraine, EIA is required for 'works located in nature reserves, spas and other protected areas' (UNECE 1996) and in Finland, EIA is required for any 'permanent alteration to natural forest, peatland or wetland over what can be considered a unified area above 200 ha in size, by carrying out new ditching or by draining unditched peatland and wetland areas, by removing the tree stock permanently or by

Table 2.3 Examples of ecological indicative thresholds for environmental impact assessment (EIA).

Country	Proposed activity subject to EIA
Austria	Forest clearing areas > 20 ha
Czech Republic	Forest clearing areas > 5 ha (Ministry of the Environment) Landscape interventions which may cause fundamental changes in the biological diversity and in the structure and function of ecosystems (District Offices)
Estonia	Changing of sanitary conditions within recreation areas and of natural conditions within nature conservation areas Establishment and zoning of nature conservation areas (national EIA) International and inter-regional schemes of nature management and conservation
Finland	Permanent alteration to natural forest, peatland or wetland

> **Box 2.4** 'Sensitive Areas' requiring consideration of the need for EIA on a case-by-case basis in the UK.
>
> ---
>
> 1 Sites of Special Scientific Interest (SSSIs) including:
>   • Ramsar Sites designated under the Convention on Wetlands of International Importance
>   • Special Protection Areas (SPAs) and candidate SPAs under the EC Directive on the Conservation of Wild Birds
>   • Special Areas of Conservation (SACs) and potential SACs under the EC Directive on the Conservation of Natural Habitats
>   • National Nature Reserves (NNRs)
> 2 World Heritage Sites and Scheduled Ancient Monuments
> 3 National Parks (and the Broads and the New Forest Heritage Area)
> 4 Areas of Outstanding Natural Beauty, National Scenic Areas and Natural Heritage Areas

replanting the area with species of trees not indigenous to Finland' (UNECE 1996).

It would be better if eligibility for EIA was based on the condition of ecological resources rather than simply registering the presence of formally designated areas. Otherwise important ecological effects may be missed, particularly trans-boundary effects of actions undertaken outside designated areas which nevertheless impinge on them (wetlands affected by pollution of water courses upstream, for example). Furthermore, the automatic exclusion of some projects from EIA means that potential cumulative impacts on critical natural capital may be neglected.

2.3.3 Consideration of alternatives

The ability to consider alternative options, whether these are policies, sites or processes, is fundamental to environmental planning. Systems that preclude effective consideration of alternatives cannot be used proactively and do not merit definition as planning systems in the strict sense. In particular, effective consideration of alternatives is essential to allow options causing least ecological impact to be explored. Some countries require information on alternative solutions to be provided as part of the EIA process. In others, only minimal information is required, full EIA being carried out only for one or a limited number of development options. Although some countries (the Netherlands, for example) do require relatively full discussion of alternative solutions to be put forward in project ESs, consideration of alternatives is constrained under project-EIA. It becomes a more realistic proposition under SEA and systematic consideration of alternatives is often one of the main motivations for introduction of SEA legislation.

Where there is no consideration of alternatives, scope for avoidance of adverse ecological impacts is considerably reduced and there tends to be an over-reliance on mitigation to 'fix' damage. As mitigation is rarely 100% successful, avoidance of impacts at source should always be the priority. Without formal requirements for consideration of alternatives, the information supplied is rarely adequate for siting and design to be optimized with respect to ecological constraints, making it impossible to identify those options likely to cause least ecological damage.

2.3.4 Public consultation and participation

In the US, there has been a tendency for consultation and negotiation with the public concerning potential environmental impacts to be introduced at earlier and earlier stages in the decision-making process. It is generally acknowledged that early and effective public consultation can be instrumental in avoiding costly delays and expensive public inquiries at a later date. In many countries, however, public participation in the EIA process has remained negligible. The 'Five Year Review' of EIA implementation in the EU (CEC 1993) concluded that the 'minimum legal rights of citizens to consult the ES and comment meaningfully on its contents are insufficiently safeguarded in law'. Public unrest at the prospect of development without formal opportunity for participation in the decision-making process is, increasingly, responsible for expensive delays. In the UK, policing of anti-road protests has cost millions of pounds.

With respect to ecological impacts, meaningful dialogue with the public is hampered by the complexity of ecology as a science and its lack of clear guiding principles. Those habitats and species with acknowledged 'public appeal' dominate the public debate and it is not always easy to put the case for less charismatic elements of the biota when they are threatened by proposed actions. Public concern about wildlife can therefore have both positive and negative effects on EcIA scope and standards. However, it is only through engaging the interest and concern of people that most wildlife resources will be safeguarded in the long term. EcIA carried out for purposes of EIA can do a great deal to promote interest and awareness in environmental issues. For some habitats and species, EIA has provided a vehicle for intensive study of species habitat requirements, which might have been difficult to implement otherwise, for example the detailed studies of the needs of the spotted owl *Strix occidentalis* in the northern forests of the US.

2.3.5 Reviewing the EIA process

Independent review by ecological professionals helps to ensure the adoption of best practice. In many countries, EIA is carried out by the proponents of development actions without any independent review. This is true of the UK, Spain and many of the other Member States of the EU. This has tended to result in EIAs with restricted scope (in both spatial and temporal terms), which lack objectivity and rigour. Countries that have mechanisms for independent review include the US and Canada, the Netherlands, Wallonia and Italy. In these countries, independent review (usually by some form of review commission) is considered to be vital in promoting best practice and achieving quality control.

The Dutch EIA legislation has made provisions for independent review of the EIA process by setting up an 'EIA Commission'. This comprises independent experts who can be called upon to comment on the quality of ESs and their compliance with statutory requirements and official guidelines. In the Netherlands, the competent (or decision-making) authority therefore has access to independent and unbiased advice concerning the quality and reliability of the information contained in ESs. The EIA Commission can also be called upon to assist the competent authority during the scoping phase of the EIA process in actually formulating the guidelines that proponents of a development are required to follow. The Commission does not make any judgement concerning the acceptability or otherwise of a proposal. Its advice is restricted purely to the quality and reliability of information about environmental aspects of the proposed project.

One advantage of the Dutch system for review is that the composition of review panels is flexible and can be structured on a case-by-case basis to ensure that the individuals selected to make up the panel have relevant expertise in the areas covered by each EIA. Decision-making authorities can therefore be kept abreast of new knowledge or state-of-the-art techniques, receiving input from specialist advisors at a stage in the process where it is still possible to influence the scope of impact assessment studies. Receipt of specialist and unbiased advice also makes it easier for decision-making or planning authorities to put conditions on development consent.

In the US, the Environmental Protection Agency (EPA) carries out review of ESs to check both the format of ESs and their adequacy in terms of content and coverage. It is widely accepted that some form of review of the quality and coverage of environmental impact statements is necessary to maintain the overall quality of the EIA process, but in some countries, including the UK, for example, there has been considerable resistance to the idea at government level.

2.3.6 Monitoring

Monitoring is essential to strengthen baseline knowledge and improve the efficacy of ecological mitigation. Few countries have established a legislative framework for follow-up or post-project development monitoring. This has hampered the development of reliable predictive techniques. Ecology is a relatively 'young science' and has often been under-resourced. Effectiveness of EcIA and its value as a tool for environmental regulation will remain limited unless monitoring is introduced (Hollick 1981). In the short term, monitoring may seem expensive, time-consuming and unnecessary, but in the long run it will make environmental prediction more efficient and reliable. It is becoming more common for monitoring to be made a planning condition, the developer being required to fund and implement a suitable programme. Some arrangements have been made for monitoring to be carried out by environmental trusts set up by the developer. As well as enhancing ecological prediction, formal follow-up would also enable separate assessments to contribute to the overall knowledge base. The need for ecological monitoring is considered in more depth in Chapter 8.

2.3.7 Guidance

Lack of independent review may be compounded by a shortage official guidance on appropriate impact assessment methodologies. This has implications for practitioners, proponents, consultants and decision-making authorities as well as for the overall quality of the process. Lack of official guidance has contributed towards the marked inconsistency in approach which characterizes EcIA in many countries and has hampered the ability to 'learn from experience' or define 'best practice methodologies'.

Furthermore, moves to define professional standards for practitioners in EIA and EcIA are only just getting underway in most countries. The professional qualifications of practitioners who undertake EcIA are sometimes questionable.

2.4 Legislation for international and trans-boundary effects

Just as it has become obvious that project-level assessments can miss important ecological impacts, there are also a number of environmental problems and ecological phenomena that can only be tackled effectively through concerted international action. Species other than man cannot be expected to recognize or respect national boundaries and their effective conservation cannot be achieved solely through national action. Many migratory bird species, for example, are affected by influences at the flyway level, which cannot be tackled independently by those countries in which they are temporarily resident. For threatened species, integrated action is required to ensure that populations are safeguarded in all countries. The role of pollution from industrialized countries such as the UK in producing acid rain and contributing to the demise of Scandinavian lake ecosystems is another well-known example, the effective management of which demands a trans-boundary approach (Fig. 2.1). There is more recent evidence that Antarctic seabirds are accumulating high levels of toxic organic chemicals generated by industrial plants thousands of kilometres away (Bewley 1997). Van den Brink (1997) measured levels of hexachlorobenzene (HCB) (released from industrial incinerators and other plants) in the preen oil of Antarctic bird species including the Adelie penguin *Pygoscelis adeliae*, the southern fulmar *Fulmaris glacialis*, the snow petrel *Pagodroma nivea* and the Antarctic petrel *Thallascica antarctica*, and found that they were hundreds of times higher than levels found in preen oil taken from common terns *Sterna hirundo* on the Dutch Waddenzee. Van den Brink postulates that volatile chemicals like HCBs undergo a 'cold condenser effect', rising as a vapour to the upper atmosphere where they are transported round the globe. On reaching the poles, they condense and drop, becoming a particular threat to animals nearer the top of food chains, in which they

Fig. 2.1 Didcot Power Station, UK.

can bioaccumulate. Similarly, many of the species traditionally hunted for food by the Inuit of northern Canada, like beluga and narwhal whales, have become increasingly toxic with accumulated polychlorinated biphenyls (PCBs), mercury, cadmium and hexachloro-cyclohexanes (HCHs). Many of these chemicals have been banned from use in Canada for many years and are probably derived from sources many thousands of kilometres away (Pearce 1997).

Global warming is another environmental problem with international dimensions and a diversity of causes or contributory factors. It is particularly difficult to tackle because of the considerable uncertainties that surround the nature, scale and timing of possible outcomes and because of the complexity of its mechanisms. While scientists might take a holistic view which considers the problem in terms of 'a single, highly complex, global-scale system', the actual managerial response required to regulate it is more likely to be 'local' (Hood *et al.* 1992).

While much of the impetus to enact legislation for assessment of trans-boundary impacts has come from recognition of global problems with multiple causes, like global warming, it is equally possible for single developments to generate trans-boundary impacts. Atmospheric, aquatic or marine pollution from a single development might well cross an international boundary, for example. In recent years, a number of international agreements have been reached in an attempt to tackle global environmental problems collectively. For example, there has therefore been growing pressure for EIA to be applied as a tool in the regulation of trans-boundary effects and to help achieve global conservation of biological diversity and environmental quality. The Convention on Environmental Impact Assessment in a Transboundary Context was drawn up under the auspices of the United Nations Economic Commission for Europe (ECE) and adopted at Espoo, Finland on 25 February 1991 to provide a framework for the environmental assessment of activities causing trans-boundary impacts in Europe. Twenty-nine countries have signed up to it and the Convention has entered into force in 11 of them (Albania, Austria, Bulgaria, Finland, Italy, Luxembourg, Netherlands, Norway, Republic of Moldova, Spain and Sweden). With respect to impacts on flora and fauna, a number of international instruments are of potential value in determining the likely significance of the trans-boundary effects of a proposed activity (UNECE 1996). These are listed in Box 2.5.

The Bern Convention, for example, carries an obligation to conserve the habitat of wild plants and animals, especially those listed in the Convention as endangered or vulnerable. This Convention also lays particular stress on areas that are important to migratory species. The EC Council Directive on the conservation of natural and seminatural habitats of wild fauna and flora (the Habitats Directive) was agreed by

Box 2.5 International instruments relating to trans-boundary assessment of ecological effects.

- Convention on Wetlands of International Importance Especially as Waterfowl Habitat, 1971 (Ramsar Convention)
- Convention on Fishing and Conservation of the Living Resources in the Baltic Sea and Belts, 1973
- Convention on International Trade in Endangered Species of Wild Fauna and Flora (CITES), 1973
- The Bern Convention on the Conservation of European Wildlife and Natural Habitats, 1979
- Convention on the Conservation of Migratory Species of Wild Animals, 1979
- United Nations Convention on the Law of the Sea, 1982
- Convention on Biological Diversity, 1992
- Various EC Directives, for example the 'Birds' and 'Habitats' Directives

the European Council in December 1991. This has a part to play in the international conservation of biological diversity by requiring Member States to take measures to maintain or restore natural habitats and wild species at a favourable conservation status in the EC as a whole. The Directive requires designation of Special Areas of Conservation (SACs), which are of acknowledged Europe-wide importance. With respect to ecological impacts, the Convention on Biological Diversity has perhaps had the most significant influence on the integration of international concerns through national law. The Convention on Biological Diversity also makes explicit reference to the potential role of EIA in conserving biological diversity.

2.4.1   The Convention on Biological Diversity

The Convention on Biological Diversity was agreed at the Earth Summit in Rio. Thirty governments had ratified the Convention by early October 1993 and 41 by the time it actually came into force in December 1993. The Convention is important not least because it represents the first time a large majority of the World's States have collectively agreed to a binding legal instrument intended to conserve biological diversity and promote the sustainable use of biological resources. The Convention 'articulates a series of national and international biodiversity-related rights and obligations' and 'sets broad goals which parties must fulfil at the national level' (Burhenne-Guilmin & Glowka 1994).

While loss of biological diversity can be regarded as a global concern, responsibility for its conservation is considered to rest with individual nations. The Convention therefore emphasizes national action. In a similar way to the EC Directive on EIA, it takes a result-orientated approach, which leaves signatories considerable flexibility in deciding how to achieve those results. The Convention contains 42 Articles, which place a requirement on signatories to identify their biological capital, to develop and implement a national biological diversity strategy, to conserve biological diversity and to use biological resources sustainably. The Convention also specifies some possible techniques by which these goals might be achieved. These include the integration of biological diversity concerns into national decision-making, use of a combination of command/control and incentive measures, controlling processes and activities that threaten or cause the local loss of biological diversity and also influencing decisions through EIA (Burhenne-Guilmin & Glowka 1994).

EIA has evolved primarily as a tool for environmental planning and management, rather than as a mechanism for the conservation of biodiversity. However, by regulating development and its impacts on ecosystems, EIA has a potential role in ensuring that the aims of the Convention on Biological Diversity are promoted through planning and development control to ensure that ecosystem viability is preserved. In Article 14 of the Convention, EIA is singled out as a potential implementation mechanism, where it is suggested that EIA should be carried out 'for proposed projects likely to have significant adverse effects on biodiversity'. Most national EIA frameworks appear to provide a statutory basis for governments to meet the requirements of Article 14 of the Convention. Although biological diversity may not be mentioned explicitly, its importance is usually implied (Sadler 1996). Certainly, EcIA is an obvious tool for addressing potential impacts on biological diversity. At the same time, the development of the 'biodiversity information base' has the potential to strengthen the scientific basis for EcIA. Biodiversity can be measured in various ways, providing objective and defensible evaluation criteria. International obligations to address the national status of wildlife habitats and species have also given impetus to national monitoring and inventory, thereby improving baseline knowledge and understanding and making it possible to evaluate local impacts in a wider context.

In practical terms there appear to be a number of barriers to implementing the requirements of the Convention on Biological Diversity through EcIA. For example, consistent approaches are needed to derive criteria for assessing the significance of impacts on biological diversity. In many countries, existing environmental quality standards are largely defined in terms of ambient, effluent and emission concentrations for surface freshwater, marine waters and air (Dahuri 1994), not in terms of uptake by species or the absorptive capacity of ecosystems. For many habitats and species it has proved very difficult to define 'favourable conservation status' or to predict the consequences of gradual genetic impoverishment. Maintenance of intraspecific diversity is generally acknowledged to be important for species adaptation and survival, but it has proved difficult to measure and quantify the minimum levels of diversity required for long-term viability. Because there are no accepted environmental standards relating to biological diversity in most countries, it is difficult to establish what constitutes a significant impact on biological diversity. This is something that is considered in more depth later in the book.

Under the Convention on Biological Diversity, scientific and policy concerns have tended to focus on threatened plant and animal species. Guidance on the incorporation of biological diversity considerations in the NEPA process issued by the US Council on Environmental Quality (CEQ) in 1993 referred to the inadequacy of attention paid to 'non-listed' species and non-protected areas. In many countries, the listing of species as 'endangered' and the implementation of appropriate protective measures lags behind the rate of species decline. Early warning signals are therefore needed to detect impending endangerment and extinction at an early enough stage for something to be done. The scientific basis for some of these is explored in Chapter 6.

Key factors contributing to declines in biological diversity in most countries are listed in Box 2.6.

Sadler (1996) identifies several principles which can help development proponents to take better account of biological diversity considerations when undertaking EIA:

Box 2.6 Key factors contributing to declines in biological diversity. (After US CEQ 1992, 1993.)

• Physical alteration as a result of resource and land conversion to more intensive uses (causes habitat destruction, fragmentation and simplification)
• Pollution (possible immediate direct, indirect and cumulative effects on species and habitats)
• Over-harvesting of fish, wildlife and other plant and animal species (can reduce populations below levels at which they can recover or indirectly affect other species with which they interact)
• Introduction of exotic species (can lead to elimination of natural species through predation, competition, genetic modification and disease transmission)
• Disruption of natural processes by intensive resource management activities (may alter ecosystem dynamics with further effects on community composition and succession)
• Global climate change if it occurs (may affect the natural ranges of many species and ecosystems)

- take a synoptic or ecosystem perspective;
- identify unique, rare ecologically sensitive areas and important patterns and interconnections (e.g. wildlife corridors);
- avoid or minimize impacts on these features through location, design and mitigation measures;
- prepare impact compensation and rehabilitation plans for unavoidable damages in accordance with the 'no net loss' criterion.

In the longer term, the application of a regional ecosystem approach will be necessary to address issues of biological diversity conservation fully. This can be illustrated by reference to the US Endangered Species Act of 1973, which requires all Federal agencies to undertake pro-grammes for the conservation of endangered and threatened species. The Act also prohibits Federal agencies from 'authorizing, funding, or carrying out any action that would jeopardize a listed species or destroy or modify its critical habitat'. In a 1978 amendment, a stipulation was made that 'critical habitat' must be designated concurrently with the listing of a species. There is a clear role for EcIA here, in ensuring that proposed development actions are compatible with the safeguard of classified species and their habitat. In the context of conserving biological diversity, however, it is important to note that a 'species' is considered to include any species or subspecies of fish, wildlife or plant, but also any variety of plant of distinct population segment of any vertebrate species that interbreeds when mature. Recently, the US Fish and Wildlife Service and Marine Fisheries Service have adopted a policy to clarify their interpretation of a 'distinct population segment' for the purposes of listing, de-listing and re-classifying species under the Endangered Species Act. The concept of 'species' is having to be re-visited to ensure that biodiversity is conserved 'across the board'. Conservation of genetic material in its full variety is not necessarily achieved by measures applied at the species level: subspecies and distinct population units may be just as important. EcIA can play an important part in ensuring that proposed development actions are compatible with the conservation of endangered and threatened species, but it will not be effective in conserving them across their ranges unless it is possible to take a strategic approach under which local variation can be safeguarded.

Despite such 'teething problems', in countries which take their obligations under the Convention on Biological Diversity seriously, scope for effective EcIA has increased dramatically. The Convention on Biological Diversity has forced governments to focus on the conservation status of species and habitats, to consider the adequacy of existing knowledge about spatial and temporal trends in their distribution and to examine the cross-sectoral implications of development. Not only has investment in survey and monitoring improved, but generally accepted

Box 2.7 Guiding principles for conservation of biodiversity. (After Wynne 1993.)

- Biological resources must be used sustainably
- Non-renewable resources must be used wisely
- Conservation policy and practice must stem from a sound knowledge base
- Biodiversity conservation must be an integral part of all Government programmes, policy and action at national and local levels
- The precautionary principle must guide all decisions that could cause environmental damage
- Environmental appraisal and economic appraisal must be widely carried out at both a strategic and project level
- Regulation that controls matters such as site protection and pollution must be used to prevent biodiversity losses below critical levels
- Economic measures may be preferred to regulations to achieve higher environmental standards within sustainable limits
- Subsidies that lead to a loss of biodiversity must be removed
- Biodiversity conservation requires the care and involvement of individuals and communities as well as government process
- The lessons of the past must be learned. There are enough examples of ecological damage caused by human actions to avoid a repetition of past mistakes

standards are emerging for key species and habitats, making it possible to evaluate the significance of local losses in a national context. Box 2.7 summarizes 11 underlying principles for the conservation of biodiversity identified in 'Biodiversity Challenge', a report produced by a consortium of UK voluntary nature conservation bodies to assist the Government in formulating its Biodiversity Action Plan (Wynne 1993).

These are worthy principles, but they raise a number of ecological questions, which are currently difficult to answer. For example, what constitutes a 'critical level' for biodiversity loss, what are 'sustainable limits' for the exploitation of ecosystems and how should we develop the knowledge base that will permit us to monitor the status of threatened habitats and species?

2.5 Regulation of industrial activity

Environmental legislation relating to the regulation of industrial operation is in place in many countries and creates an additional need for EcIA techniques. There is legislation that applies to consent or approval procedures for the siting of some industrial activities, control of industrial pollution and also to the control of accidents and hazards, including major pollution incidents.

2.5.1 Control of industrial hazards

Industrial accidents can pose major hazards to both human and eco-logical 'health'. In many countries, specific regulatory procedures are in place to deal with the consequences of major industrial accidents in order to plan for the mitigation of adverse health effects. The EC, for example, issued a Council Directive on the major-accident hazards of certain industrial activities, generally referred to as the 'Seveso Directive' (EEC 1982). This was enacted in the UK in the form of the CIMAH or 'Control of Industrial Major Accident Hazard Regulations, 1984'. Under these regulations, a major accident is defined as 'an occurrence (such as a major emission, fire or explosion) resulting from uncontrolled developments in the course of an industrial activity, leading to serious danger to persons … or to the environment, and involving one or more dangerous substances'. It is important to note that 'environment' in this context is intended to include both the 'natural' and man-made environment, so that, in terms of assessment and accident prevention, all major accidents have equal status whether their effects are on man or on the 'natural' environment (Treweek *et al.* 1993). An incident is considered to be a major accident if it causes long-term or permanent damage to particular unique, rare or otherwise valued components of the man-made or natural environment, or if there is widespread environmental loss or damage. The criteria used to determine 'value' are considered below.

The Directive created a particular need for techniques of environ-mental risk analysis that could be used to predict the possible impacts of major industrial accidents or hazards on the environment and to estimate their likelihood of occurrence. More specifically, it generated a need for techniques that could be used to characterize potentially affected ecosystems, to evaluate their likely responses to defined stresses, for example through exposure to fire or chemical spillages, and to predict the risks and consequences of any irreversible damage that might occur. In other words, the CIMAH regulations also require some form of EcIA or ecological risk analysis.

Official UK guidance (Department of the Environment 1991) sum-marizes those aspects of the 'natural' environment that should be taken into account when assessing the likely effects of major industrial accidents. With respect to nature conservation, accidents likely to affect designated sites (National Nature Reserves and Sites of Special Scientific Interest, for example) are automatically taken very seriously. As established methods and criteria have been used in the selection and designation of such sites, it is assumed that it can be 'taken as read' that they are of nature conservation importance. A similar assumption is applied to many of the EcAs undertaken for purposes of EIA. The risk of neglecting habitats and species that are not afforded formal protection, but that are nevertheless declining or threatened, has been alluded to earlier. The regulations on CIMAH in the UK are particularly worthy of note because they do require consideration of non-designated sites if impacts are likely to exceed certain thresholds of damage. These thresholds are defined in terms of the area likely to be affected by a major accident. Under the CIMAH regulations an incident is considered to constitute a 'major accident' if permanent or long-term damage is caused to 'areas of scarce, intermediate or unclassified habitat' as follows.

• Two or more hectares of *scarce* habitat, including vegetated shingle beaches, saline lagoons, dune slacks, unimproved neutral grassland (including seasonally flooded grassland), lowland limestone pavement or other lowland basic rock less than 300 m in altitude, fens (including marsh and *Phragmites* reedbeds) lowland raised bogs, lowland heathland of southern Britain.

• Five or more hectares of *intermediate* habitat, including heathland less than 300 m in altitude, lowland limestone grassland less than 300 m in altitude, salt marsh and sand dunes.

• Ten or more hectares of more *widespread* habitat not otherwise classified.

Unfortunately, it is not always easy to interpret simple measures of 'area affected' in ecologically meaningful terms. As will become clear later in the book, minimum viable habitat sizes vary considerably between species and, in many cases, are not known.

The Regulations also require the significance of accidents to be considered in relation to particular species, using 'thresholds of significance' for individual species, which are determined in relation to the known or estimated British population. Thus, a major accident is considered to have occurred when an incident causes 'death, or inability to reproduce in a significant percentage of the known or estimated British population of a particular species, whether caused directly or indirectly'.

Again, it is not easy to determine what actually constitutes a 'significant percentage' for many species, but at least the Regulations emphasize the need to consider the effects of localized incidents in a wider context, determined by the national status and distribution of habitats and their associated species. The official guidance (Department of the Environment 1991) suggests that 'loss (or inability to reproduce) of 1% of any species would be considered significant'. However, this does not take into account the variation between species in terms of reproductive rates, or in the ability of their populations to recover following impacts. Lack of knowledge about the distributions of many species also makes it difficult to evaluate the regional significance of impacts.

> **Box 2.8** Techniques needed to address ecological impacts under the CIMAH regulations.
>
> ---
>
> - Describe/identify environmental components
> - Measure susceptibility and vulnerability to defined impacts
> - Assess risk (probability) of impact occurring
> - Identify environmental components that can be considered to have 'high value' ('valued ecosystem components')
> - Predict the likely extent of loss/damage should an incident occur
> - Estimate potential of affected ecosystem components to recover
> - Estimate recovery times
> - Assess the importance of residual loss/damage

For accidents affecting the natural environment, the time taken for unassisted recovery to a state close to the original is an important factor. It will depend on the type, susceptibility, diversity, abundance, colonizing ability and population processes of the species involved. 'A state close to the original' in this context denotes not only that particular species have returned to a site, but also that their respective age/size distributions and community structures are more or less as they were prior to the incident.

In other words, in theory, it would be necessary to consider both the vulnerability of habitats and species to damage and their resilience following damage. The need to consider capacity for 'recovery' implies that, in the context of CIMAH regulations, scarce habitats that are slow to recover from damage should be given higher priority than widely distributed and resilient habitats. This emphasizes the need for ecological risk assessment methods that incorporate some measure of functional attributes, and do not only measure static characteristics of the environment.

Techniques required to carry out EcIA under the CIMAH regulations are listed in Box 2.8.

There is an initiative currently underway to develop some of these techniques, which are also common to EcIA in many other contexts. Meanwhile, they emphasize the need for considerable background knowledge about the distribution, status and behaviour of ecosystem components.

2.5.2 Integrated pollution control

Procedures for IPC emerged from a recognition that it was impossible to regulate releases of pollutants to air, water and land effectively through entirely separate control regimes. In the UK, the Fifth Report of the Royal Commission on Environmental Pollution (1976: Cmnd 6371) proposed that pollutant releases should be directed to that environmental medium where the least environmental damage would be done, and recommended that a body should be created with responsibility for ensuring that industrial wastes would be disposed of so as to minimize effects on all three environmental media and optimize the resulting environmental solution. In other words, every effort should be made to search for the 'best practicable environmental option' (BPEO). One of the main objectives of IPC as developed in the UK, is to approach pollution control by considering discharges from industrial processes to all media in the context of effects on the environment as a whole (Department of the Environment 1990). IPC is now the responsibility of the UK's Environment Agency. The legislative framework for IPC rests on the prescription of processes and substances to which IPC applies. Prescribed processes cannot be operated without formal authorization. The need to consider the cost implications of pollution control has led to a situation where the BPEO is to be achieved using the 'best available techniques not entailing excessive cost' (BATNEEC). Much less clear are the environmental quality standards that these techniques are intended to deliver. For this reason, there is a clear role for ecological assessment techniques that could be used to establish acceptable environmental quality standards and to ensure that the potential cumulative effects of multisource pollution can be tackled effectively. In this context, there is a considerable need for research on the combined toxic effects of contaminants on organisms at the population level. In other words to characterize the actual exposure of organisms to contaminant 'cocktails'. To date, the majority of ecotoxicological information has been collected for captive organisms and has involved tests on single chemicals. This makes it difficult to estimate the likely ecological effects of pollutants that have accumulated slowly over time, or that interact synergistically with others released into ecosystems. This is a source of considerable uncertainty in ecological risk assessment. Knowledge is equally scanty concerning the possible population-level effects of chronic pollution. This is an area where EcIA and ecological risk assessment need to develop a common ground for research and application.

2.6 EcIA's role in sustainable development

Increasingly, environmental legislation seeks to ensure a more holistic approach to the management and regulation of the environment. In most cases, such an approach rests on the concept of sustainable development and the acknowledged need for conservation of biological diversity. Prior to its launch in 1996, for example, the UK's Environment Agency undertook 'to take a holistic approach to the protection and enhancement of the environment'. Its objectives were based on the premise that 'conserving and where practicable enhancing biodiversity and protecting the natural heritage is an essential element of sustainable development' (Environment Agency 1996). The Environment Act (1995) required the Environment Agency to contribute towards the achievement of sustainable development. When considering proposals affecting sites that are important for biodiversity or nature conservation, the Agency undertook to pay particular attention to its statutory obligations with respect to conservation, including international obligations associated with the Biodiversity Convention. Specific reference was also made to the need to conserve 'those non-designated sites which are nevertheless important for nature conservation' (Environment Agency 1996).

With respect to land-use planning, the Agency made it clear that it expected to become involved through (*inter alia*) responding to 'consultations by the Planning Authority under the Town and Country Planning (Assessment of Environmental Effects) Regulations 1988 (SI 1199)', by responding to requests for information from developers and by responding to requests for input to Development Plans. In its intended remit, therefore, the UK's Environment Agency makes explicit links between the conservation of biodiversity, sustainable development and environmental assessment. This is directly in line with the Biodiversity Convention, where EIA is given an explicit role in the implementation of national sustainable development strategies (NSDS) (Sadler 1993). In such a context, EcIA must be expected to have a key role.

If implemented through EIA legislation, EcIA is a potential mechanism for considering the ecological implications of development in any context and at any scale, ensuring that ecological factors are included in multidisciplinary appraisals of economic, social and technical issues. If used effectively in this way, EcIA should ensure that human activities (whether individual or collective) are designed and planned to avoid ecological damage 'at source', thereby reducing the need for operational constraints and controls and the dependency on mitigation to 'fix' damage. In other words, EcIA should embody 'the type of anticipatory approach to environmental issues under which an attempt is made to determine the likely nature and cost of environmental problems in advance of their occurrence, so that appropriate measures can be taken to prevent the impoverishment of the future' (Hughes 1992). This predicates a 'precautionary principle' being an integral feature of all projects and is an essential prerequisite for the achievement of sustainability (Pearce *et al.* 1989). The overall aim is to anticipate and prevent, rather than react and cure, and in so doing, to:
• conserve the variety and range of species;
• ensure that landscape and wildlife conservation is given full weight in policies for other economic sectors;

- ensure that populations of commercially exploited wild species are managed in a sustainable way.

Nevertheless, it remains difficult to reconcile conflict between human material need, development and the environment. Advocates of the concept of 'sustainable development' envisage a 'win–win' situation in which all potential conflicts are removed and development proceeds in harmony with the long-term viability of natural resources. Sceptics tend more to the view that 'sustainable development' is a contradiction in terms and that, as a general rule, 'the environment' bears the brunt of compromise.

All economic activity depends, ultimately, on a finite environmental resource base. It is generally acknowledged that imprudent use of the environmental resource base might reduce the capacity for generating material production in the future and this reduction might be irreversible (Arrow *et al.* 1996). Technical improvements in the management of resources together with resource-conservation measures might delay the point of 'no return', but 'perpetual growth is only conceivable if signals of scarcity are generated and acted upon within the economic system' (Arrow *et al.* 1996). It is presumably for this reason that the EC's Habitats and Species Directive (92/43/EEC) requires Member States to maintain or restore natural habitats and wild species at a 'favourable conservation status', but recognition of a 'favourable conservation status' requires definition of measurable attributes that can be used to derive thresholds.

One possibility is to express 'sustainability' in terms of environmental 'carrying capacities' (Arrow *et al.* 1996), sustainability being achieved if carrying capacities are not exceeded (Pritchard 1993). Alternatively, measures of ecosystem 'resilience' might be used to define the limits of the ability of ecosystems to absorb the effects of human exploitation. However, carrying capacities are not straightforward to define or measure: not only are they contingent on the availability of resources, which is a dynamic property, they can also be difficult to calculate in situations where population trends are unknown and the current condition of ecosystems is unclear (in other words it is not known whether a system is currently in an optimal or suboptimal state). Ecological resilience is also difficult to measure and may vary between systems and kinds of disturbance (see section 6.3.11). The use of concepts like 'carrying capacity' and 'resilience' to provide benchmarks for regulation of development would demand measurement and monitoring of the state of resources, knowledge of current and possible future uses or threats and of the effectiveness of mitigation measures available. In short, a strategic approach to EcIA is required with a firm grounding in resource monitoring if it is to play an effective part in efforts to strive for sustainable use of natural resources.

2.7 Recommended reading

The following texts are recommended as further reading to pursue the themes explored in this chapter.

Ecology and nature conservation in impact assessment

English Nature (1996) *Strategic Environmental Assessment and Nature Conservation*. English Nature, Peterborough.

Royal Society for the Protection of Birds (1995) *The Treatment of Nature Conservation in Environmental Impact Assessment*. RSPB, Sandy.

Thompson, S., Treweek, J.R. & Thurling, D.J. (1997) The ecological component of environmental impact assessment: a critical review of British environmental statements. *Journal of Environmental Planning and Management* 40 (2), 157–171.

Treweek, J. (1996) Ecology and environmental impact assessment. *Journal of Applied Ecology* 33, 191–199.

Westman, W.E. (1985) *Ecology, Impact Assessment and Environmental Planning*. John Wiley and Sons, New York.

Principles and theory of impact assessment

Sadler, B. (1996) *Environmental impact assessment in a changing world: evaluating practice to improve performance*. Final Report of the International Study of the Effectiveness of Environmental Impact Assessment. Canadian Environmental Assessment Agency and the International Association for Impact Assessment, Ottawa.

Vanclay, F. & Bronstein, D.A. (eds) (1995) *Environmental and Social Impact Assessment*. John Wiley and Sons, Chichester.

Review of EIA systems and legislation

Gilpin, A. (1995) *Environmental Impact Assessment. Cutting Edge for the Twenty-First Century*. Cambridge University Press, Cambridge.

Wood, C. (1995) *Environmental Impact Assessment. A Comparative Review*. Longman, Harlow.

Strategic environmental impact assessment

Therivel, R., Wilson, E., Thompson, S., Heaney, D. & Pritchard, D. (1992) *Strategic Environmental Assessment*. Earthscan, London.

Sustainable development

Smith, L.G. (1993) *Impact Assessment and Sustainable Resource Development*. Longman, Harlow.

There are also a number of journals that contain relevant material, including:
The Environmental Professional
Impact Assessment and *Project Appraisal* (Now merged to form *Impact Assessment and Project Appraisal*)
International Journal of Sustainable Development and World Ecology
Journal of Environmental Management
Journal of Planning and Environmental Management

Also, the EIA newsletters produced by the EIA Centre at the University of Manchester.

3 Scoping

3.1 Introduction

Scoping is all about ecological impact assessment (EcIA) design. This chapter outlines the factors that should be taken into account when setting study limits and deciding what methods to use. The importance of scoping cannot be over-emphasized: get it wrong, and important ecological components and effects may be missed from the EcIA entirely, or only discovered when it is too late to do anything about them. There is no one, clear recipe for effective scoping. To a large extent, your approach will depend on what is already known about a proposal and its receiving environment.

Information about 'what is where and why it is there' provides the basis for all subsequent exercises in prediction. In an ideal world, we would be in possession of full knowledge about the distributions of ecological components and about the factors explaining that distribution. This would enable us to characterize baseline conditions straight away and to predict ecosystem conditions in the absence of any proposed action. We would then have a detailed first layer for EcIA against which different impact scenarios could be analysed. In fact our knowledge is invariably limited and patchy. One of the first main tasks in EcIA is to ascertain the limits of knowledge about potentially affected ecological receptors and to design studies that will plug some of the gaps in our baseline knowledge of the receiving environment.

In summary, what is required is:
• an interpretation of the proposal and its associated sources of ecological stress or disturbance ('stressors');
• information about potentially affected ecological 'receptors' (their spatial and temporal distributions).

The information needed to establish the spatial and temporal distributions of potentially affected ecological receptors may come from a variety of sources. In some cases, existing knowledge may be adequate. In others, it may be necessary to initiate new surveys or inventories just to get enough information to ensure that all relevant aspects of potentially affected ecosystems are taken into account in the EcIA. Considered together, information about the proposal and ecological receptors in the 'receiving environment' provides the raw material for impact screening, carried out to establish the range of likely interactions between potential stressors and ecosystem components (see section 3.5). The results of impact screening are an important input into the focusing procedures that form the subject of the following chapter. These are used to select the valued ecosystem components (VECs), which merit more detailed study and will form the focus of ecological impact assessment and evaluation (see Chapter 4).

3.2 Deriving EcIA study limits

Appropriate study limits are the product of effective scoping. EcIA studies must be structured to take account of the full range of potential stressors and ecological outcomes. Deriving suitable study limits is invariably complicated and it is generally necessary to review and refine them as knowledge is gathered. The type and intensity of study required will differ in every case, as will the VECs selected for detailed study, but regardless of whether field surveys, literature searches or remote sensing are being undertaken, it will be necessary to take certain constraints into account when deriving appropriate study limits for EcIA. These are likely to include the following (after Beanlands & Duinker 1983).

- Proposal-driven constraints: activities associated with the proposal, their scale and distribution in time and space.
- Ecological constraints: distributions of habitats, species and physical factors, ecological linkages and relationships, locations of key features, scales of variation, vulnerability and likely exposure to proposal activities, etc.
- Technical constraints: deriving from limits in our abilities to predict and measure change, effectiveness of survey and sampling techniques, availability of suitable models, etc.
- Administrative constraints: social, economic or political factors, for example timeframes for determination of development consent, county boundaries, funds available for impact assessment studies, etc.

In addition, it is always important to bear in mind the evaluation criteria that are to be used and to consider how ecological information will ultimately be used in the decision-making process.

Ideally, all constraints should be considered together to derive appropriate boundaries or limits for ecological study and both spatial and temporal factors should be considered. Clearly, to some extent the scope of EcIA is predetermined: it depends to a considerable degree on the spatial and temporal limits of a proposed action and the likely distribution and scale of resulting stressors. Beanlands and Duinker (1984) preferred a 'resource-based' or 'ecosystem-up' approach, starting with identification of VECs in a general area of search and working back towards the proposal to search for possible impacts. At the end of the day, however, both the characteristics of the proposal and the characteristics of the receiving environment must be known and used to derive appropriate study limits.

If study limits are determined solely on the basis of administrative and technical constraints, important ecological impacts are likely to be missed, as administrative boundaries, and those of the proposal(s) are unlikely to coincide with ecological boundaries. Ideally, ecological assessment should transcend such artificial limits and reflect the range and limits of ecological processes. Similarly, timescales for determination of development consent are often incompatible with those of ecological study. Repeated surveys may be needed to establish patterns and limits of seasonal variation in population numbers, activity or habitat use. The timing and duration of key stages in the implementation of a proposal should therefore be considered in relation to the periodicity of ecological change and key life-history stages of any species that may be affected. While a certain amount of pragmatism will always be necessary, the timeframe of EcIA should not be determined solely on the basis of administrative requirements, nor should it be constrained within schedules determined for other impact assessment studies.

While 'ecosystem approaches' to EcIA are often recommended, the concept of the ecosystem is of no value unless ecosystem components and limits are clearly defined. An 'ecosystem' can be defined simply as an interdependent body of living organisms and their physical environment and the term is often used very loosely. One way to delimit ecosystems is to identify boundaries that correspond with barriers to resource flow. Merriam *et al.* (1993) cite the example of a river, which 'may act as a barrier to the dispersal of small mammals and therefore delimit a boundary to the flow of genetic variation between populations'. It is not always easy to delineate meaningful ecosystem boundaries, however. Most barriers and boundaries are like membranes in organisms, which vary in their permeability or resistance to flows (Wiens *et al.* 1985). Watershed boundaries, for example, can be crossed by some species of wildlife or by atmospheric pollutants. It will never be possible to account for all cross-boundary flows or to delineate ecosystems perfectly, but every effort should be made to 'capture' or encompass as large a proportion of essential ecosystem processes as possible. In simple terms, the study area for EcIA should include all those areas where biological changes related to the proposal under study are expected to occur (US Fish and Wildlife Service 1980).

Resources that might be considered in deriving 'ecologically appropriate' study limits include:
• environmental/physical media (water, air, geological strata, topography, climate);
• soils, vegetation;

- biotopes/ecoregions (landscape 'units');
- habitats;
- species, distinct genotypes;
- communities;
- populations;
- individuals (e.g. for protected species).

It is important to take account of scales or levels of ecosystem organization. For larger vertebrates, for example, consideration of home-range size can be useful in setting spatial limits for study areas. As a simple example, a study area of 10 ha would be inadequate to address potential impacts on the black bear *Ursus americanus*, but might be more than adequate to study effects on red-backed voles (Stiehl 1994). Some wildlife species range over vast areas. The threatened species of monkey, the Yunnan snub nose *Rhinopithecus bieti*, in the Baimaxueshan Nature Reserve in Yunnan Province, China, has a home range of about 25 km^2 and covers about 1500 m or so each day to collect the fruticose lichen *Bryoria*, which forms its diet (Small 1997). Study areas would therefore have to be at least 25 km^2 in size, and would probably need to be larger to take account of potential indirect effects. (A home range can rarely be regarded as a 'closed system'!) An even more extreme example was provided by Purves *et al.* (1992) who concluded that recolonization of Banff National Park in Canada by wolves *Canis lupus* appeared to depend on an ecological 'neighbourhood' extending from Kluane in the Yukon, to Montana. Wildlife habitat use often transcends national and administrative boundaries: many migratory birds, for example, use habitat in different countries.

Resources for EcIA are often limited, resulting in superficial surveys and the limited use of appropriate technology, such as geographical information systems (GIS), which might assist in deriving appropriate spatial boundaries for EcIA or predicting landscape-scale effects. Technical constraints may also limit the effectiveness of EcIA. For some species there are no known effective sampling techniques and there is insufficient existing knowledge to be able to predict their population responses with any degree of confidence, but this is not a valid reason for excluding species from a study.

The process of deriving study limits should always be sufficiently flexible and iterative to take account of any new issues or information that emerge during an EcIA, particularly in situations where information about ecological and proposal parameters is limited at the outset. Often a phased approach is necessary: where existing ecological information is limited, preliminary field surveys may be undertaken simply to evaluate the need for further, more detailed studies. EcIA design will also require adaptation in the light of which VECs are selected (see section 4.1). This influences not only the geographical limits of any studies carried out, but the level or intensity of study required to permit effective prediction of impacts.

3.3 Characteristics of the proposal

Reliable information about the impact-generating aspects of a proposal is crucial and it is important that this information should be provided (to ecologists and decision-makers alike) in such a way that its potential ecological implications can be understood. Accurate information is therefore required about the nature of a proposed action, its location, its size or extent, its intended lifetime and so on. Review of UK environmental statements (ESs) suggests that this information may not always be taken into account when identifying potential ecological receptors. In a review of proposed road developments, for example, only 3% of ESs actually specified the width and length of the proposed scheme (Treweek *et al.* 1993) and cited it as a factor in the derivation of study limits, yet it is clearly impossible to choose an appropriate width of survey corridor, or to determine which ecological receptors are likely to be affected, if the dimensions of a proposed road scheme are unknown.

Information likely to be required for effective EcIA is summarized in Box 3.1, where it can be seen that it will not be enough to know how much land will actually be occupied by infrastructure. Information will also be needed about the likely range, magnitude and duration of associated activities, emissions and disturbances. Often, it is necessary to

---

Box 3.1 Information about a proposal required for effective EcIA.

- Location(s)
- Size/areal extent
- Spatial organization
- Timetable and proposed duration of each phase (e.g. construction, operation and decommissioning)
- Potential sources of impact
- Emissions (type, volume, range)
- Receiving environment for emissions (water, soil, air)
- Disturbances: extent, magnitude, duration
- Best- and worst-case operating conditions

To take any widespread or cumulative effects into account, information on the development context will also be required, including:

- Alternative sites
- Alternative designs
- Past, current and future proposals
- Generation of other associated/connected developments

draw on numerous studies in the literature to obtain this information. For example, one study has been shown that application of salt to roads for de-icing can increase soil salinity up to 15 m away from the road itself (Jones 1981). Other research carried out in the Netherlands has established 'effect distances' for reductions in density of breeding birds in woodland and grassland habitats adjacent to roads (Reijnen *et al.* 1995b). For all species studied, the 'effect distance' and associated density reduction were 46 m and 34% in wooded areas and 710 m and 39% in grassland. The largest effect distances were found in open areas and were close to 1000 m for some species. To identify the causal factors for density reduction, Reijnen *et al.* (1995b) studied the range of possible causal factors to identify those most likely to be responsible for observed

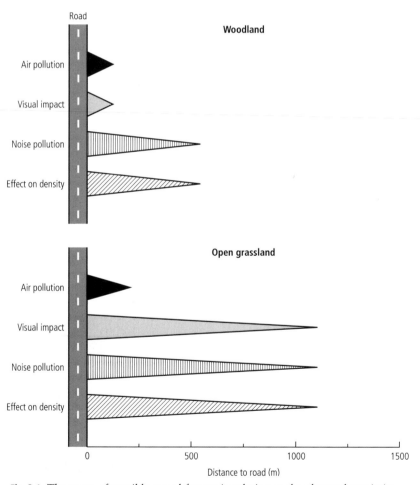

Fig. 3.1 The range of possible causal factors in relation to the observed maximim effect distance for reductions in density of breeding birds. (After Reijnen *et al.* 1995b.)

density reductions. Figure 3.1 shows how these ranges compare with the observed maximum effect distance. They concluded that noise was the key factor linking traffic levels and density of breeding birds, although visual disturbance was a contributory factor in more open areas. Similar studies are needed in other countries to characterize the relationship between levels of road traffic, associated disturbance and effects on birds and other species in the vicinity. Likewise, there are many examples of studies on the barrier effects of roads, but it is not yet possible to quantify barrier effects for all species in all situations.

Study boundaries should be drawn to include any areas of direct impact and also any areas of secondary or indirect impact. For major roads, for example, the zone of direct impact could extend up to 1000 m on either side of the carriageway. If indirect effects are also taken into account (e.g. the effects of displaced individuals on the occupancy of alternative habitat), then the 'effect zone' could be considerably greater. For a reservoir development it would be necessary to include the area of flood basin actually occupied by the proposed reservoir, any down-stream stretches of rivers that might have altered flow or increased turbidity (Stiehl 1994) or other areas where land-use changes might occur as a result of dam construction. Areas affected indirectly will not necessarily be contiguous with areas of direct impact. Figures 3.2 and

Fig. 3.2 Reservoir upstream of Canadian dam constructed for generation of hydro-electric power.

Fig. 3.3 Channelization of river and disruption of river flow downstream of Canadian dam constructed for hydro-electric power generation.

3.3 show the area flooded by a dam in Canada for hydro-electric power generation and some of the downstream impacts of channel alteration and reduced flow. The overall area affected is considerable and includes land far removed from the reservoir itself. There are cases where it may be difficult to establish the full extent of potential impacts prior to development. It makes sense to err on the side of caution and incorporate more than might be necessary into the study area so that field studies can be relied on to establish areas of impact and 'no impact'. Likewise, the timeframe of EcIA studies should be set to take account of the full schedule of development, including preparation, implementation, operation, decommissioning and follow-up.

Although information on proposed locations, designs, processes, duration, etc. is essentially an external input to EcIA, segregating the processes of development planning and EcIA is a recipe for inflexible approaches to design, impact assessment and mitigation. Effective dialogue between engineers and ecologists should take place as early as possible in the design and development of a proposal, to ensure mutual understanding of constraints. All too often, delayed awareness of potential ecological constraints means that, by the time they are acknowledged, design implementation has progressed too far for cost-effective modification to be possible. Ecological mitigation measures then become necessary. These are less likely to guarantee the conservation of natural resources and can be expensive to implement. On the other hand, early consideration of options for mitigation improves the chances of implementing an effective mitigation scheme. Neglect of key issues in the early stages of impact assessment can also result in expensive delays to development programmes and the gathering of data that is of little diagnostic value.

3.4 Characteristics of the receiving environment

Before any consideration of potential impacts is possible, a certain amount of background knowledge is required about the ecosystems and ecosystem components likely to be affected in the 'receiving environment'. A variety of information sources may be drawn upon to obtain the necessary information about potential ecological receptors, from existing literature sources and databases to new surveys. The desired output is an ecological characterization of the receiving environment that will permit:
- mapping of distributions of potential receptors in space and time;
- mapping of ecologically relevant units for impact assessment and/or monitoring;
- derivation of ecologically appropriate study limits;
- impact screening and the identification of valued environmental components.

In situations where existing information proves inadequate, a phased approach to characterization of the receiving environment will be necessary, preliminary studies being used to:
- identify survey needs to fill current knowledge gaps;
- provide a suitable framework for survey/inventory;
- determine the level or intensity of study required to characterize the environment as a basis for impact assessment.

The first requirement is to establish simply which resources, ecosystems or ecosystem components occur within the 'impact zone' and to derive provisional study limits based on knowledge of their ecology and approximate distributions. Ecosystems can be inventoried and studied using either 'bottom-up' or 'top-down' approaches. 'Bottom-up' approaches start with inventory of the abiotic factors that influence distribution patterns, such as climate, surficial geology, topography or air quality. A top-down perspective, on the other hand, starts by studying the behaviour and dynamics of 'target', 'key' or 'evaluation' species. For example, a top-down approach might be used to address 'the control exerted by top predators on the population dynamics and distributions of species below them in the food chain' (Power 1992). Merriam *et al.* (1993) recommended the use of both in parallel to characterize the ecological 'state' of Canadian national parks. It is then possible to link physical environmental attributes with the behaviour of organisms in a meaningful way.

The starting point for EcIA in practice is generally the collation of existing information to gain an appreciation of the range of variation present and to find out what is already known about the ecology of the receiving environment in the notional study area. This may involve use of existing ecological databases and maps, the scientific literature, consultation with specialists, local anecdotal knowledge or the results of preliminary field surveys or reconnaissance. The following types of ecological information are likely to prove useful in characterizing the receiving environment and structuring EcIA studies:

- ecosystem/ecoregion/biotope/biogeographic classifications;
- species/habitat/ecosystem relationships or affinities;
- species distributions and status;
- historical distributions or management.

It is always dangerous to exclude any possibility without verification in the field, but existing knowledge of ecological associations and distributions enhances efficiency by providing consistent, objective frameworks for EcIA and reliable baselines of basic information. Once the biological systems found in an area have been classified and inventoried, it is possible to gain some indication of the associations that can be expected between organisms and particular sets of physical conditions, biogeographic regions, biotopes or even countries. In some countries, such as the UK, there are relatively well-developed national databases of species distribution that can be drawn on by EcIA practitioners, although they do not always summarize distributions at the required scale. Finally, ecosystems are dynamic and must be understood in terms of both past and present states and their current and historical management. This is fundamental to the establishment of baseline conditions.

3.4.1 Ecosystem classifications

A robust system for classifying land into biologically relevant categories is required. Various approaches have been used, but most systems for ecosystem classification differentiate classes or types using information on climate, geology, topography and vegetation. Common to all systems for ecosystem classification and mapping is the integration of information about abiotic variables with information about the associations and distributions of organisms.

In some countries and regions, considerable effort has been invested in ecosystem classification and the establishment of databases of ecological information that can be drawn upon by EcIA practitioners. Characterization of the receiving environment is considerably easier where consistent, formal methods for ecosystem mapping are well established. Ecosystem mapping is a natural follow-on from ecosystem classification, which makes it possible to stratify the landscape into homogeneous map units based on ecological criteria. It is discussed in more detail in section 9.2.2. The US Forest Service (Bailey 1980) and the province of British Columbia (Demarchi 1993) both use the concept of 'ecoregions' to identify geographical areas with similar biological characteristics. British Columbia's ecoregion classification is intended to provide a systematic view of broad biogeographic relationships within the province (Demarchi 1993). It evolved from recognition that soils and vegetation are a function of factors like climate, geology, relief and time as well as the activity of organisms (Jenny 1941; Major 1951) and it is based on the identification of permanent landscape units that characterize or determine the distribution of plant communities at the landscape level. These landscape units are defined using information on geology, terrain, topography (slope and aspect), moisture regime and nutrient regime. There is a three-level ecosystem classification hierarchy (Table 3.1), including ecoregion units, biogeoclimatic units and ecosystem units (Ecosystem Working Group 1998). The parameters used to define these levels of ecosystem classification are summarized in Table 3.2.

Often, the concept of 'habitat' is relied on to 'make the link between species and the physical environment' (Beanlands & Duinker 1983). The US habitat evaluation procedure (HEP: US Fish and Wildlife Service 1980), for example, uses 'habitat' as the basic unit of assessment for predicting impacts on wildlife (see section 5.4.4) and a phased habitat survey approach has been developed by the UK's Joint Nature Conservation Committee (1993). The prevalence of habitat-based approaches to characterization of the receiving environment derives largely from their relative ease, efficiency and cost-effectiveness. It is generally much more straightforward to describe and quantify distributions of habitat than it is to inventory or count populations, or to

Table 3.1 Summary of the hierarchy of ecosystem classifications used for ecological mapping in British Columbia. (After Ecosystem Working Group 1998. Copyright of the Province of British Columbia.)

Ecoregion units	Biogeoclimatic units	Ecosystem units	Map scale
Ecodomain			1 : 30 000 000
Ecodivision			1 : 7 000 000
Ecoprovince			1 : 7 000 000
Ecoregion			1 : 2 000 000
Ecosection	Zone		1 : 2 000 000
	Subzone		1 : 250 000
	Variant	Broad ecosystem units	1 : 250 000
	Phase	Site series groups	1 : 250 000
		Site series	1 : 50 000
		Specific ecosystem units	1 : 5000–1 : 20 000

Table 3.2 Examples of physical and biological factors considered when classifying ecosystems in British Columbia. (After Demarchi & Lea 1989. Copyright of the Province of British Columbia.)

Classification level and map scale	Parameters used to define ecosystem units				
	Climate	Landform and geology	Soil	Vegetation	Associated fauna
Specific ecosystem units 1:5000– 1:20000	Specific microclimates	Specific landforms and parent materials	Soil series (many classes)	Plant communities (e.g. successional stage, including stand structure)	Specific habitat use as influenced by distribution, range and social behaviour
General ecosystem units 1:50000– 1:100000	Detailed level mesoclimates	Local landforms, including topography (slope, aspect) and parent materials	Soil subgroups (few classes)	Plant communities (potential structural stages, including climax)	Units of potential and current habitat use
Broad ecosystem units 1:250000	General level mesoclimates	General landforms (including topography, slope, aspect)	Soil 'great groups'	Broad plant communities (potential structural states, including climax)	Broad units of potential and current habitat use
Ecoregion/ biogeoclimatic units 1:250000	Climatic regimes, macroclimates	Subdivision of regional physiography to represent groups of local landforms	Soil 'great groups'	Climatic climax communities	Broad distributions, e.g. belts of seasonal habitat use by migratory species

study the population dynamics of a species directly. In effect, the concept of 'habitat' bridges top-down and bottom-up approaches to ecosystem study, by linking abiotic factors to the behaviour, distributions and dynamics of associated species. In countries where habitats have been classified using consistent systems and their associations with species have been well studied and inventoried, it is possible to infer a great deal about potential species distributions from habitat maps.

For terrestrial ecosystems, descriptions and classifications of habitat are predominantly vegetation based. Plants and their communities ('vegetation') form mosaic patterns made up of distinct patches or phases in community type and development, which evolve through responses to both abiotic factors and biotic interaction. As plants respond directly (and often predictably) to abiotic factors they are often an effective descriptor for large-scale biogeographic or landscape patterns. One warning note, however: many plant species have not occupied all their potential environment, so it can be misleading to draw conclusions about environmental conditions based on the absence of a given species (Billings 1952).

Most systems of ecosystem mapping use vegetation distributions as a key factor in discriminating between ecosystem units. Not only is vegetation relatively static and therefore easier to survey, but new information technologies are making it relatively easy to map vegetation patterns directly using remotely sensed data. In general, vegetation forms the 'best understood, most general and repeatable patterns on any landscape and in any biogeographic region in response to abiotic factors (Hunter & Price 1992 in Merriam *et al.* 1993). Biogeographic zones reflecting broad-scale responses to climate, for example, are primarily defined on the basis of characteristic vegetation types. Vegetation also constitutes a significant part of the habitat of many species. Other trophic levels depend on primary production by plants and 'hierarchy theory suggests that since vegetation patterns are large scale, they will constrain ecological functioning at smaller scales' (Allen & Starr 1982 in Merriam *et al.* 1993). As well as indicating something about abiotic conditions, therefore, vegetation can also be used to predict possible associations with other species. Species that rely on vegetation for habitat, cover or food are affected by the spatial organization of vegetation (types and amounts), sometimes in predictable ways.

In some cases the spatial configuration of habitat may have more impact on population dynamics than the total amount of habitat available (Lefkovitch & Fahrig 1985; Pulliam & Danielson 1991; Taylor *et al.* in press). The science of landscape ecology can be used to study the relationships between spatial configuration of habitat in the landscape and the ecological patterns and processes that determine the dynamics of associated species.

In terms of habitat provision at the landscape level, diversity can be measured in terms of the number of different vegetation patch 'units' and their spatial arrangement or organization. Predicting the effects of qualitative and quantitative changes in the vegetation mosaic (caused directly by removal or indirectly through altered ecosystem processes) on the identity, abundance and distribution of associated species is an important part of EcIA, and this is an area where 'landscape ecology' has much to offer. However, it is important not to neglect the importance of factors and processes that may not be easily detected, recognized or classified at a 'landscape level'. For many species, vegetation structure is just as important as vegetation composition in the provision of 'habitat', and for smaller species such as invertebrates, broad-scale mapping of vegetation is unlikely to reveal very much about the identity or abundance of associated species. Vegetation-based habitat mapping is generally easiest and most effective for larger vertebrates. This is partly because broader classifications are simpler and easier to map using remotely sensed data and partly because knowledge of the habitat needs of these species is more extensive and consistent.

3.4.2 Species–habitat relationships

Relationships between organisms and their 'habitat' may be inferred directly from observed associations or estimated by drawing on published information about species' needs or life requisites. Combined approaches may also be used to model potential distributions.

Published information (or expert knowledge) can sometimes be used to derive 'rules of thumb' about the likely occurrences of species in certain areas or habitats. In the UK, there has been concern for some time about the potentially adverse effects of afforestation with non-native conifers on the wildlife of upland moorlands. Afforestation has been particularly widespread in Scotland and adverse impacts on breeding bird populations have been reported. Harris (1983) drew on information derived from a number of bird censuses carried out in coniferous plantation woodlands in England, Wales and Northern Ireland in an attempt to characterize the use of coniferous plantations by birds. These censuses had been carried out using a number of different methods, so direct and detailed comparisons were difficult. Nevertheless, it was possible to detect certain general patterns in the use of coniferous woodland by birds. For example:
• some species never use coniferous woodland either for feeding or breeding;
• woodlands at lower altitude generally hold more species than those at high altitude;
• woodlands in the south hold more species than those in the north;
• woodlands in the east hold more species than those in the west;
• the widest variety of species is found where there is diversity of habitat;
• in plantations and woods of comparable size, shape and isolation, densities are highest in mixed coniferous/deciduous woods, followed by oak, birch, spruce, larch and planted pine;
• species richness is highest in mature woodland and in woods at the 'thicket' stage.

Without field survey, it was therefore possible to identify that pool of bird species most likely to be lost from newly afforested areas or to colonize them. It was also possible to draw general conclusions about the likely species richness or diversity of afforested areas compared with open moorland. Afforestation of moorland in the uplands is clearly likely to have most impact on those species that are restricted to moorland habitat and are unable to colonize forested areas. Again, drawing on published data, Harris was able to suggest which bird species might be able to escape the effects of afforestation because they breed largely above the tree-planting limit: species such as the snow bunting *Plectrosphenax nivalis*, for example. Some of the information on habitat use and autecology required to predict which species would be able to remain and breed on afforested land and which would be displaced was therefore available in the published literature and could be used to provide a framework for a more refined approach.

In an attempt to generate a national database of information on the associations between species and habitat, the UK's Natural Environment Research Council constructed a 'Biotopes Occupancy Database' (Eversham *et al.* 1992). The database was compiled using existing British published sources, information from the literature being supplemented and validated by relevant taxonomic experts. The database was also evaluated using species lists from known biotopes. The approach was applied to butterflies and it was confirmed that the initial assignment of species to biotope, using the taxonomic literature, was effective in most cases (Loder 1992). The Biotopes Occupancy Database was constructed to contain all the Red Data Book species for which adequate distribution data existed, all the nationally scarce vascular plants, all the breeding birds, mammals, reptiles and amphibians, a wide range of invertebrates (including butterflies, grasshoppers and crickets, dragonflies and macro-moths) and a selection of bryophytes and vascular plants characteristic of certain, well-defined natural habitats. The intention was to derive measures of species fidelity to defined biotopes. Some species show a very high fidelity to biotope, whereas others are more 'generalist'. Clearly such information is immensely valuable as an aid in evaluating the importance of a biotope to a species and the relative seriousness of its destruction or loss.

By compiling published information and existing knowledge about re-lationships between species and habitat, the usefulness of reconnaissance surveys can be greatly increased, and survey methods can be selected that are likely to detect the species generally associated with the habitats that have been identified, even if they are not detected following a first look. For purposes of EcIA, knowledge of theoretical biotope affinities is therefore particularly useful if it can be callibrated for available local or site-specific data on the location and distribution of biotopes. In other words, if biotopes can be classified at a scale appropriate to the use of interpreted aerial photography, satellite imagery or field survey for mapping habitats.

For example, British Columbia's ecosystem classification combines ecosystem mapping with use of published information to derive pre-dictive algorithms that clearly establish the expected relationship between mapped ecological parameters and species' needs or measurable life requisites. The system is used as a basis for wildlife interpretations, primarily to estimate habitat potential ('capability') and its actual or current 'suitability' for associated fauna. To some extent the approach mirrors that used to develop habitat suitability indices (HSIs) for use in the US Fish and Wildlife Service HEP (see section 5.4.4). Relationships between life requisites and mapped habitat attributes can be used firstly to predict species presence or absence. If habitat or life requisite relationships are sufficiently well defined and mapping detail is adequate, it may also be possible to predict relative abundance. Assigning ratings to the ecological units according to their relative importance for associated species provides a basis for predicting the likely implications of various management activities or stressors in a study area.

Species' habitat value ratings can be derived that reflect:
• the ability of the habitat (ecosystem unit) in its current condition (including any anthropogenic influences or effects) to support a given species ('suitability');
• the ability of the habitat in its optimal condition to support a given species (its capability).

As for HEP (US Fish and Wildlife Service 1980), ratings are based on habitat potential to support a particular species and compare habitat (whether defined for ecosections, biogeoclimatic units or ecosystem units) to the best available for that species in the province. They are also expressed in terms of carrying capacity, in this case based on a standard density measure: the number of individuals using $1\,km^2$ of habitat for a month.

The development of GIS technology has made it considerably easier to combine information about habitat preferences derived from the literature, descriptions of habitat provision (often derived from satellite imagery) and predictive modelling. Some examples are given in section 9.2.

3.4.3 Species distribution data

Direct knowledge of species distributions makes it easier to:
• predict which species are likely to occur in a given area of search;
• estimate species status for EcIA evaluation;
• place local impacts in a wider context;
• design appropriate EcIA studies;
• select suitable survey methods.

Figures 3.4 and 3.5 are national distribution maps for two species of sedge found in lowland wetlands in the UK: *Carex elata* and *Carex aquatilis*, respectively (plotted by the Biological Records Centre (Institute of Terrestrial Ecology) on the basis of data collected in national surveys of aquatic and scarce species (Stewart *et al.* 1994; Preston & Croft

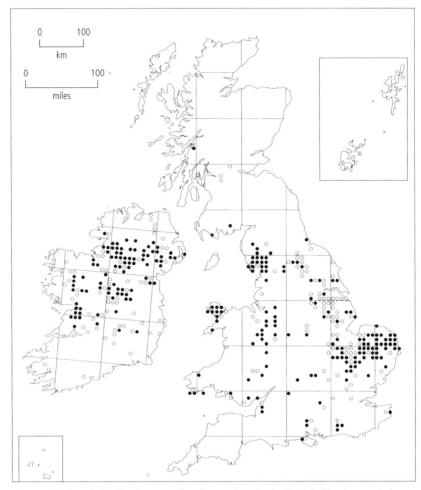

Fig. 3.4 National distribution map for *Carex elata* in the British Isles (open circles represent records up to and including 1969, filled circles refer to records from 1970 onwards). Plotted by the Biological Records Centre (Institute of Terrestrial Ecology) on the basis of data collected for national surveys of aquatic and scarce species (Stewart *et al.* 1994; Preston & Croft 1997).

1997)). Using such maps, it is possible to screen out, at a glance, those areas of the country where these species are highly unlikely to be found. The slender tufted sedge *Carex elata* is found in marshes and beside water bodies predominantly in England, Wales and Ireland, with only a few isolated records in Scotland. The water sedge *Carex aquatilis*, on the other hand, is found predominantly in Scotland, with only a few records in northern England and some in Wales and Ireland. This reflects their world range, as *Carex aquatilis* is a boreo-arctic species and *C. elata* a temperate species (*sensu* Preston & Hill 1997). The distribution of both species has declined over the years. If reasons for decline are

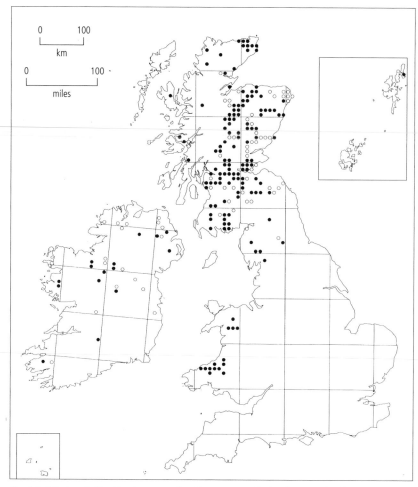

Fig. 3.5 National distribution map for *Carex aquatilis* in the British Isles (open circles represent records up to and including 1969, filled circles refer to records from 1970 onwards). Plotted by the Biological Records Centre (Institute of Terrestrial Ecology) on the basis of data collected for national surveys of aquatic and scarce species (Stewart *et al.* 1994; Preston & Croft 1997).

known, information about historic locations for a species can be a useful guide to the likely availability of potential sites for restoration (often an important factor in mitigation planning).

The Convention on Biodiversity has given particular impetus to the development of national species-recording schemes. Without reliable information on species distributions and status, rational approaches to the conservation of biological diversity are impossible. However, the quality and coverage of data on species distributions varies considerably between countries and locally within them. Investment in comprehensive, national, up-to-date biological recording is often limited, despite its proven benefits. There is also very uneven coverage of different groups and species, with an almost universal bias towards the more charismatic. The frequency of mention of bird species in UK ESs, for example, greatly exceeds references to slugs and nematodes! It is therefore important to establish the extent of any under-recording when using existing species distribution data, particularly when making population estimates. This is especially the case in countries where national biological recording is under-resourced and a large number of species may not even be officially identified and named (Natural Environment Research Council 1992). Estimates of the total number of species in the world fall between 3 and 30 million (May 1990), of which only about 1.8 million have been identified (Stork 1988). Invertebrates are particularly 'under-recorded'. Shortage of time and resources for comprehensive new surveys makes it essential that all relevant existing information should be collated. Awareness of what information does exist is therefore always important. Directories of ecological information sources can be a valuable guide, such as the county-by-county directory produced by Donn and Wade (1994) in the UK. A directory of record repositories in the UK was also published (Anon. 1979) but has not been updated recently.

Always bear in mind that information drawn from existing literary sources, from contacting local conservation groups or from biological records centres may have limitations with respect to EcIA: it has often been gathered for other purposes and may not be up to date.

3.4.3.1   *Geographical representation or range*
There are clear differences in geographical representation or range for most groups and species. Any field surveys must therefore be designed to detect species characteristic of the area affected by a proposed action. Taking account of inherent natural variation in species distributions is also important in planning mitigation, when, for example, the use of plant species or genetic strains of local provenance to restore damaged areas can play a part in maintaining locally characteristic vegetation.

In the UK, many invertebrate species have highly localized distributions. Several rare butterflies are confined to southern England where they are effectively at the edges of their European distribution. Some even have disjunct distributions, for example being found in north-west Scotland and in southern England, but nowhere in between (Ravenscroft & Young 1996). There are geographical trends, not only in the distributions of individual species but also in the numbers of species likely to be encountered (Schall & Pianka 1978). Geographical gradients have been observed in the species richness of most groups of organisms in the UK, running from south-east to north-west (Lawton *et al.* 1994), related to major environmental gradients of temperature, rainfall, geology and topography (Roy & Eversham 1994). Dragonflies, for example, show an

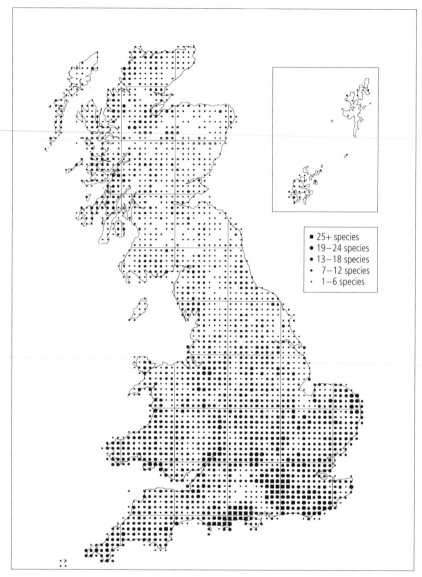

Fig. 3.6 Map illustrating northward and westward decline in dragonfly species richness in Britain (Institute of Terrestrial Ecology, Biological Records Centre).

almost uniform northward and westward decline in species richness (Fig. 3.6). The precise reason for this is unclear, although insolation may play a role (Hamilton-Wright 1983).

In common with most other parts of the globe, the British flora and fauna also show a strong (but not universal) decline in species richness at higher latitudes (Schall & Pianka 1978). Although general geographical trends such as these can be observed, there will always be exceptions. In the UK, for example, liverworts and ferns have a higher alpha-diversity in the north and west where rainfall is highest (Hill *et al.* 1994). Knowledge of which species have a random distribution in the UK and which show significant geographical gradients in their distribution can be very valuable in designing sampling or survey strategies in cases where existing site-specific knowledge is scanty.

Knowledge of national species distributions not only gives an indication of the areas in which a species is likely to be found, it can also be very valuable for estimating the relative importance of different areas in maintaining national populations. The golden plover *Pluvialis apricaria*, for example, is at the southern limit of its breeding range in Wales and on Dartmoor, in the south-west, but is relatively common as a breeding bird further north. Likewise the main breeding population of the twite *Carduelis flavirostris* is in Scotland. It is at the southern limit of its range in the Peak District, so nesting areas in this region are regarded as particularly important. 'Edge-of-range' considerations must be tempered by knowledge of whether populations are acting as sources or sinks. Clearly, a population that is at the edge of a species' range and is also acting as a source (providing colonists from a net population 'surplus') may be more valuable than populations that are acting as a sink (absorbing individuals 'over-produced' elsewhere in the species' range).

Information on the national status of a species should also be considered relative to other countries. For example, 90% of the north-west European population of the black-tailed godwit *Limosa limosa* is to be found in the Netherlands (Osiek 1986). The species is relatively common there, but very rare elsewhere in Europe. Maintaining the status of the Dutch population might therefore be crucial to the overall survival of the species. Similarly, the great crested newt *Triturus cristatus* is more common in the UK than elsewhere in continental Europe.

3.4.3.2 *Seasonal patterns*
For some groups there may be marked seasonal differences in distribution that should be taken into account. This is certainly the case for many bird species, which migrate between wintering and breeding habitat, and also for many plant species. In temperate regions, for example, there is little point in surveying for the ground flora of deciduous woodlands in the depths of winter.

3.4.3.3 *Sources of information*
Quality of information about species distributions varies considerably between groups. There are a number of very well-researched bird atlases, for example, which summarize historical and current information on the distributions of species, although often at a relatively coarse scale.

The European Bird Census Council's Atlas of European Breeding Birds (Hagemeijer & Blair 1997) summarizes distributions on the basis of 50-km squares, but at least provides a baseline of knowledge compiled using consistent guidelines. Compilation of reliable species atlases is a laborious and expensive task, often drastically under-resourced. The data used to produce the *Atlas of European Breeding Birds*, for example, are quite old (some collected in the 1970s) and have been collected over a number of years during which many further species declines have taken place. Nevertheless, increasing availability of 'atlas' data in digital form will greatly enhance the ability of EcIA practitioners to make meaningful judgements about the likely coincidence of proposal-related stresses with species, populations or communities.

3.4.4 Historical distributions or management

In most countries, it is impossible to ignore the impacts of past and current human activities. Anthropogenic effects can override many other determinants of ecological process and function and can have a major impact on the development of ecosystems. In a number of countries, including many of those in western Europe, wildlife habitats are predominantly 'seminatural'. They have not only evolved under anthropogenic influence, but depend on consistent management intervention for their continued existence. In such cases, characterization of current conditions can be greatly assisted by an understanding of past activities and their effects. In particular, knowledge of the historical circumstances in which ecosystems have developed can assist in determining whether an ecosystem is likely to be in a stable or fluctuating state under current circumstances or those likely to pertain following implementation of a proposal.

Ecologists may therefore benefit from prefacing study of any particular community or site with an historical review of how its present composition, character and function developed (Sheail 1983). The first priority is to find out what relevant information is available for a particular system or locality. Obviously the availability of recorded historical information varies considerably. For example, hardly any large-scale maps were compiled in England before 1500 (Sheail 1983). In the UK, the search for what is available is best started in the local studies sections of the principal public libraries and the appropriate county record office (Anon. 1979). University theses can also be very valuable. Occasionally, inventories of useful sources are published. It is a shame that the resources are rarely available to update these more regularly. For example, an inventory of all the natural history manuscript resources preserved in the British Isles outside the Public Record Office (PRO) was published by Bridson *et al.* in 1980, but has not been updated.

When a suitable source has been identified, it is important to establish the reasons for its original compilation (and preservation), so that you can decide how much reliance can or should be placed on the facts it records. In particular, it is important to gain some indication of the tolerance given to inaccuracies during compilation. The coverage and detail of recorded historical information is patchy and variable. Not only has much passed unrecorded, but 'what is known cannot be submitted to trial and experimental proof' (Sheail 1983). Nevertheless, knowledge of the past can assist in understanding present conditions and predicting how they might develop in the future.

In some countries, recorded histories go back a very long way. The first British survey with national coverage, for example, was the Domesday Survey of 1086, which includes information on land use and to some extent on land cover. The first national Land Utilization Survey was directed in the 1930s by Dudley Stamp and repeated in the late 1950s. The published maps are usually too small in scale to reveal sufficient information for detailed ecological study of land-use change (for example) with time, but such detail can sometimes be found in the manuscript sheets actually compiled in the field. These took the form of Ordnance Survey (OS) 1 : 10 560 sheets on which land use was recorded. Those maps of the First Land Utilization Survey which survive can be consulted in the British Library of Political and Economic Science. Despite their limitations, these surveys provide an invaluable baseline against which to compare general trends in land use for the whole country (Sheail 1983).

The importance of recording new information in such a way that it can be interpreted effectively in the future cannot be over-emphasized. New methods of digital storage of data have opened up unprecedented opportunities for consistent recording of environmental features over time. This can be of benefit in the short and medium term as well as the longer term, reducing the expense of large-scale monitoring considerably and making it possible to detect and quantify overall trends. Historical aerial photographs, for example, have proved particularly useful for detecting longer-term trends in vegetation cover. Increasingly, repeated satellite surveys are available and there are some projects underway to repeat national surveys originally carried out to classify satellite imagery, which can now be used to establish baseline trends, for example, in land cover. A notable example is the UK's Countryside Survey, which was carried out in 1990 and is to be repeated for the year 2000.

Where reliable historical data are available, they can be invaluable in helping practitioners to decide whether ecosystem components are stable in the study area, or undergoing prolonged declines. Clearly, ecosystem components that are acknowledged to be rare or threatened at present and that also appear to have been undergoing a steady decline are likely

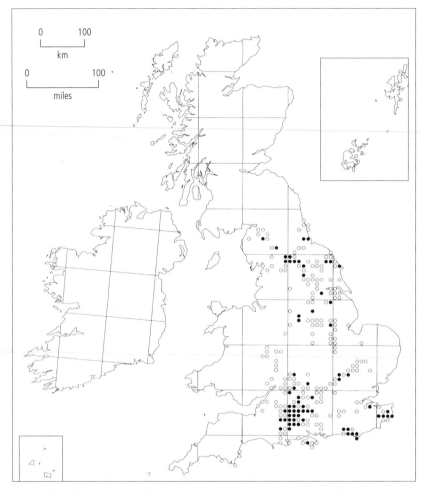

Fig. 3.7 Distribution of *Orchis ustulata* (open circles represent records up to and including 1969, filled circles refer to records from 1970 onwards). Plotted by the Biological Records Centre (Institute of Terrestrial Ecology) on the basis of data collected for a national scarce species survey (Stewart *et al.* 1994).

to merit careful study and to be selected as VECs (see section 4.1). Knowledge of historical species distributions can therefore be very useful. Figure 3.7 shows how the distribution of the burnt orchid *Orchis ustulata* declined from 1970 onwards (represented by the filled circles), disappearing from a number of its previously known locations in England (plotted by the Biological Records Centre (Institute of Terrestrial Ecology) on the basis of data collected for a national scarce species survey (Stewart *et al.* 1994)). *Orchis ustulata* has declined mainly as a result of adverse agricultural practices (such as ploughing, failure to maintain suitable grazing regimes, and the use of artificial fertilizers and herbicides) but also as a result of encroachment by building and destruction of sites through quarrying. Due to its continuing decline and the fact that it is restricted to calcareous sites with short grassland (which are themselves limited), burnt orchid would be an obvious choice for selection as a VEC in any case where it was known to be potentially affected.

3.4.5 Preliminary studies

In most cases, there will be insufficient site-specific and up-to-date information available to form the basis for an effective EcIA, and it will be necessary to carry out some sort of field survey or original study. Preliminary surveys may be needed simply to establish enough about the habitats and species that are present to derive suitable study limits for more detailed surveys to be undertaken at a later date. In these circumstances, it is important to ensure that there are clear criteria for interpreting results so that 'triggers' for more detailed investigation can be identified. The UK's Institute of Environmental Assessment, for example, produced guidelines for baseline ecological assessment

---

Box 3.2 Criteria for seeking additional information following preliminary field survey for amphibians and reptiles in the UK. (After Institute of Environmental Assessment 1995.)

Detailed information on amphibians and reptiles should be sought under the following circumstances
When preliminary studies indicate that a site contains (or has contained in the past) protected species, good assemblages of species or species on the edge of their range

For amphibians this means any of the following
- Sites with great crested newts *Triturus cristatus* or natterjack toads *Bufo calamita* (and the smooth newt *T. vulgaris* in Northern Ireland)
- Sites with four or more amphibian species
- Sites in an area where an amphibian is rare or at the edge of its geographical range

For reptiles this means
- Sites with sand lizard *Lacerta agilis* or smooth snake *Coronella austriaca* and, in Northern Ireland, common lizard *L. vivipara*
- Sites with at least three reptile species
- Sites in an area where a reptile is rare or at the edge of its geographical range

Further information should also be sought if preliminary field survey identifies apparently suitable habitat for protected reptile and amphibian species lying within the known range of the species, but for which no records exist

(Institute of Environmental Assessment 1995), which include criteria for deciding when detailed site surveys are likely to be warranted. Box 3.2 summarizes survey criteria for amphibians and reptiles in the UK. These reinforce the need for reliable information about the distributions and ranges of species that can be used to assess habitat suitability. Such an approach is useful because it supplies practitioners with consistent 'rules of thumb', which they can use to decide when to intensify their searches.

Sometimes (rarely), one-off surveys will be sufficient to provide all the information necessary for impact prediction. It is not uncommon for EcIA budgets to stretch to little more than a 'walk over' in which species lists are drawn up from chance encounters, rather than from any sort of systematic survey. There is actually relatively little information available about survey efficiencies for different techniques, habitats and species, and more research is required to establish what 'return' can be expected from a given level of investment or input. Otherwise important ecological components may be missed in preliminary surveys and omitted from subsequent study. Gaston (1996b) gives a number of examples

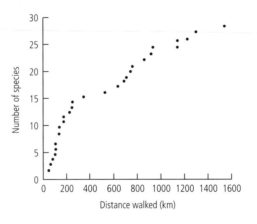

Fig. 3.8 Species accumulation curve for the number of snake species encountered by one observer in the INPA-WWF reserves near Manaus, Brazil (sampling effort expressed in terms of number of kilometres walked). (From Zimmerman & Rodrigues 1990.)

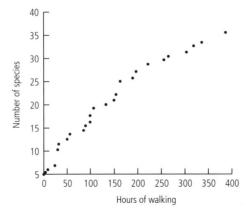

Fig. 3.9 Cumulative number of frog species recorded by one observer in the INPA-WWF reserves near Manaus, Brazil (sampling effort expressed in terms of hours of walking). (From Zimmerman & Rodrigues 1990.)

where increased survey effort has resulted in higher numbers of species being recorded. Figures 3.8 and 3.9 demonstrate how the distance walked by a recorder, or the hours spent walking, can result in marked increases in the numbers of species recorded (snakes and frogs, respectively).

A paper by Block *et al.* (1987) described how eight study sites were surveyed repeatedly until no new species were detected. One hundred and thirty person days were spent in total and the number of surveys needed per site ranged from 10 to 28. It is clear that 'one-off' surveys are prone to miss the majority of species present in an area of search. As well as suggesting that more rigorous surveys should be adopted as the norm, this emphasizes the importance of comprehensive national monitoring data that can be used to provide reference lists of species and habitats that are likely to occur or known to have been recorded in the area of interest.

In some circumstances, particularly for EIAs with a wide geographical coverage (for proposed new roads or pipelines, for example), it may be worth investing in new aerial photographs or satellite images. Use of remotely sensed data is considered in more detail in section 9.1.2. However, decisions to use remotely sensed data are best made early on in the EcIA process. Aerial photographs or satellite images can then be used to derive base maps, which not only benefit the presentation of results, but can help you to visualize spatial relationships between the proposal and ecological receptors or to structure field surveys so that relevant ecological processes are 'captured' at an appropriate scale.

3.5 Impact screening

3.5.1 Introduction

Just as it is important to ensure that study limits will permit the effective 'capture' of relevant biological changes, the ecosystem components selected to form the focus of study should provide useful information about those changes. Some form of 'impact screening' is generally required to identify key linkages between a proposal and potential ecological receptors so that the approaches taken to impact assessment and evaluation are relevant both to the proposal and to affected ecosystems. Impact screening requires 'decomposition' of proposals and the affected environment into their component parts to identify potential interactions between them. In other words, given what we know about the nature and distributions of proposed activities and potential ecological receptors, how and where are they likely to interact? There is little point in focusing the EcIA on ecosystem components that are unlikely to be vulnerable to proposal-related stressors, despite appearing to fall within their zone of impact.

Having identified 'what is where', it is possible to identify the extent to which the ranges and distributions of stressors and ecological receptors will overlap. This process of 'spatial screening' can be greatly assisted by use of GIS (see section 3.5.5). To predict ecological impacts, however (the subject of Chapter 5), it is not enough to know how ecosystem components are distributed and therefore which may be affected. It is also necessary to ask why things are where they are and to characterize the potentially impacted environment in terms of 'function', rather than just describing its key elements.

The complexity of possible interactions between ecosystem components is immense (Peters 1991) as demonstrated by a graphic representation of possible interactions between a plant (P) and its environment (Fig. 3.10). These environmental factors act on the plant and on each other, resulting in a vast range of potential relationships to address. Because it is never practicable to study the entire range of potential interactions in depth for purposes of EcIA, it is necessary to sift through the pool of potential stressors and ecological receptors to determine which are actually likely to interact with each other and to ensure that the range of potential interactions has at least been identified. This is often a conceptual exercise: the first step in narrowing down the limits of study to a more manageable subset of stressors, receptors and relationships. Chapter 4 outlines methods that can be used to refine this selection further, screening not only for stressors and receptors that will interact, but for

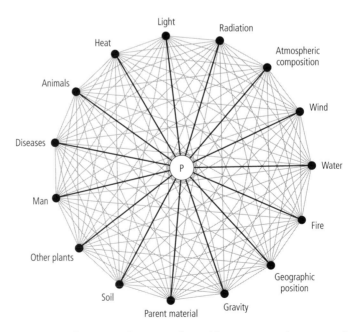

Fig. 3.10 Diagram illustrating the range of possible interactions between a plant (P) and a representative range of environmental factors. (Modified from Peters 1991.)

Table 3.3 Selected ecological stressors and impacts.

'Stressor' or source	Possible consequences
Direct removal or destruction of vegetation cover, e.g. prior to construction	Vegetation cover is lost. Flora is destroyed and associated species are displaced, temporarily or permanently
Soil is removed, e.g. for construction or prior to opencast mining	Soil/plant associations break down, perhaps irreversibly. Profound effects on soil flora and fauna (especially mycorrhizal associations). Any restored vegetation likely to undergo alternative succession and develop new associated fauna
Soil structure is altered, e.g. through compaction or erosion as caused by heavy machinery	Any restored vegetation will be different, often less productive
Soil compaction or introduction of hard cover results in increased run-off, hydrological disruption and sometimes pollution of watercourses	Terrestrial cover is removed. Soil biota is restricted. Water quality and aquatic flora and fauna may be affected
Soil erosion (caused by removal of vegetation cover) results in sedimentation in watercourses	Water quality (e.g. oxygen availability) and associated flora and fauna affected
Soil contamination (may occur during construction or operation)	Effects depend on contaminants and susceptibility of receptors. Short- or long-term, reversible or irreversible effects are possible. Profound effects on vegetation and associated fauna are possible
Hydrological change (lowering or raising of watertable)	Can have implications for vegetation composition and for composition of associated fauna. In extreme cases, whole wetlands can be lost
Hydrological change: altered watercourses, canalization of watercourses, in-filling of water bodies	Aquatic ecosystems altered or removed. Note possibly important 'downstream' consequences
Introduction of linear infrastructure, which acts as barriers to species movement	Can reduce genetic diversity and population viability. Can also reduce overall carrying capacity of regions
Introduction of infrastructure that alters animal behaviour, e.g. through visual or noise disturbance	For example, a road across an estuary might disrupt flocking behaviour in bird species that require 'open-vista' habitats to aggregate
Operation of linear infrastructure, e.g. traffic on roads	Can cause mortality of individuals, altered population sizes and age structures. A significant factor in causing barrier effects, which may affect regional and local biodiversity. Can also alter behaviour patterns, e.g. roads may reduce densities of breeding birds by interfering with territorial behaviour
Operational infrastructure disrupts access or traditional nomadic routes, resulting in reduced ability to maintain 'traditional' management	Isolated habitat areas alter due to lapsed 'traditional' management, e.g. scrub encroaches on abandoned grassland, or increased grazing pressure on reduced area causes botanical change

Continued on p. 76

Table 3.3 (*Continued*).

'Stressor' or source	Possible consequences
Pollution of air	Can have widespread effects, often chronic and/or cumulative. May disrupt breeding success or simply reduce the vigour and viability of individuals. Can have indirect effects on soil, e.g. nutrient enrichment due to nitrogen deposition
Climate change	Widespread and various effects, difficult to 'pin down'
Replacement of characteristic, local 'habitat' with 'poor replicas' (often a consequence of inappropriate mitigation)	Reduced habitat diversity, e.g. at the landscape level (associated with reduced biological diversity at other levels in organization hierarchy)

ecosystem components that can be regarded as 'valuable' or important. These, including the processes or functions that are key to ecosystem viability, are generally referred to as 'valued ecosystem components' or VECs. The processes of impact screening and VEC selection really need to be integrated. A certain amount of iteration will generally be required to narrow down the ranges of impacts and receptors that are actually studied in detail. Examples of possible stressors and impacts are given in Table 3.3.

3.5.2 Checklists and matrices

Where a large number of potential stressors and receptors are involved, the complexity of potential interactions between them can make it difficult to reduce impact assessments to manageable proportions. *Checklists and matrices* can be very useful for impact screening, providing effective means of summarizing and presenting large quantities of disparate information. They are generally used simply to alert practitioners to the full range of potential impacts and to help refine impact assessment frameworks.

Checklists are simple lists of factors or issues that should be taken into account when undertaking an impact assessment. They may be generated on a case-by-case basis or issued in the form of generic guidance. The South African Department of Environment Affairs, for example, issued a 'Checklist of Environmental Characteristics' in 1992 (Department of Environment Affairs 1992). This 'identifies environmental characteristics which may potentially be affected by proposed development actions, or which could place significant constraints on a proposed development'. This checklist is not intended to be exhaustive, but simply indicates the major 'characteristics and linkages' that should be taken into account in most situations. Box 3.3 shows those factors listed with respect to the 'ecological characteristics of the site and its surroundings'.

Box 3.3 Ecological characteristics of the site and its surroundings: a checklist of factors to be considered. (After Department of Environment Affairs 1992.)

Vegetation
- Survival of rare or endangered plant species
- Diversity of plant communities
- 'Stabilizing' vegetation such as that found on sand dunes
- Vegetation communities of conservation or scientific importance
- Conservation of vegetation communities of particular recreational value
- Introduction or spread of invasive alien seeds and plants
- Natural replenishment of existing species
- Frequency of veld/bush fires
- Degree of trampling/disturbance
- Degree of grazing/exploitation
- Genetically engineered organisms

Animals
- Survival of rare or endangered animals
- Diversity of animal communities
- Animal communities of particular scientific, conservation or educational value
- Non-resident or migrant species
- Alien species (including domestic and invasive species)
- Survival of animal communities due to frequency of fires, poaching, disturbance, restricted mobility, over-exploitation
- Genetically engineered organisms

Natural and seminatural communities
- Local, regional or national importance (e.g. economic, scientific, conservational, educational)
- Survival or persistence of 'natural' communities
- Appropriateness of conservation methods to be employed
- Ecological function, for example in relation to physical destruction of habitat, reduction in effective size of community, altered predator–prey relationships, availability of food, barriers to animal movement or migration
- Ecological function in relation to water availability and quality, for example quality and flow of groundwater, quality of standing or flowing water, oxygen content of water, salinity, turbidity, flow rate, temperature, level of chemical and other forms of pollution, eutrophication, toxins, siltation patterns
- Ecological function in relation to air quality (including dust pollution and deposition)
- Ecological function in relation to disturbance, including recreation pressure, altered grazing pressure, altered fire regimes, etc.

Matrices can be used to identify possible interactions between identified project actions or activities and ecological components. Simple 'interaction matrices' as developed by Leopold *et al.* (1971) display project actions or activities along one axis with appropriate environmental factors listed along the other (Canter 1996). Leopold *et al.* (1971) constructed a generic interaction matrix based on about 100 categories of action and 90 categories of 'environmental item'. To use the 'Leopold' matrix, each action and its potential for creating an impact must be considered for each environmental item or receptor. Where an impact is anticipated, the matrix is marked with a diagonal line in the appropriate cell. It is also possible to indicate the possibility of beneficial or detrimental impacts using + or − signs, respectively. Leopold *et al.* recommended going on to describe interactions in terms of their magnitude and importance, assigning numerical values of 1 (small magnitude or low importance) to 10 (large magnitude or high importance).

Matrices of this kind are particularly useful for preliminary screening of impacts, making it easier to generate a firm conceptual framework for impact assessment. For ecological considerations, for example, proposal activities might be set against VECs to identify possible ecological impacts. With respect to ecological considerations, the 'Leopold' matrix lists the following relevant 'environmental items' (Box 3.4).

Some of these categories are so broad that they are of relatively little diagnostic value. For example, identifying that there may be effects on 'insects' as a result of 'urbanization' (one of the generic 'action' types) is not all that helpful with respect to clarifying study limits and selecting appropriate ecological survey methodologies. In most cases therefore it is preferable to construct matrices that are specific rather than generic, to ensure that the full range of anticipated actions and potential ecological receptors is considered in any particular case.

One of the disadvantages of conventional two-dimensional matrices is that they hamper identification of the secondary and higher order impacts, which characterize most ecosystems. This led to the development of more complex matrix methodologies, including the construction of 'stepped' or 'cross-impact' matrices, which can be used to address secondary and tertiary impacts of 'actions' or stressors. The principles of stepped matrices are described comprehensively by Canter (1996). A stepped matrix is one in which environmental factors are set against other environmental factors so that the consequences of initial changes can be displayed. This is illustrated in Fig. 3.11 (after Canter 1996) in which action 3 affects environmental factor D, which then has consequences for factors A and F. Tertiary impacts occur as changes in factor A has further repercussions for factors B and I and changes in factor F effect changes in factor H. Stepped matrices of this kind lend themselves well to impact identification and assessment in EcIA because

Box 3.4 Biological conditions to be incorporated in a 'Leopold matrix'. (After Canter 1996.)

Biological conditions
1 Flora
   (a) Trees
   (b) Shrubs
   (c) Grass
   (d) Crops
   (e) Microflora
   (f) Aquatic plants
   (g) Endangered species
   (h) Barriers
   (i) Corridors
2 Fauna
   (a) Birds
   (b) Land animals, including reptiles
   (c) Fish and shellfish
   (d) Benthic organisms
   (e) Insects
   (f) Microfauna
   (g) Endangered species
   (h) Barriers
   (i) Corridors

Ecological relationships
   (a) Salinization of water resources
   (b) Eutrophication
   (c) Disease and insect vectors
   (d) Food chains
   (e) Salinization of surficial material
   (f) Brush encroachment
   (g) Other

they permit impacts to be traced through more complex systems, rather than attempting to reduce complex ecological systems to simple, two-dimensional models of reality. However, they can become unwieldy and difficult to follow in cases where a large number of potential stressors and receptors are involved.

In an attempt to overcome the limitations of conventional matrix techniques, Shopley *et al.* (1990) developed a 'component interaction technique' specifically to tackle secondary ecological effects. They modelled the environment as a list of components ranked in order of their ability to initiate secondary impacts, based on known chains of dependence between them. They used computerized matrix powering procedures to structure the data and facilitate investigation of the secondary impact potential of affected ecosystems. In their method, dependencies

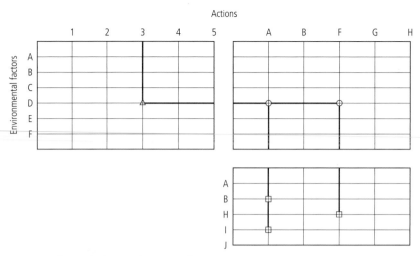

Fig. 3.11 Example format for stepped matrix. (From Canter, L.W. (1996) *Environmental Impact Assessment*, 2nd edn. McGraw-Hill, New York; reproduced with permission of The McGraw-Hill Companies.)

were scored as simply present or absent in order to measure the extent to which ecosystem components were linked. This enabled the degree of disruption likely to be caused by removal of any one component to be traced through its associated linkages and helped to identify critical ecosystem components that might merit further study. Such approaches demonstrate the benefits of modern data handling capabilities, but they also rely on a thorough understanding of how key components of ecosystems do actually interact with each other: an understanding that is often limited.

3.5.3 Networks

Networks are widely used to represent possible chains of events for impacts of activities on ecosystems. Again, Canter (1996) gives a number of examples. The development of networks can provide a useful way of visualizing systems and clarifying the linkages between ecosystem components and specific activities. In an example provided by the US Soil Conservation Service (1977), the creation of an impoundment was considered in terms of the basic resources affected (land, water or air), the resultant changes in cover type or land use, potential physical and chemical effects (such as downstream water quality changes), associated

Fig. 3.12 (*Opposite*) US Soil Conservation Service network diagram. (From Canter, L.W. (1996) *Environmental Impact Assessment*, 2nd edn. McGraw-Hill, New York; reproduced with permission of The McGraw-Hill Companies.)

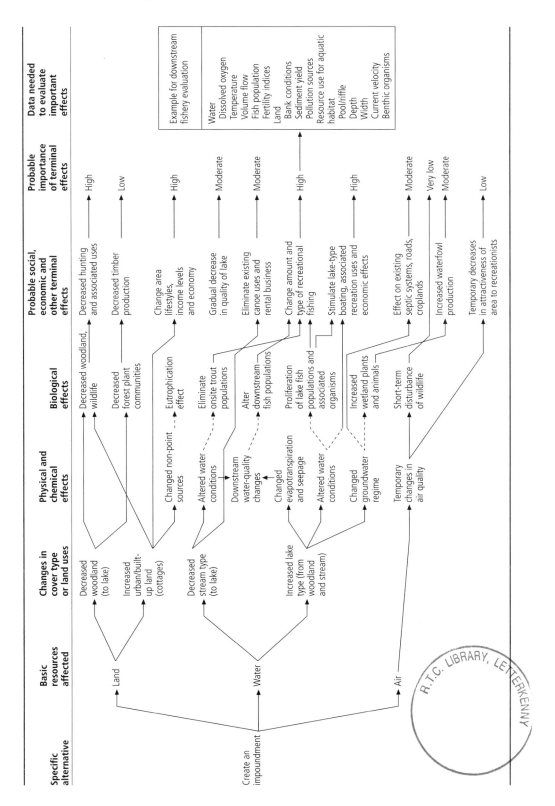

biological effects (such as altered downstream fish populations) and their probable importance (high, medium or low). The resultant network not only identified key linkages between physico-chemical and biological parameters, but also provided a framework for identifying current knowledge gaps and therefore the information that would need to be collected to evaluate important effects (Fig. 3.12).

3.5.4 Conceptual models

Conceptual models generally consist of written descriptions or visual representations of possible responses by ecological receptors to specific activities or impacts. Their complexity may vary considerably, for example primary, secondary or tertiary linkages between actions and potential receptors may be represented. While complex conceptual models may be developed to assist in quantitative impact prediction, simple models can be useful for identifying possible linkages between activities and receptors in order to identify those which should be investigated in further detail. Early conceptual models therefore tend to be broad in scope, identifying as many potential relationships as possible (US Environmental Protection Agency 1996). As more information is gathered throughout the EcIA process, models can be refined to help clarify the relative plausibility of impact scenarios. Approaches taken to subsequent analyses can then be justified. The use of conceptual models is particularly common in ecological risk assessment to clarify possible sources or hazards, potential receptors and the pathways by which they may be connected.

Clearly, conceptual models may be constructed using a variety of approaches. They may be based on matrices or networks, or even on simple pictoral representations of stress-generating activities and affected systems (Fig. 3.13). In some cases, these models may play a part not only in summarizing the environmental factors to be considered in the analysis, but also in impact prediction.

3.5.5 Geographical information systems

Increasingly, GIS are emerging as tools for the identification and quantification of potential ecological impacts (see section 9.2). Linked to comprehensive databases on the distributions of abiotic and biotic variables (as might be constructed in the early phases of EcIA to store information on the identity and distributions of VECs), they offer powerful techniques for addressing 'where?' and 'what if?' questions about the location and likely magnitude of interactions between ecosystem components and stress-generating activities, whether for one proposal or for a number of possible scenarios (Treweek 1995). In

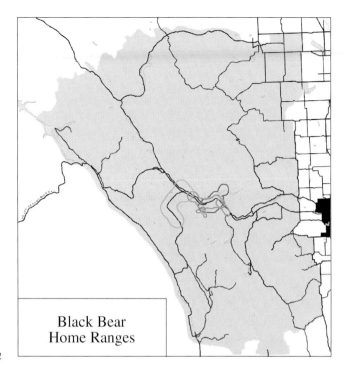

— Black Bear 11
— Black Bear 13
— Black Bear 14
— Black Bear 18
— Black Bear 19
— Black Bear 192

Black Bear
Home Ranges

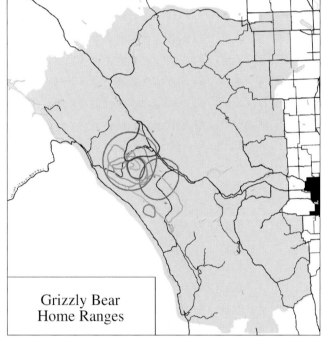

— Grizzly Bear 1   — Grizzly Bear 8
— Grizzly Bear 2   — Grizzly Bear 9
— Grizzly Bear 4   — Grizzly Bear 10
— Grizzly Bear 5   — Grizzly Bear 11
— Grizzly Bear 7   — Grizzly Bear 23

Grizzly Bear
Home Ranges

Plate 3.1 Home ranges for black bear, grizzly bear and wolf (*Overleaf*). (Sources: black bear, 75% home-range contour from Kansas *et al*. (1989) and analysis on Banff National Park Warden Service GIS; grizzly bear, 75% home-range contour from Raine and Riddel (1991) and Purves *et al*. (1992); wolves, 75% home-range contour from Purves *et al*. (1992). © Komex International Ltd 1994.)

[*Facing p. 82*]

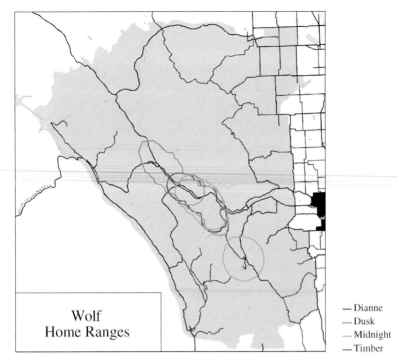

Wolf
Home Ranges

— Dianne
— Dusk
— Midnight
— Timber

Plate 3.1 (*Continued*)

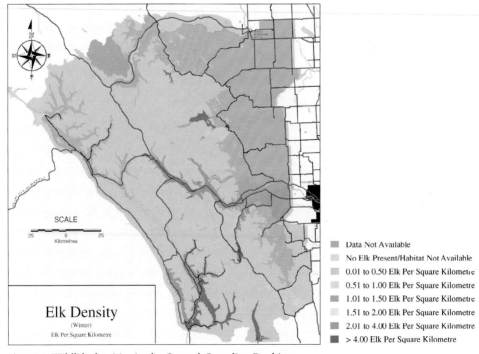

SCALE

25   0   25
Kilometres

Elk Density
(Winter)
Elk Per Square Kilometre

Data Not Available
No Elk Present/Habitat Not Available
0.01 to 0.50 Elk Per Square Kilometre
0.51 to 1.00 Elk Per Square Kilometre
1.01 to 1.50 Elk Per Square Kilometre
1.51 to 2.00 Elk Per Square Kilometre
2.01 to 4.00 Elk Per Square Kilometre
> 4.00 Elk Per Square Kilometre

Plate 3.2 Wildlife densities in the Central Canadian Rockies
Ecosystem: elk, moose and grizzly bear (*Opposite*). (© Komex
International Ltd 1994.)

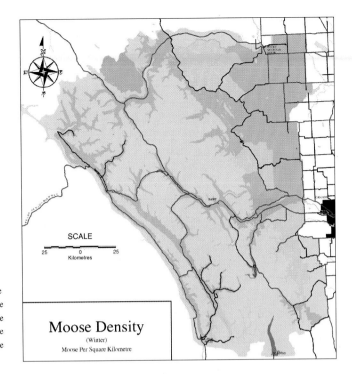

Data Not Available
No Moose Present/Habitat Not Available
0.01 to 0.50 Moose Per Square Kilometre
0.51 to 1.00 Moose Per Square Kilometre
1.01 to 2.00 Moose Per Square Kilometre
2.01 to 4.00 Moose Per Square Kilometre
> 4.00 Moose Per Square Kilometre

SCALE
25   0   25
Kilometres

Moose Density
(Winter)
Moose Per Square Kilometre

Data Not Available
No Bear Present/Habitat Not Available
0.001 to 0.012 Bear Per Square Kilometre
0.013 to 0.035 Bear Per Square Kilometre
>0.035 Bear Per Square Kilometre
— Primary Road
— Secondary Road

SCALE
25   0   25
Kilometres

Grizzly Bear Density
(Summer)
Bear Per Square Kilometre

Plate 3.2 (*Continued*)

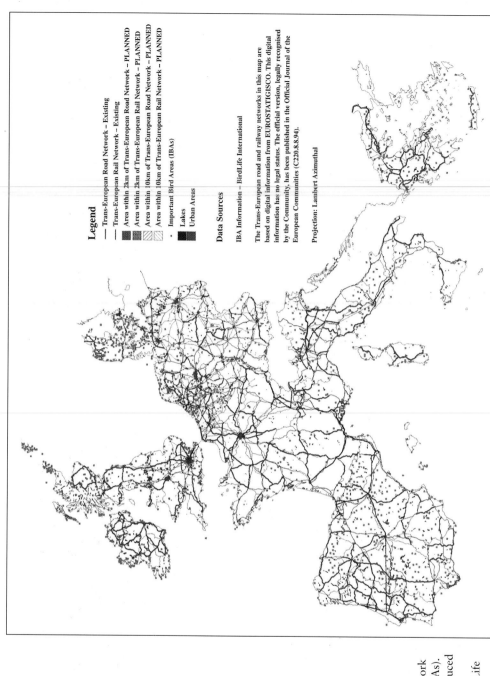

Legend

—— Trans-European Road Network – Existing

—— Trans-European Rail Network – Existing

Area within 2km of Trans-European Road Network – PLANNED

Area within 2km of Trans-European Rail Network – PLANNED

Area within 10km of Trans-European Road Network – PLANNED

Area within 10km of Trans-European Rail Network – PLANNED

· Important Bird Areas (IBAs)

Lakes

Urban Areas

Data Sources

IBA Information – BirdLife International

The Trans-European road and railway networks in this map are based on digital information from EUROSTAT/GISCO. This digital information has no legal status. The official version, legally recognised by the Community, has been published in the Official Journal of the European Communities (C220.8.8.94).

Projection: Lambert Azimuthal

Plate 9.1 Map of the trans-European road and rail network and important bird areas (IBAs). (From Bina *et al.* 1997; produced by World Conservation Monitoring Centre and BirdLife International.)

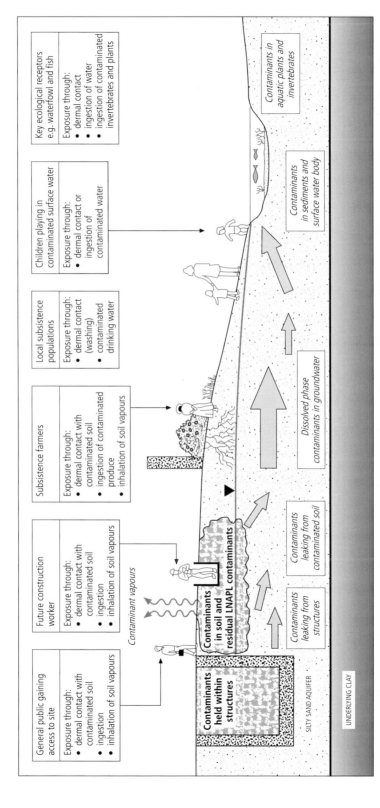

Fig. 3.13 Pictorial conceptual diagram to illustrate potential impacts of groundwater pollution. **Bold** type signifies sources, *italic* type signifies pathways and Roman type signifies receptors. LNAPL, light non-aqueous phase liquid.

particular, simple overlaying techniques can be used to identify (and measure) areas where there is likely to be coincidence (or an overlap) between an activity and an important ecological receptor (referred to as 'constraint mapping'). They can also be used to explore the consequences of altering the location, timing or extent of proposed activities or to modify proposals to avoid particularly sensitive receptors. More complex operations can be carried out to establish or research functional relationships between key ecosystem components and possible explanatory variables. For example, data on the distributions of wetland plants might be combined with data on soil watertable (on one or many occasions) to explore the soil water requirements of different species. Once such relationships are established (possibly through associated statistical analysis) they can be used predictively in impact assessment (e.g. to predict the possible consequences of watertable change for selected species of interest).

In summary, with respect to impact identification and assessment, GIS can play a useful part in:

- locating and measuring areas of coincidence between key receptors and defined activities ('exposure assessment');
- targeting detailed field surveys on appropriate areas or receptors;
- constraint mapping (to avoid impacts on particularly sensitive or important areas or receptors);
- comparing impact scenarios for different proposal 'designs';
- developing a predictive framework.

When constructing GIS for purposes of impact screening or constraint mapping, the conceptual basis for building the system remains fundamental. It is important to include the right thematic layers in the system and to ensure that the range of potential stressors and receptors can be mapped at an appropriate scale from the data available. In other words, the ability of the system to perform the task demanded of it will be constrained by the input data. Thinking about the availability and suitability of data while refining conceptual frameworks for EcIA is useful in itself, however. It can help to ensure that limited resources are invested in data that are most likely to be of diagnostic value.

3.6 Exposure assessment

For an impact to be expressed, ecological components must be 'sensitive' (i.e. respond readily) to a stressor and must also be exposed to it. 'Sensitivity' of response is closely dependent on the mode of activity or expression of the stressor in question (US Environmental Protection Agency 1996) and the specific characteristics of receptors. For example, sensitivity to a potentially toxic chemical pollutant is influenced both by the characteristics of the chemical itself and by the physiology of receptor organisms. Similarly, the likelihood of exposure to impact depends not only on the range of stressor activity, but also on the 'behaviour' and activity of receptors. Individual and community life-history characteristics, for example, can play a part, such that species with long life cycles and low reproductive rates are more likely to be vulnerable to extinction if mortality rates increase than those with short life cycles and high reproductive rates (US Environmental Protection Agency 1996).

There are several parameters that can be used to describe the distribution and abundance of ecological receptors in relation to the spatial and temporal extent of a proposed action or set of activities and to estimate their susceptibility to exposure. It may be necessary to take a number of these factors into account when screening potential receptors to select VECs or when determining assessment endpoints. They include:
- range or distribution (relative to stressor range);
- abundance or population density;
- population dynamics;
- social organization and behaviour;
- seasonality of presence or activity;
- mobility with respect to availability of alternative habitat;
- resource dependence and habitat specificity;
- interdependencies (linkages).

3.6.1 Home-range size

Home-range size affects the proportion of time for which an individual animal or group of animals is likely to be exposed to impacts of known spatial influence. Home range is defined as the geographical area encompassed by an animal's activities (excluding migration) over a specified time. Other related measures include 'foraging radius', or the distance that an animal will travel to find food sources. In homogeneous habitats, it may be possible to assume that home ranges are roughly circular, but in some environments animals may use a variety of resources and have very irregular home ranges. Some examples of wildlife home ranges are illustrated in Plate 3.1 (facing p. 82) for black bear *Ursus americanus*, grizzly bear *U. horribilis* and wolf *Canis lupus* in the Canadian Central Rockies Ecosystem (Komex International Ltd 1995). The value of such information when designing surveys and assessments can be seen clearly. If stored digitally it can also be incredibly useful for ecological constraint mapping.

Home-range size for individuals within a population can vary with season, latitude or altitude as a consequence of changes in the distribution and abundance of food or other resources. It tends to vary with body size and age because of differences in the distribution of preferred forage or prey. Home-range sizes also depend on habitat quality. Poorer habitats are associated with sparse populations and larger home ranges. Finally, home ranges are influenced by sex and season. For example, females with young may forage 'close to home' while males range more widely. It cannot be assumed that home ranges will remain constant either. Many animals have to adjust their behaviour and home ranges to adapt to changing food supplies over time. For example, there is evidence that changes in salmon distribution in the American north-west influence the viability and distribution of grizzly bear populations, which rely on salmon as nutrient-rich food to reproduce successfully and prepare for hibernation (Levy 1997).

For non-territorial animals there may be considerable overlap of activity areas among neighbouring individuals or groups. For example, several individuals or mated pairs may share the same geographical area, but use signalling behaviours to ensure there is some degree of temporal separation. Other species are strongly territorial and will defend mutually exclusive areas, with individuals, breeding pairs or family units actively advertising identifiable boundaries and excluding neighbouring individuals or groups. All aspects of activity should be taken into account when determining home ranges. Areas should not be excluded simply because they are only visited rarely: they may still be crucial for survival.

3.6.2 Population density

Population density (number of individuals per unit area) influences how many individuals (or what proportion of a local population) might be exposed within a zone of impact. For strongly territorial species it is sometimes possible to infer population density from territory size, but in other cases, particularly where home ranges overlap, this is not possible and direct counting may be necessary: something that is carried out surprisingly little in EcIA. Plate 3.2 (facing p. 82) shows wildlife densities for three species (elk *Cervus canadensis*, moose *Alces alces* and grizzly bear *Ursus horribilis*) in the Central Canadian Rockies Ecosystem (Komex International Ltd 1995).

3.6.3 Social organization

Social organization influences the way in which animals of various ages and sizes are likely to be distributed and therefore their likelihood of exposure. In some species, for example, individual home ranges do not overlap. In others, all individuals are likely to use the same home range. In between these two extremes, home ranges may be shared with mates, offspring or extended family groups (US Environmental Protection Agency 1993).

Often, young animals are more sensitive than adults. For example, Pacific salmon (*Oncorhynchus* spp.) eggs and fry are more sensitive to sedimentation than adults (US Environmental Protection Agency 1996). The adverse effects of a particular activity may be important only to some age classes, or only at critical life stages, both of which are influenced by social organization.

3.6.4 Population dynamics

Factors such as the annual fecundity and mortality rates for a species can influence the abilities of populations to recover following impact. The longevity of individuals influences the potential for cumulative deleterious effects, for example long-lived individuals are more likely to accumulate toxic concentrations of chemical pollutants than those whose exposure is limited by virtue of their short life. Whether populations are density dependent or independent has an important bearing on how they are likely to respond either to imposed stresses or to mitigation efforts (see section 4.6.2).

3.6.5 Seasonal patterns of use or activity

In selecting suitable species for inclusion in survey or monitoring programmes, it is always important to consider all potential habitat users, for example passage migrants and occasional feeders as well as resident breeding species. Many life-cycle attributes affect animals' activity and foraging patterns in time and space. For example, many species of bird are present in the northern hemisphere only during the winter months, or move seasonally between the northern and southern parts of continents. Some species of mammal, reptile and amphibian hibernate, or spend a dormant period during winter months. Likewise species from all other groups may show similar, seasonal variations in presence or activity, which must be taken into account when screening for impacts or designing field schedules for EcIA. These include plant species as well as animals.

Other temporal variations in patterns of activity may also occur and influence the exposure of a species to impacts. Characteristic diurnal patterns are very common. Some species are nocturnal and are difficult to detect during the day. Others vary their activity during the day. Many songbirds in temperate regions, for example, show characteristic patterns of singing activity. The Dartford warbler *Sylvia undata* in the UK, for example, shows most singing activity in the early to mid-morning period, with a later peak in the late afternoon, whereas foraging behaviour is much more consistent throughout the day. Just as it is inappropriate to survey for species out of season, it is also important to

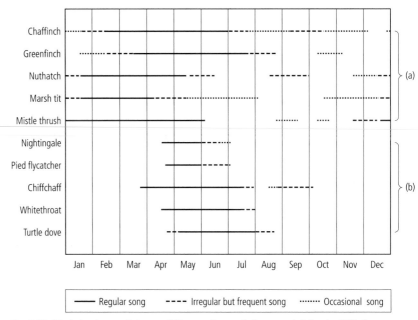

Fig. 3.14 Bird singing activity at different times of the year. (a) In the UK the resident species sing early in the spring. (b) The resident species have all but finished singing before African migrant species arrive and begin to sing. (Reprinted from Bibby, C.J., Burgess, N.D. & Hill, D.A. (1992) *Bird Census Techniques*, by permission of the publisher Academic Press Limited, London.)

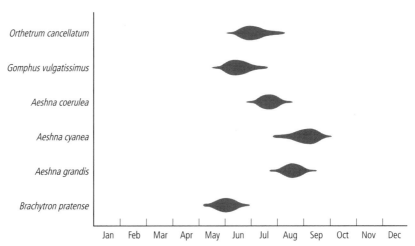

Fig. 3.15 Gant diagram showing flight periods for selected species of dragonfly. (From Institute of Environmental Assessment 1995.)

ensure that field surveys are timed to coincide with peak periods of specific animal activity.

Gant diagrams or charts can be very useful for highlighting peak periods of activity or occurrence for different species. Figure 3.14 shows variation in the singing activity of bird species in the UK, while Fig. 3.15 was included in guidelines for baseline ecological assessment by the UK's Institute of Environmental Assessment (1995) to illustrate how flight periods can vary between dragonfly species. Of course, exposure to impact is not necessarily more likely during active periods. Some species may be more vulnerable during dormancy because of reduced mobility and ability to evade impacts. Some seasonal activities influence vulnerability to impact because of their effects on the energy levels of species. The energetic demands of migration, for example, are so great, that individuals may become susceptible to a whole range of impacts, which they might otherwise be able to withstand (US Environmental Protection Agency 1996).

3.6.6 Mobility

Mobility is a major factor affecting the ability of species to evade or avoid exposure, not just to direct habitat destruction, but also to the effects of habitat fragmentation and isolation. However, mobility is not a straightforward property. Species that are theoretically mobile (many birds, for example), may nevertheless be unable to respond to habitat redistribution if access to replacement habitat makes unacceptable demands on their energy resources. In other words, there might be energy costs associated with avoiding stresses or disturbance. Mobility might confer a short-term ability to avoid impact, but not a longer-term ability to maintain populations. Some species are also very habitat-faithful or tenacious.

3.6.7 Resource dependence and habitat specificity

Species with highly discerning habitat requirements are less likely to be able to find replacement habitat if displaced, than species with more 'catholic tastes'. These species are also most likely to be those of high nature conservation importance or 'rarity value'. Mobility and the ability to avoid exposure is of little benefit if no replacement habitat can be found.

3.6.8 Interdependencies (linkages)

Sensitivity to impacts can be increased by the presence of other 'natural' and anthropogenic stresses (synergistic effects), or effects on one species can have knock-on effects on associated species. In generating exposure scenarios, it is therefore essential to consider all possible ecological linkages, whether between species or between different life stages of the same species.

3.7 Recommended reading

EIA techniques

Canter, L.W. (1996) *Environmental Impact Assessment*, 2nd edn. McGraw-Hill, New York.
Canter, L.W. & Sadler, B. (1997) *A tool kit for effective EIA practice — review of methods and perspectives on their application.* A Supplementary Report of the International Study of the Effectiveness of Environmental Impact Assessment. Environmental and Ground Water Institute, University of Oklahoma, USA; Institute of Environmental Assessment, UK; International Association for Impact Assessment.

Ecological study design for impact assessment

Institute of Environmental Assessment (1995) *Guidelines for Baseline Ecological Impact Assessment.* E. and F.N. Spon, London.
Beanlands, G.E. & Duinker, P.N. (1983) *An Ecological Framework for Environmental Impact Assessment in Canada.* Institute for Resource and Environmental Studies, Dalhousie University, Halifax; in cooperation with the Federal Environmental Impact Assessment Review Office, Canada.

Species distributions and information about protected species

Hagemeijer, W.J.M. & Blair, M.J. (1997) *The European Bird Census Council's Atlas of European Breeding Birds.* T. and A.D. Poyser, London.

UK: Red Data Books
US: US Fish and Wildlife Service web pages (include endangered species listings and action plans)

4 Focusing procedures

While survey and inventory of all ecosystem components might be desirable to achieve a full characterization of baseline conditions, limited time and funds for surveys and analysis invariably make it necessary to restrict studies to a more limited range of species or ecosystem parameters. It would be impossible to establish comprehensive, long-term monitoring programmes for all ecosystem components (Treweek 1995) and the 'count everything' approach tends to result in superficial survey and the collection of a great deal of material with little predictive value.

The need to restrict the range of ecosystem components for survey has resulted in the development of 'focusing procedures', which can be used to identify key biological components and processes and to rationalize the subsequent impact assessment. The outcome of focusing should be the selection of suitable 'valued ecosystem components' (VECs) to provide the focal point for ecological impact assessment (EcIA). Focusing therefore depends quite heavily on a form of evaluation, not of impacts (as discussed in Chapter 6) but of the value of ecosystems or ecosystem components themselves. This value might be measured in terms of usefulness as indicators, economic worth, inherent 'value' or societal benefit. Criteria for selection of VECs are considered in more detail in the remainder of this chapter.

Focusing procedures are not always formalized and there is considerable inconsistency in the methods used. This aspect of impact assessment is fraught with difficulties. Attempts to rationalize subsequent studies by focusing only on preselected ecological components can result in considerable bias due to ill-informed assumptions about 'importance' or 'value' and the tendency to exclude 'unfashionable' or under-recorded species (Treweek 1995). The effectiveness of all focusing procedures rests on knowledge of the ecological characteristics and behaviour of the environmental components represented. If there are gaps in this knowledge, then it is important always to make some provision for the detection of the unexpected. Also, to give good reasons both for the selection of the species or other parameters to be studied in detail and for the exclusion of the remainder. Selection of VECs is the first step in selecting 'assessment endpoints' that are ecologically relevant. It is subsequently necessary to consider the specific susceptibility (sensitivity and exposure) of these VECs (see section 4.8), to consider their relevance to societal or management goals and to ensure that they are actually measurable (in other words, that measurement endpoints can be found).

The following section considers the theoretical basis for selecting VECs. The site-specific information to which focusing procedures will be applied may be derived either from pre-existing data sources ('raw' or 'interpreted') or from preliminary survey as discussed earlier in section 3.4.5.

4.1 Valued ecosystem components

VECs are, by definition, ecosystem components that are considered to be important or valuable and that merit detailed consideration in the EcIA process. The methods used for selecting VECs are important, because the results of an EcIA can be greatly prejudiced if important ecosystem components are inadvertently screened out at an early stage. Ideally, the criteria used to 'screen out' or 'select in' components should be objective, consistent, transparent and defensible. In practice, the choice of subjects for detailed analysis in EcIA is often completely arbitrary. Attributes often selected as VECs include those listed in Box 4.1.

Suitable attributes will vary in each case and it is not necessary to select VECs from each of the categories listed in Box 4.1 in every one. Often, EcIA studies are primarily species centred. For example, Gibeau (1993) used the grizzly bear *Ursus horribilis* as a VEC for assessing the cumulative effects of development in the western mountain parks of Canada.

However, it is always important to consider the questions that will have to be addressed in the EcIA and the nature of the decision that will have to be reached. 'Species' will only be suitable as VECs in EcIA if suitable measurement and assessment endpoints can be found. This

Box 4.1 Attributes which may be selected as VECs.

- Abiotic components or environmental media such as water or air
- Biogeographic units, landscape units or 'ecoregions'
- Habitats
- Species
- Populations or communities
- Individual organisms (especially if protected species are affected)
- Functional groups of species (guilds)
- Ecosystem functions
- Special sites (e.g. protected sites)

might require data on, for example, population dynamics, environmental carrying capacity or behavioural responses to disturbance. It is worth remembering that impacts are likely to be expressed sooner at other levels of ecosystem organization than they are at the species level. Furthermore, there may not be enough information available about status and distribution at the species level for meaningful interpretation of data. For these reasons, while it is common for species to form the ultimate focus of interest, other categories of attribute tend to be selected as VECs so that meaningful measurement endpoints can be found. For example, measures of habitat distribution and quality might be used, or population-level studies carried out only for guild indicators, to reduce the number of species for which detailed analysis is required. There may be cases where it is possible to identify important ecological functions or characteristics, the modification of which is likely to have significant knock-on effects for other components, whether these are protected species or ecosystem functions valued by people. Geppert *et al.* (1984), for example, used water quality as a VEC in a study of the cumulative environmental effects of timber harvesting, and Horak *et al.* (1983) focused a study of the effects of coal mining on fish stocks and 'wildlife'. Where sites, habitats, species (or individual populations) are the subject of formal designation or protection they are, by definition, 'valuable' and merit detailed consideration.

The VECs eventually selected may be a 'mixed bag' of species, habitats, ecological and economic functions. For example, Kalff (1995) explains how the 12 VECs identified for a cumulative ecological assessment of Kouchibouguac National Park in Canada included special landscape features, habitats and ecosystems as well as species like the nationally endangered piping plover *Charadrius melodus* and the provincially endangered osprey *Pandion haliaetus* (Box 4.2). Selection of VECs in this case was influenced by the goals of the park, the goals of Parks Canada as a whole (Environment Canada, Canadian Parks Service 1993) and on the requirements of international agreements like the Convention on Biological Diversity (UNEP 1992).

When VECs have been selected, it may then be necessary to establish general trends and patterns in their abundance, areal extent or concentration. Information on rates of change is useful as this can have a bearing on vulnerability: a VEC with rapidly deteriorating status is likely to be more at risk than one which is changing more slowly. At a later stage, when it comes to evaluating impacts (section 6.3), proximity to VEC thresholds of viability should be considered: does the impact tip the VEC over its threshold or bring it dangerously close, or is there a big safety margin? This comes into play when impacts are evaluated, but it is worth noting here with respect to the need for some sort of measurement endpoint. VECs for which no reliable measurement or assessment endpoint can be found, are not generally worth including.

> **Box 4.2** VECs selected for ecological assessment of Kouchibouguac National Park, Canada. (After Kalff 1995.)
>
> - Land system 16
> - Salt-marsh ecosystem
> - Lagoon ecosystem
> - Barrier dune
> - Rare and functionally important dune vegetation
> - Piping plovers
> - Waterfowl
> - Osprey
> - Soft-shelled clams
> - Bank swallows
> - Woodland jumping mouse
> - Grey seals

4.2 Criteria for selecting species as VECs

Criteria for selecting species for further study have been based on (*inter alia*) their commercial value, their rarity (or 'endangerment'), their role in supporting other 'important' species, their importance for continued ecosystem function, and so on (e.g. Eberhardt 1976). In many circumstances it will be appropriate for a variety of selection criteria to be used. The 'habitat evaluation procedure' (HEP) (US Fish and Wildlife Service 1980), for example, identifies four main categories of 'evaluation species':

1 species with high public interest, economic value or both;

2 species known to be sensitive to specific land-use actions and that might serve as 'early warning' or indicator species for an affected wildlife community;

> **Box 4.3** Criteria for selecting species as VECs.
>
> - Public visibility/appeal (charismatic and emblematic species)
> - Economic importance
> - Protected status
> - Rarity
> - Endangerment/conservation status
> - Susceptibility and/or responsiveness to defined impacts (indicators)
> - Representativeness of responses of guilds (guild indicators)
> - Umbrella species
> - Important ecological role (e.g. position in food chain, keystone species)
> - Availability of consistent survey methods
> - Expediency/tractability for survey

3 species that perform a key role in a community because, for example, of their role in nutrient cycling or energy flows;

4 species that represent groups of species that utilize a common environmental resource ('guilds').

Commonly used criteria for selecting species as VECs are summarized in Box 4.3.

4.2.1 Public appeal (charismatic and emblematic species)

Species with public appeal or 'charisma' are often selected as evaluation species for EcIA because they can play a significant part in mobilizing public support. The path of the Newbury bypass in the UK impinged on the habitat of two protected species: the dormouse *Muscardinus avellenarius* (Fig. 4.1) and Desmoulin's whorl snail *Vertigo moulinsiana*. The dormouse undoubtedly did more to engage public sympathies than did discovery of the less obviously charismatic or cuddly Desmoulin's whorl snail (although in this case even the appeal of the dormouse was not enough to safeguard its habitat).

For whatever reason, there is considerable bias in the species selected for study in EcIA. In the UK, review of environmental statements (ESs) suggests that birds are sampled much more than invertebrates and that the microflora and fauna are almost never surveyed (Spellerberg & Minshull 1992; Treweek *et al.* 1993; Thompson 1995). The selection of species for EcIA on account of their public appeal alone can mean that limited resources are diverted away from the survey and assessment of species that are considerably more valuable as indicators of ecological condition or change. In some cases it can also mean the neglect of species that are more vulnerable and threatened.

4.2.2 Economic importance

Species of commercial importance are often selected as VECs and they are relatively easy to use in evaluation because they can be given a

Fig. 4.1 Dormouse *Muscardinus avellenarius*. (From Corbet & Harris 1996.)

straightforward monetary 'value'. For EcIA carried out in countries where wildlife is still widely 'harvested', important species for hunting or fishing can be identified that have a clear economic value and are often selected as VECs. Game species may also constitute an important economic resource for ecotourism (African 'safari' animals, for example).

4.2.3 Protected status

Most environmental impact assessment (EIA) legislation requires information on the presence of protected species. While protected status invariably triggers some level of EcIA, it brings no guarantee that habitat will actually be preserved and rarely seems to act as a significant deterrent to applications for development consent. In fact, it is very difficult to determine how often the presence of a protected species has resulted in the redesign of a proposal or denial of development consent. Anecdotal evidence suggests that it is unusual. Legislative requirements to consider impacts on protected species can result in EcIAs that are biased towards protected species at the expense of others, which might actually be more valuable as indicators of environmental impact. While legal protection is often conferred following calculations of importance, rarity, threat and so on, it cannot always be assumed that fully objective methods will have been used. Furthermore, the status of species is not static. There are numerous instances where species once regarded as common have declined towards endangered status before mechanisms for their protection have been implemented. Rates of decline generally exceed the rate of bureaucratic response. How can we know when a once common species is likely to reach that threshold of rarity at which it will become endangered if we cannot measure its status and detect trends? This problem emphasizes the need for regularly updated and comprehensive national datasets, which permit regular checks on the viability and status of species.

4.2.4 Rarity

Some species are inherently rare by virtue of extreme ecological specialism. However, species that are rare, or restricted in their distribution, are more vulnerable to extinction than those which are more widespread and less likely to be able to recover from any damage or loss attributable to a proposed action (see section 6.3). For this reason alone, 'rarity' is an appropriate criterion to use to select species for further study. It is important to establish whether or not the area affected by a proposed action is fundamental to the continued existence of rare species. In countries with well-established legislation for nature conservation, rare species will also tend to be protected. However, as mentioned above, legislation can lag behind the need and scrupulous practitioners should take a precautionary approach. In addition, 'rarity' is a property that can be measured, making it possible to rank species using objective and consistent methods.

In situations where rare species are likely to be affected by a proposed action, it may be useful to try and seek out the reasons; for example, is rarity a consequence of historical accident or a clear response to biogeographic forces (Roy & Eversham 1994)? Explaining the causes of rarity is particularly important with respect to the replaceability of species in alternative locations and is therefore an important consideration when planning mitigation.

4.2.5 Endangerment or conservation status

A distinction should be drawn between 'rarity' and 'endangerment', which is a combined measure of rarity, vulnerability and threat (see also section 6.3.16). Species that are rare or restricted in distribution, vulnerable to the specific impacts associated with a proposed action and threatened throughout their range, are endangered and should be regarded as prime targets for detailed study.

International legal obligations to conserve the world's biological diversity have spurred governments on to inventory national resources and to assess the conservation status of their native species. Various systems are in use that rank species according to different criteria. 'Endangerment' is one of the criteria in common use. It is impossible to assess conservation status or endangerment without comprehensive, long-term, national monitoring data. This is more likely to be available in countries that have a long history of wildlife conservation and management.

Any species (or its 'critical habitat') as listed under the US Endangered Species Act, for example, would constitute a VEC in the context of EcIA. Species are classified as 'endangered' when they are in danger of extinction within the foreseeable future throughout all or a significant portion of their range, or as 'threatened' if they are likely to become endangered within the foreseeable future throughout all or a significant portion of their range. Species may be determined to be endangered or threatened because of any or a combination of the following factors:
• present or threatened destruction, modification or curtailment of its habitat or range;
• over-utilization for commercial, recreational, scientific or educational purposes;
• disease or predation;
• the inadequacy of other existing regulatory mechanisms; or
• other natural or man-made factors affecting its continued existence.

Note that recent amendments to the Act and current discussions are focusing on the need to recognize distinct genetic units that are effectively below 'species level', so that genetic diversity can be safeguarded in its full variety (see section 2.4.1).

The conservation statuses of the most threatened species of plants and animals in the UK have been assessed by the authors of Red Data Books (RDBs), which contain information on those species believed to be in danger of extinction (Roy & Eversham 1994). Four British RDBs have been published to date, covering vascular plants (Perring & Farrell 1977, 1983), insects (Shirt 1987), other invertebrates (Bratton 1991) and birds (Batten *et al.* 1990). Although RDBs have yet to be published for vertebrates other than birds, RDB categories for reptiles, amphibians and mammals may be derived from other publications (e.g. Anon. 1989a,b; Harris & Jefferies 1991). The categories included in the RDBs are defined as shown in Table 4.1.

Knowledge about the number of RDB species in each taxonomic group can be useful in alerting practitioners to the potential conservation status of species they encounter in EcIA. Roy and Eversham (1994), for example, summarized the numbers of RDB species in each taxonomic group and their RDB status categories. All species of bat found in the UK are included in Schedule 5 of the Wildlife and Countryside Act 1981 and were included in RDB categories 1, 2 or 3. Further survey would obviously be indicated in any situation where adverse impacts on British bats was suspected.

A high proportion of British butterflies and dragonflies are also RDB species. These groups appear to have declined more than most other invertebrate groups in the UK. Four out of 68 butterflies and three out of 44 dragonflies have become extinct as breeding species since recording began and many other serious declines have been documented (Roy & Eversham 1994). Therefore, in the UK, species of butterfly and

Table 4.1 Red Data Book (RDB) categories.

RDB1	Endangered: likely to become extinct if causal factors continue
RDB2	Vulnerable: likely to move into the Endangered category in the near future
RDB3	Rare: taxa that are not at present Endangered or Vulnerable, but that are at risk
RDB4	Out of Danger: formerly meeting one of the above criteria, but now relatively secure due to effective conservation measures or the removal of threats
RDB5	Endemic: not thought to occur naturally outside the British Isles
RDB app	Formerly native, but thought to have become extinct before 1900
RDBK	Insufficiently known: believed to fall into one of RDB categories 1–3, but with insufficient information to determine which category

dragonfly are often potential candidates as VECs meriting detailed
investigation for purposes of EcIA.

The International Union for the Conservation of Nature and Natural
Resources (IUCN) has devised criteria for estimating threat to native
species. Using these criteria for Britain, 21% of fish species, 11% of invertebrate species, 13% of seed plants and 17% of mosses can be considered to be 'threatened'. To derive indicators of sustainable development
for the UK (Fig. 4.2) the Department of the Environment (1996) also
included species that are nationally scarce to estimate the numbers of
native species 'at risk'. Using these figures, over a third of the UK's 2700
native species of mosses, liverworts and lichens are threatened or nationally scarce, about a quarter of the 2300 seed plants, ferns and related
plants, and about a quarter of the 15 000 invertebrate species (Fig. 4.2).

When considering the conservation status of candidate species for
EcIA, it is important to take account of any trends (is the species stable,
increasing or declining) as well as the possible explanation. Is a species
declining because of specific human actions, or because of other reasons,
for example because it is at the geographical limits of its natural range
and susceptible to climatic variation? Species that are both rare and
declining should be included in any EcIA for proposed actions that might
pose a further threat to their survival.

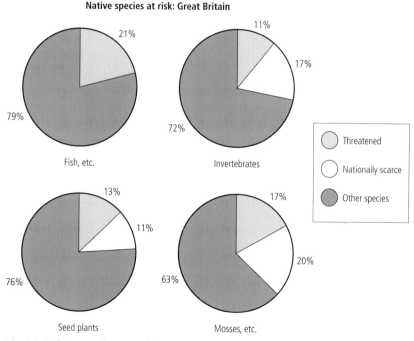

Fig. 4.2 Indicators of sustainable development for the UK. (From Department of the
Environment 1996. Crown copyright is reproduced with the permission of the
Controller of Her Majesty's Stationery Office.)

In conclusion, the status of species should be considered in both spatial and temporal terms: is their geographical range/distribution shrinking and how is it faring relative to historical records? The importance of repeated national recording (monitoring) in this respect cannot be over-emphasized.

4.2.6 Indicator species

Individuals, populations or communities of species, by their presence, their productivity or their condition may act as indicators of habitat conditions, pollution, habitat disturbance, habitat age or continuity or associated biological diversity.

There is considerable potential for the application of 'indicator' theory to EcIA, particularly in selecting candidates for in-depth survey. There are many species that, by their presence alone, reveal a great deal about the environment, ecosystem or habitat under study. For example, all plants can be said, simply by their occurrence, to indicate specific combinations of soil and climatic conditions (Clements 1920). The presence of some species may also imply the likely presence of others, as many species (both plants and animals) form predictable associations.

Many species have been used as indicators in EcIA, but most of these share certain properties or attributes. In general, good indicators:
- respond measurably or detectably (are sensitive);
- respond rapidly or promptly;
- respond reliably or predictably.

Box 4.4 Desirable qualities for invertebrate indicator groups. (After Brown 1991.)

- Taxonomically and ecologically highly diversified (many species in each locality or system)
- Species have high ecological fidelity
- Relatively sedentary
- Species narrowly endemic, or, if widespread, well differentiated (local or regional)
- Taxonomically well known, easy to identify
- Well studied (genetics, behaviour, biochemistry, ecology, biogeography)
- Abundant, non-furtive, easy to find in the field
- Damped fluctuations (always present)
- Easy to obtain large, random samples of species and variation
- Functionally important in ecosystem
- Response to disturbance predictable, rapid, sensitive, analysable and linear
- Associates closely with and indicates other species and specific resources

Table 4.2 Indicator values for invertebrate groups according to Brown (1991).

Collembola	19
Odonata (dragonflies and damselflies)	18
Isoptera (termites)	20
Hemiptera: Coreidae, Pentatomidae, Cygaeidae, Tingidae, Myridae	14
Homoptera: Membracidae, Cercopidae	13
Coleoptera: Carabidae, Cicindellidae, Elateridae, Cerambycidae, Chrysomelidae, Curculionidae	19
Diptera: Asilidae, Tabanidae	16
Hymenoptera: Formicidae	21
Hymenoptera: Apoidae, Vespidae, Sphecidae	18
Lepidoptera: Sphingidae, Saturnoidae	13
Lepidoptera: Arctiidae	15
Lepidoptera: Papillionidae, Pieridae	16
Lepidoptera: Morphinae, Satyrinae	16
Lepidoptera: Bait-attracted Nymphalinae	14
Lepidoptera: Heliconiini, Ithomiinae	21

Good indicators are also likely to be species for which there are reliable survey methods. Box 4.4 lists desirable qualities for invertebrate indicator groups in ecology and biogeography (Brown 1991). Brown (1991) classified the various groups of insects according to their suitability as indicators of biodiversity by scoring them with 0, 1 or 2 for these criteria. The total value of each group as an indicator according to this system (out of 24) is summarized in Table 4.2.

The US Environmental Protection Agency (US EPA), the United Nations Economic and Social Council, the United Nations Environment Programme (UNEP) and other agencies have drawn up lists of indicators suitable for use in environmental quality assessment (see Spellerberg 1991).

4.2.6.1 *Indicators of pollution*

Spellerberg (1991) distinguishes between the following main types of pollution indicator.

• Sentinels: sensitive organisms that can be introduced deliberately as early warning devices or as pollution 'meters' (e.g. phytometers).

• Detectors: species occurring naturally in the area of interest which show measurable responses to environmental change.

• Exploiters: species whose presence indicates the probability of disturbance or pollution.

• Accumulators: organisms that take up and accumulate chemicals in measurable quantities.

• Bioassay organisms: selected organisms used as laboratory reagents to detect the presence and/or concentration of pollutants, or to rank them in order of toxicity, for example.

In EcIA, most use is likely to be made of 'detectors' and 'exploiters'. Well-known examples are lichens, which are useful as 'detectors' because of their sensitivity to atmospheric pollution and their longevity. Pollutants like SO_2 affect the algal component of lichens, causing a breakdown in the symbiotic relationship between algae and fungus. Lichens have been used in ecological assessment for many years (Hawksworth & Rose 1976; Skye 1979) and there are a number of guides to their use in monitoring air pollution. Hawksworth and Rose (1976), for example, developed a method of estimating mean winter levels of atmospheric SO_2 based on the presence of certain indicator taxa of lichen, lichen species composition then being used to derive maps of pollution zones. One problem with the use of lichens as pollution indicators for monitoring for the purposes of EcIA is their relatively slow response time: often years, rather than weeks or days (Spellerberg 1991).

Numerous other plant species are good detectors and exploiters. Baker (1987) refers to their use in studying levels and types of metal contamination in mine workings. Metalliferous mine spoils often support floras that are very distinct from surrounding vegetation on uncontaminated soil. Metallophytes (found only on metal-contaminated soils) and pseudometallophytes (occurring on both contaminated and 'normal soils') can be recognized. Plant community composition appears often to be determined by a mixture of metal tolerance and interspecific competition for other resources, notably nitrogen, phosphorous and potassium. *Psychotria douarrei* is a New Caledonian species of the genus *Psychotria* (Rubiaceae) which grows only on ultramafic soils, which are typically infertile (having low nitrogen and potassium status), have unfavourable magnesium/calcium ratios and are often enriched with nickel, cadmium and sometimes cobalt.

4.2.6.2 *Indicators of disturbance*

It is well known that there are some species particularly vulnerable to disturbance and others that are more robust and able to exploit disturbed areas readily. Disturbances can therefore affect both habitat quality and availability. The mud snail *Lymnaea glabra* (Fig. 4.3), for example, is an amphibious snail, which occurs in shallow seasonal pools in the UK, primarily in areas of unimproved (relatively undisturbed) agricultural pasture that flood in winter and do not dry out again until early summer when the snail breeds. It is unlikely to be found in pastures that have been drained or treated with herbicides or pesticides (B.C. Eversham, personal communication 1998). Similar relationships can be observed for other groups or species. For example, some bird species are tolerant of noise or disturbance by people, whereas others are elusive and shy.

Fig. 4.3 Mud snail *Lymnaea glabra*. (Courtesy of Brian Eversham.)

Different forms of disturbance can also be recognized through characteristic and predictable changes in species composition. This can be illustrated clearly by reference to the UK's National Vegetation Classification, which recognizes zonations and successions between distinct vegetation communities that are mediated by management (a special form of disturbance!). For example, *Arrhenatherum elatioris* grassland communities develop in mown but ungrazed areas. In damper areas, introduction of winter grazing is often associated with a transition to *Alopecurus pratensis–Sanguisorba officinalis* flood meadow, while introduction of grazing in both summer and winter can be associated with a transition to a *Lolio–Cynosuretum*. Many such characteristic successions can be recognized which are associated with certain patterns of disturbance.

4.2.6.3 *Indicators of age or continuity*
The presence of some species, or certain combinations of species may, in some cases, indicate the continuous presence of undisturbed habitat in a location for long periods of time. In the UK, for example, certain components of the ground flora are found to be present almost entirely in woods known to have been in existence on the same site for long periods (referred to as 'ancient woodland indicators') (Peterken 1974). Ancient woodlands are defined as those over 400 years old and some are believed to have been present continuously since the 'wild wood' occupied most of the British landscape after the last ice age. Interestingly, the reliability of some plant species as ancient woodland indicators varies regionally. Species like yellow archangel *Lamiastrum galeobdolon* can nearly always be regarded as an ancient woodland indicator. The bluebell *Hyacinthoides nonscripta*, on the other hand, is largely confined to ancient woodlands in the south, but is able to colonize newer plantations in the uplands (Buckley & Fraser 1998). The lemon slug *Limax tenellus* (Fig. 4.4) is among the most stenotopic ancient woodland

Fig. 4.4 The lemon slug *Limax tenellus*. (Courtesy of Brian Eversham.)

indicators (B.C. Eversham, personal communication 1998), although I am not aware of any example where it has been used as an indicator in EcIA. Unlike many other potential ancient woodland indicators it occurs on a wide range of soils and in a wide range of woodland types: Caledonian pinewoods as well as broadleaved lowland woodlands. It seems to be incapable of colonizing recent plantations, even those which are two or three hundred years old, perhaps because it feeds on fungi and lichens, which are only found in ancient woodland.

Just as some species appear to indicate 'age' or lack of historical disturbance, community composition can also be influenced by 'age', sometimes in predictable ways. Hooper's (1970) surveys of hedgerows in Britain suggested that they could actually be dated on the basis of their species composition. The rule of thumb derived from his surveys was that a 30-yd (27.4-m) hedge would contain approximately one woody species for every 100 years of its existence (Pollard *et al.* 1974). Subsequent studies have demonstrated that Hooper's 'rule' does not always apply. Other factors must also be taken into account, including the management of the hedgerow and its location (Willmot 1980 in Spellerberg 1991), but the 'rule' remains useful if used with caution.

'Older' and more 'continuous' habitats are often rich in species and also have a species composition that differs significantly from that of 'younger', more 'disturbed' habitats. For this reason, it is important to be able to recognize them and to ensure that they are protected. Ancient woodlands, by definition, cannot be replaced without a very long wait, and their destruction should be avoided if at all possible. Kirby (1992) cites the examples of the New Forest, Windsor and Sherwood Forests in the UK, which support invertebrate faunas endangered throughout Europe. If such sites were lost, their associated characteristic flora and faunas would probably be lost forever. In the UK, even isolated individual ancient trees can provide important habitat for particular invertebrate species and communities rarely found on younger trees, but these faunas are unsustainable unless suitably mature trees are available in the locality for the time when the original host tree dies (Kirby 1992).

4.2.7 Guild indicators

Root (1967) introduced the guild concept in a synecological study of five species of birds that foraged in the same general fashion. He defined a 'guild' as a 'group of species that exploit the same class of resources in a similar way'. However, he also showed that, although the species obtained resources in a similar fashion, they differed in both the actual resources they exploited and in the methods used to obtain them (Block *et al.* 1987). One advantage of using an approach based on guilds is that it allows habitat resources to be assessed despite changes in species composition (Short & Burnham 1982). Species assemblages are not fixed, but new species may adapt to use resources similar to those of the original complement of species. Using guilds as a basis for assessment therefore encourages a resource-based approach.

A guild-indicator species is a member of an ecological guild that acts as an indicator for all species from the guild to environmental change (Block *et al.* 1987). The concept is only valid if species in a guild use similar resources so that, if the impact of environmental change is determined for one species, the other species present can be assumed to be similarly affected (Severinghaus 1981). Species must therefore be sympatric and have very similar patterns of resource use. As a general rule, the more restricted the membership of the guild (in terms of numbers of species), the more likely it is that species within the guild will be ecologically similar (Mannan *et al.* 1984). However, there are probably few cases where the habitat suitability or population status of a guild indicator will exactly parallel those of other guild members. Block *et al.* (1987) therefore recommended that the use of guild indicators should be restricted to species that are closely related ecologically and are constrained within the same vegetation type and seral stage. For endangered species, there is a risk that its status may be jeopardized if it is monitored indirectly using a guild indicator, so guild indicators should not be used to substitute for monitoring of other VECs.

One problem with applying this concept to EcIA, is that a considerable amount of knowledge is needed to establish the presence of guilds in an area and to estimate the significance of differences between guild members. The concept should only be used for species that have been well studied and for which detailed autecological information is available.

The US Fish and Wildlife Service (1980) has developed HEP, which relies heavily on the concept of guilds for the selection of suitable 'evaluation species', categorizing vertebrate species according to 'feeding' and 'reproductive' guilds. For HEP, feeding guilds are defined in terms of feeding mode (e.g. 'carnivore', 'herbivore' or 'omnivore') and the vegetation stratum where feeding occurs (e.g. 'forest canopy', 'shrub layer' or 'surface'). Reproductive guilds are defined only in terms of the stratum where reproduction occurs (US Fish and Wildlife Service 1980). Guilding can be structured to take account of different levels of detail. For wetlands, for example, there are three levels of 'locational descriptor'. The first, broadest level of descriptor is simply 'water within cover type'. 'Level 2' descriptors include the 'water surface', 'water subsurface' and 'water bottom', while 'Level 3' descriptors divide 'water surface' into 'protruding structures' like stumps and logs, 'emergent vegetation' and 'floating vegetation'. For purposes of EcIA, only those strata likely to be affected by a proposal need to be considered.

Use of the guild concept to select evaluation species requires the following steps:

Table 4.3 Terrestrial feeding guilds in deciduous forest in the south-central US. (US Fish and Wildlife Service 1980.)

Cover type	Feeding mode					
	Vertebrate carnivore	Invertebrate carnivore	General omnivore	Scavenger	Herbivore (fungi)	Herbivore (vascular plants)
Tree canopy		Hairy woodpecker				Fox squirrel Grey squirrel
Tree boles		Hairy woodpecker	Pileated woodpecker Carolina chickadee			
Shrub layer						White-tailed deer Eastern cottontail Eastern woodrat
Land surface	Bobcat Red-tailed hawk Red-shouldered hawk Barred owl	Nine-banded armadillo	Grey fox Racoon			White-tailed deer Eastern cottontail Eastern woodrat Golden mouse Fox squirrel
Subsurface						

1 definition of 'feeding and reproductive guild cells' (selecting appropriate descriptors);

2 selection of species from each cover type that meet guild definitions; and

3 selection of species from each guild to act as 'study evaluation species'.

Feeding and reproductive matrices are required for each main cover type. An example of a feeding matrix is shown in Table 4.3, with feeding modes entered horizontally across the top and locational descriptors (i.e. strata) entered down the left side of the matrix. Some species that are representative of the different 'guild cells' have been entered into the matrix. The categorization of species into guilds is not always straightforward, as there may be hundreds of vertebrate species in the study area. One criterion that might be used to select species might be the availability of adequate habitat information (US Fish and Wildlife Service 1980). If more than one species is entered into any guild, at least one should be selected as a species to represent the guild. Again, it can be necessary to screen species, for example according to their sensitivity to proposed activities, their ecological role and so on.

4.2.8 'Umbrella species'

'Umbrella species' are those for which targeted conservation management will also benefit other species using the same habitat. They include species such as the Florida scrub jay *Aphelocoma coerulescens coerulescens*, which is restricted to scrub oak habitats on the Florida peninsula, which are also inhabited by a number of other rare species of reptiles, insects and plants. Many of these species, are restricted to these scrub habitats and benefit from management of land carried out to protect the Florida scrub jay. Similarly many, but not all, of the species requiring old-growth temperate rain forest benefit if habitat is protected for the benefit of the spotted owl *Strix occidentalis*. The most useful umbrella species are those whose habitats harbour numerous rare, endemic species. Umbrella species should therefore be given priority in proportion to the number of other endemic, rare species that co-occur with them. However, it is important to recognize that all species could probably be considered to be an umbrella species at some scale. Furthermore, most endangered species have very exacting and subtly different requirements, so even when a suitable umbrella species exists, the specific ecological needs of other community members must also be taken into account. Nevertheless the inclusion of umbrella species as VECs for purposes of EcIA often makes a great deal of sense, ensuring that limited resources are allocated to study of species known to indicate something about the quality and provision of habitat or a number of other, associated species.

4.2.9 Ecological role: keystone species

Some species appear to play a particularly important role in ecosystem function, with other species depending on them for their survival. Removal of such species from an ecosystem is likely to result in substantial alterations in ecosystem function.

'Keystone species', for example, are 'controlling' species, which have a significant role in ecosystem function, which is often disproportionately large relative to their abundance (Power & Mills 1995) and whose removal from a system can therefore lead to significant and often disproportionate changes (Paine 1974). Holling (1992) suggested that all terrestrial ecosystems are controlled and organized by a small set of key plants, animals and abiotic processes, and, not surprisingly, keystone species or processes are often selected as VECs in EcIA. However, note that keystone components of ecosystems may be important only for restricted periods, or have long-term periodicity (e.g. fire in boreal forests), which is difficult to characterize on the basis of limited surveys.

Working out which species are keystone species can be difficult, because the importance of a species in an ecosystem is not necessarily proportional to its size, abundance, ease of sampling or charisma! For example, tiny fig wasps and African elephants are both keystone species. There is no theoretical reason why organisms at any trophic level in an ecosystem should not act as 'controlling species' (Hunter & Price 1992). Spencer *et al.* (1991) documented a case where an introduced species, the opossum shrimp *Mysis relicta*, significantly reduced populations of native zooplankton. This reduction cascaded up the food chain and had negative effects on the top predators in the system, bears and eagles. The opossum shrimp therefore appeared to be a keystone species.

Keystone processes can also be identified; for example, the habitat modification carried out by beavers is a biotic keystone process (Merriam *et al.* 1993). By creating ponds and eventually beaver meadows, beavers both add new elements to the landscape and control processes that determine the evolution of the landscape mosaic. Traditional agronomic theories about limiting growth factors reflect the fact that abiotic factors can also act in a 'keystone' fashion. Water in a drought-prone system has an overriding influence on ecosystem structure and function compared with other factors because it is so limited in supply (Merriam *et al.* 1993). The concept of 'key limiting factors' is well developed in agronomic science, where it is crucial to the determination of crop nutrient needs. Similarly, disturbance is a key process in many systems. Often it is initiated by humans, and associated with proposed development actions. The 'keystone' concept has even been extended to whole ecosystems, 'keystone ecosystems' being defined as ecosystems that contain or support a high number of species or have distinctive species compositions (e.g. Stohlgren *et al.* 1997).

4.2.10 Availability of consistent survey methods

Those environmental attributes selected to form the focus of the impact assessment should be those for which follow-up monitoring is a realistic option. There is little point in going on to invest time and resources in the one-off survey of attributes that cannot subsequently be monitored. By failing to institute a formal requirement for monitoring, much of the current EIA and related legislation effectively sanctions 'one-off' surveys, which are of little predictive value and do not contribute to the overall knowledge base. In the UK there are national monitoring schemes for butterflies (the 'Butterfly Monitoring Scheme') and for birds (the 'Breeding Bird Survey' (BBS)), for example. The BBS is based on ten 100-m transects within a 1-km survey square. Such schemes make it possible to record species using consistent survey and recording methods, thereby facilitating subsequent monitoring and follow-up. Where consistent survey methods can be used there is considerably more scope for comparative analysis, whether between sites or between years. EcIA-related surveys can then also contribute to national monitoring, thereby playing a part in the recording and conservation of native biodiversity. Figure 4.5 shows population indices for two bird species between 1981

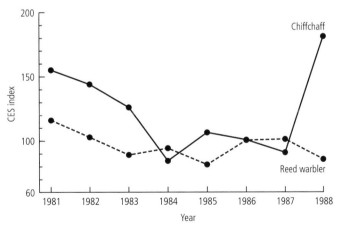

Fig. 4.5 The UK Constant Effort Sites (CES) Scheme population indices for the chiffchaff *Phylloscopus collybita* and the reed warbler *Acrocephalus scirpaceus*. The base year of the index is 1986, and chiffchaff and reed warbler indices are shown from 1981 to 1988. Reed warbler populations have remained stable and chiffchaff populations have increased, for unknown reasons. (Reprinted from Bibby, C.J., Burgess, N.D. & Hill, D.A. (1992) *Bird Census Techniques*, by permission of the publisher Academic Press Limited, London; using data from Peach & Baillie 1989.)

and 1988, derived using data from the British Trust for Ornithology's (BTO's) Constant Effort Sites (CES) Scheme. This scheme requires constant net sites to be visited 12 times between May and August. Results are used to investigate population change (between-year changes in numbers of adults captured), productivity (the ratio of juveniles to adults captured late in the breeding season) and survival (based on between-year recaptures of ringed birds). The same ringing and netting sites are used each year and the same length and type of net (Bibby *et al.* 1992a). The scheme ensures that a reasonably reliable baseline of information is available for some bird species and allows the results of other surveys to be compared with known trends.

4.2.11 Expediency

Often, it is nothing more than expediency that is the deciding factor. Certainly, EcIAs for purposes of EIA in the UK demonstrate a considerable bias towards those taxonomic groups which are easily recorded, cheap to survey, well known and for which effective evaluation mechanisms are available. Review of ESs has shown that higher plants (conveniently static) are the most frequently recorded group of organisms, followed by birds, mammals and invertebrates (Treweek 1996). Some other groups (the fungi, for example) are virtually ignored. Certainly, the relative investments in survey effort for the different taxonomic groups of species found in the UK fail to reflect their relative abundance. There are almost 30 000 species of invertebrate in the UK, which greatly outnumbers the total for higher plants and vertebrates combined (Kirby 1992). Not only are some of the more neglected groups of species more numerous than those generally studied, they are often more appropriate as indicators.

Often, surveys are, quite simply, focused on those groups or species which are the taxonomic favourite of in-house ecologists, or the consultants employed to carry them out. As well as the method *per se*, the way in which methods are applied (sampling regime and sampling protocols) will be of great importance in any quantitative studies. With respect to biodiversity assessment, Janzen and Hallwachs (1994) divided all taxa into four groupings according to the feasibility with which each one can be fully inventoried, and urged that there should not be an over-emphasis on the 'very feasible' (i.e. all vertebrates, higher plants, lichens, moths, molluscs, crustaceans) at the expense of the more challenging groups, including the more species-rich and less identifiable groups such as weevils, nematodes, mites and fungi.

4.3 Habitats

VECs may also be habitats. The UK's Biodiversity Steering Group identified 38 'key habitats' that are considered to be 'at risk' (such as those which have shown a high rate of decline, particularly over the last 20 years), rare, functionally critical or important for associated key species which are themselves rare or at risk. A monitoring programme is being established under the UK's Biodiversity Action Plan to measure changes in both the extent and quality of these key habitats. Quality is currently measured largely on the basis of populations of characteristic fauna and flora. In time, it may be possible to derive other indicators that can be used to assess the conservation status of key habitats. These indicators would have an important role in ecological evaluation for EcIA, making it possible to consider localized impacts in the context of national trends in the status of key habitats and associated species. In the meantime, these 38 habitats should automatically be considered to constitute VECs in the context of EcIA in the UK. A full list of the UK's Key Habitats and the Annex 1 habitat types identified for priority action in the EC Habitats Directive can be found in *Biodiversity: the UK Steering Group Report*, published by HMSO (1996). Their presence in an area affected by a proposal should dictate the need for more detailed study just as much as the presence of any site formally designated for nature conservation (see section 4.4). Similar initiatives are underway throughout the European Union.

The needs of individual species should never be considered in isolation. By focusing on habitats it is easier to consider the combined needs of different species and their interdependencies. One of the main advantages of selecting 'habitats' as VECs is the fact that the concept of 'habitat' can be used to link local and landscape-scale effects on associated species so that their individual and combined needs can be considered at different levels as necessary. Habitat distribution or extent is often also more straightforward to measure and monitor than the distributions of associated species. More importantly, observed species distributions are often meaningless without some associated information about habitat variables. An example used by Bibby *et al.* (1992a) using data from Williamson (1968) illustrates this very clearly. The map showing territories alone reveals very little. If associated habitat features are marked, however, it becomes clear that yellowhammers *Emberiza citrinella* prefer territories along field boundaries like hedgerows: few occur in the centres of fields (Fig. 4.6). Habitat-based EcIA is considered in detail in section 5.4.4.

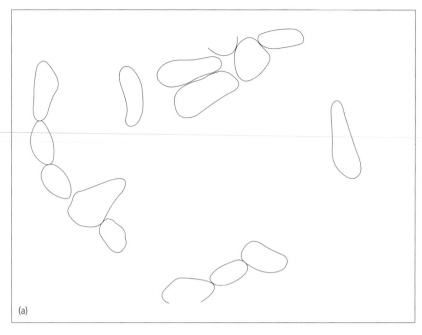

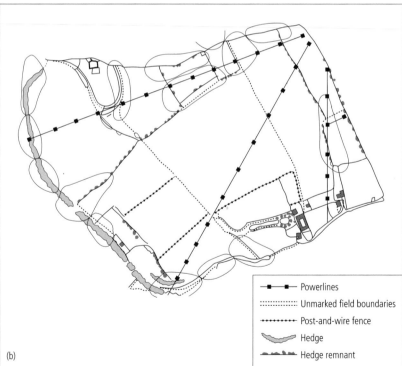

Fig. 4.6 (a) Map of yellowhammer *Emberiza citrinella* territories alone and (b) shown in association with habitat. (Reprinted from Bibby, C.J., Burgess, N.D. & Hill, D.A. (1992) *Bird Census Techniques*, by permission of the publisher Academic Press Limited, London; using data from Williamson 1968.)

4.4 Special (designated) sites

Formal designation of sites for nature conservation is an obvious trigger for further assessment. It can be taken as read that such resources are valued by society without applying any additional screening procedure. Many protected sites, or sites that support protected species, are unique or irreplaceable with current technology.

In the UK, areas of land that have national or international conservation value are designated as 'Sites of Special Scientific Interest' (SSSI) under the Wildlife and Countryside Act of 1981. In 1994 SSSIs (including those in Northern Ireland) covered about 8% of the UK's land surface (RCEP 1994). They included 304 national nature reserves, 86 areas designated as Special Protection Areas (SPAs) under the European Community's Birds Directive and 76 wetlands designated as being of international importance under the Ramsar Convention. Designation

Table 4.4 Formal designations that should be considered in ecological impact assessment. (After English Nature 1994a.)

Status	Designation
International	Special Protection Areas (SPAs)
	RAMSAR Sites
	World Heritage Sites
	Biosphere Reserves
	Special Areas of Conservation
National	National Nature Reserves (NNRs)
	Sites of Special Scientific Interest (SSSIs)
	Nature Conservation Review Sites (Grade 1)
	Marine Nature Reserves
Other	Local Nature Reserves
	County Trust Reserves*
	Local Authority Reserves, including Sites of Importance for Nature Conservation (SINCs)*
	Wildlife Trust Reserves*
	Reserves owned by charities and other non-governmental organizations (e.g. sites owned by the Royal Society for the Protection of Birds, RSPB)*
Additional designations/descriptions to consider	National Parks
	Areas of Outstanding Natural Beauty
	Heritage Coast
	Aquifer Protection Zones
	Environmentally Sensitive Areas
	Historic Parks and Gardens*
	Tree Preservation Orders (TPOs)
	Conservation Areas

NB: Sites may be the subject of more than one designation.
* Non-statutory sites.

as an SSSI was not intended to provide absolute protection against development, but the government's stated policy is to avoid them wherever possible. In practice, lack of clarity about which sites merit absolute protection, which constitute critical natural capital and which should merit replacement in kind if damaged or lost through development has meant that a number of valuable wildlife sites have disappeared. Linear developments are particularly damaging to SSSIs in terms of overall numbers affected. English Nature estimated that 150 SSSIs were potentially at risk from the UK's proposed trunk road programme as it existed in 1992 and a survey by county wildlife trusts in south-east England identified 372 wildlife sites that were under threat from road schemes, including 50 SSSIs (RCEP 1994). The trunk road programme was later reviewed and the number of SSSIs likely to be affected was reduced, but proposed new roads and road widening schemes remain a major cause of threat to protected sites and species in the UK.

For the UK, designations that should be considered in EcIA are listed in Table 4.4.

Site-based designations make it possible to identify important sites without preliminary survey, although it is important to remember that many of these sites may not be surveyed or monitored on a regular basis. For purposes of ecological impact prediction, additional information from survey may well be required.

4.5 Ecosystem structure

Measures of ecosystem structure selected as VECs might include measures of abundance or biomass, aspects of community composition, species richness, species diversity, trophic organization or spatial structure. It is not necessarily appropriate to draw a distinction between measures of ecosystem structure and function, as they may interact. For example, there is evidence to suggest that species diversity can affect ecosystem function directly. Naeem *et al.* (1994) constructed experimental microcosms with different trophic levels (decomposers, primary producers, primary consumers and secondary consumers) and three levels of biodiversity (low, medium and high). They found that treatments with higher diversity fixed more carbon (intercepted more light) and had a higher overall cover than treatments with lower diversity. Other ecosystem functions (such as nutrient retention, respiration and decomposition) did vary between treatments but showed no clear pattern with respect to species diversity. Nevertheless, these experiments do suggest that species diversity can actually influence some ecosystem functions.

Conversely, ecosystem structure can certainly affect function. There are obvious links between vegetation structure and the composition of

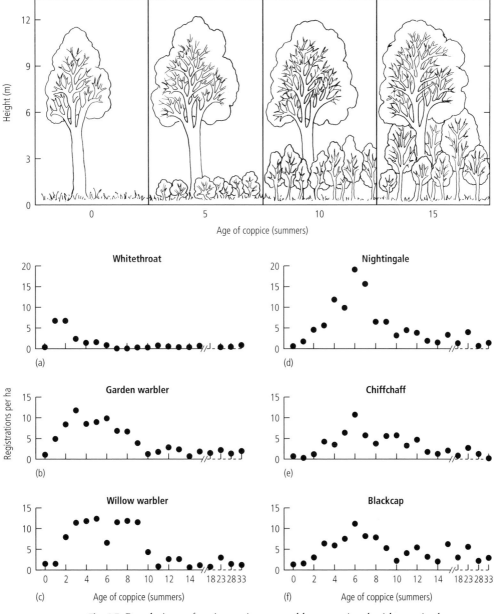

Fig. 4.7 Populations of various migrant warblers associated with coppiced woodland of different ages. Following coppicing the cut shrubs grow rapidly and there is a rapid change in the vertical vegetation structure and shrub density of the example woodland. This rapid structural change is reflected in populations of various African migrant warbler species (a–f) which select different structural characteristics (in this case age/density) of coppice. After around 10 years of growth the structure of the coppiced shrub layer is changing much more slowly and the bird community becomes more stable. Populations of the African migrant warbler species fall, and are replaced by resident birds such as tits and thrushes (not shown). The species graphs are presented (a–f) approximately in order of habitat selection with species preferring the youngest coppice coming first. (Reprinted from Bibby, C.J., Burgess, N.D. & Hill, D.A. (1992) *Bird Census Techniques*, by permission of the publisher Academic Press Limited, London; after Fuller *et al*. 1989.)

the associated fauna. Fuller *et al.* (1989) demonstrated this for coppiced woodland in the UK. Different populations of songbirds were associated with different ages and structures of post-coppice vegetation. Following coppicing the cut shrubs grow rapidly and there are associated rapid changes in the vertical structure of the vegetation and its density. Various African migrant species of warbler are associated with different ages and densities of coppice vegetation as shown in Fig. 4.7. After about 10 years, however, the vegetation structure stabilizes and resident birds move in (Bibby *et al.* 1992a).

The relationship between structure and function must also be considered with respect to the controversy over 'redundancy' in ecosystems. This has important implications for EcIA where we may wish to consider the implications of losing different species from an ecosystem under different development scenarios.

4.5.1 Community composition

Characteristic assemblages of species occur, which reflect particular coincidences of abiotic conditions and management (past and present). In countries where anthropogenic influences have a long history, many distinct communities evolved under particular traditions of land use, which decline as management practices alter. In the UK, for example, the plant communities characteristic of traditionally managed wet grassland habitats have become very restricted in extent due to land drainage for agriculture and the increased use of artificial fertilizers, herbicides and large machinery. These communities do not necessarily consist entirely of rare or threatened species, but often constitute unique or unusual groupings or assemblages of species. Analysis at the community level is more challenging than analysis of individual species' needs and demonstrates the need for holistic approaches to ecosystem study. Predicting community responses can be very difficult even when individual species' requirements are relatively well known, due to interspecific interactions. However, some characteristic community-level responses can be identified. For example, Fig. 4.8 shows well-established patterns of community change in wet grasslands under different management regimes. Where such patterns of change can be clearly identified and explained in terms of management change or the imposition of a specific stress or disturbance, general impact patterns can be identified. Communities that are relatively well understood in terms of their responses to specific stressors should be considered as VECs.

4.5.2 Species richness and species diversity

Characteristic levels of species richness and species diversity can

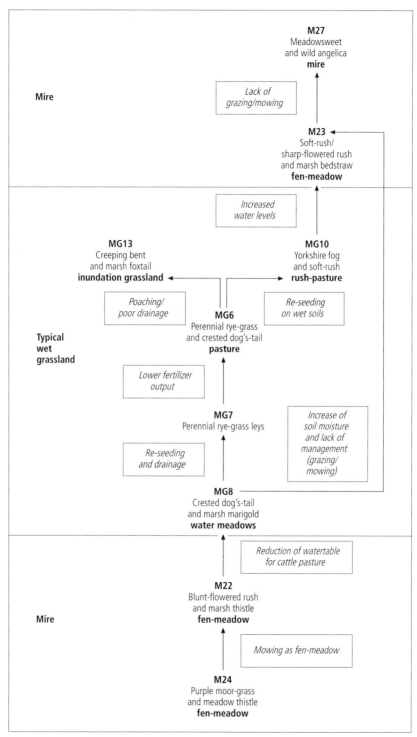

Fig. 4.8 The effect of grassland and water management on grassland communities. (After Prosser & Wallace 1995.)

sometimes be identified. The advantage of including these aspects of ecosystem structure as VECs for EcIA is that they both lend themselves to objective measurement. In other words, clear measurement endpoints can be identified to reflect assessment criteria. Declines in species richness or diversity commonly reflect adverse conditions for certain community constituents and can therefore be useful as indicators of degradation. On the other hand, there are cases where increases in species richness or diversity might actually reflect an undesirable change, for example the ingress of 'alien' or non-native species, which are not characteristic of the community affected. Qualitative aspects of changes in species richness or diversity should therefore always be taken into account. Figure 4.9 shows how chalk downland turf increased in species richness following translocation. However, the main increase was in 'weedy' and 'generalist' species rather than chalk downland 'specialists' (R. Pywell & M. Stevenson, personal communication 1998).

As demonstrated above, it is important to consider the contribution made by each species to ecosystem function and to make sure that species with unique and vital roles are protected. Maybe we should be concerned more with functional diversity than with species diversity *per se*, and should focus more closely on the actual roles of species? Furthermore, as pressure to conserve biodiversity increases, it is becoming clear that populations containing 'representative genetic diversity' may often be below the level of species (Mallet 1996).

Species richness is the total number of species present. It may be expressed as the number of species in a sample or habitat, but is more usefully expressed as species richness per unit area (Spellerberg 1991). It has been used often as a criterion in nature conservation evaluation

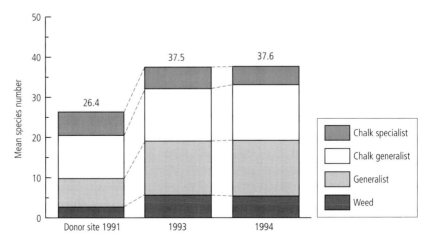

Fig. 4.9 Species composition of translocated chalk downland. (Data from Ward & Stevenson 1994.)

to rank sites and prioritize them for selection as nature reserves. For example, species richness was one of the 'biological features' used in Tans' (1974) 'Priority Ranking System', which he applied to 'natural areas' in Wisconsin, North America.

Measures of species diversity take account of the relative abundance of species, thereby giving a more accurate representation of community composition. The concepts of species richness and diversity are explored in more depth in section 6.3.14.

4.6   Ecosystem functions or processes

There are a number of 'key functional attributes' of ecosystems or ecosystem processes that should be taken into account when carrying out EcIA and that may be selected as VECs. These might include measures of energy flow, respiration or decomposition rates or nutrient cycling. These parameters are not always straightforward to measure, but ecosystem functions like nutrient cycling are often affected before impacts are expressed by other VECs, for example at the population level (Shaeffer 1991). In fact, ecosystem functions drive ecological responses, so it may be necessary to measure them directly in some cases, to be able to predict ecological responses. Some ecosystems have characteristic patterns of species succession that might be regarded as key attributes. Certainly, it is always important to remember the dynamic nature of ecosystems. Altered processes at one point in time may have numerous delayed effects on ecosystem characteristics long into the future. It is not always easy to pinpoint those functions or processes which are fundamental to the longer-term continuity of an ecosystems special characteristics.

Some of the important functions or processes that may be suitable for inclusion as 'valued ecosystem processes' (VEPs) are listed in Box 4.5.

While it is clearly important to understand ecosystem function in order to develop a rational basis for prediction, attributes of ecosystem function are not always straightforward to measure directly. For example, one way to characterize ecosystem function is to track energy flows through ecosystems. Almost any impact on ecosystems can be translated into altered energy flow. Each trophic level in an ecosystem shows measurable energy transfers, but a great deal of energy may also be locked up in inaccessible forms for long periods. Techniques for systems analysis and simulation have improved ability to measure and analyse energy flow in large systems on an experimental basis, but the large-scale models that have been developed rarely produce reliable predictions (Odum 1977).

Some changes in function can be quantified, alternatively, by measuring structural attributes of ecosystems. In other words, there may be more

---

Box 4.5 Some possible 'valued ecosystem processes'.

- Nutrient cycling (can affect system productivity and species composition)
- Energy flow (affects ability of systems to 'support' component species)
- Productivity (affects ecosystem function and species composition)
- Eutrophication (a form of increased productivity with implications for species composition)
- Succession (knowledge of patterns of succession is important for predicting community change over time)
- Colonization (can be key in maintaining populations)
- Dispersal (can be key in maintaining populations and is also important with respect to ability to recover following impact)
- Competition (altered competition has implications for species composition and patterns of succession)
- Assimilative capacity (can affect ability of a system to absorb or recover from pollution, for example)
- Various population processes (see section 4.6.1)

---

easily measurable surrogates for ecosystem function. For example, the structure of a plant community can be described in various ways, based on variables such as biomass at different trophic levels, species composition, life forms, the relative abundance of species, etc. (see section 4.5). In some cases, species composition is known to reflect the amount of energy stored or available in a system. Nutrient-rich calcareous soils in the UK, for example, support less species-rich grassland communities than 'impoverished' calcareous soils. In this example species composition is heavily mediated by interspecific competition.

4.6.1 Population processes

Understanding of the mechanisms that drive population growth or decline is very important in predicting population-level responses to particular stressors, as well as being essential for effective management of populations of endangered species. In the context of EcIA it may be important to know how populations will be affected by mortality induced directly by a development (e.g. through collisions with vehicles on a new road) or less directly through habitat loss. Understanding population responses to habitat loss or alteration is actually very limited. A series of recent papers describe research on the dynamics of bird populations following habitat disruption. Although birds may be able to move on to new habitat following disturbance or habitat loss, there are energetic costs associated with this, which can affect population dynamics in the longer term. For example, a study by Goss-Custard *et al.* (1995) used a demographic modelling approach to show that different breeding subpopulations of a migrating shorebird, the oyster-catcher *Haematopus ostralagus* sharing the same wintering area might be affected in different ways by progressive reduction of their winter habitat. Simulated responses to removal of winter habitat were initially similar in two subpopulations, which decreased in parallel. However, as habitat loss continued, the subpopulation with lower fledgling production began to be disproportionately affected. This demonstrates that population processes and dynamics can have a significant influence on the responses of different populations of the same species to the same intensifying competitive pressures. A follow-on study explored the causes of the differences observed in the modelling study (Durell *et al.* 1997) (Fig. 4.10). A simplified version of the model was used with two or four subpopulations wintering in one area. Two simple density-dependent functions were included: one for breeding territoriality and one for winter mortality, with density dependence starting only above a certain threshold density (see section 4.6.2). Each pair that acquired a breeding territory reared a constant number of fledged young per year. As the amount of winter habitat was reduced, the subpopulations with the lower reproductive rates began to decline at an accelerated rate as soon as the point was reached when all adults were able to obtain a territory and breed. Before this point, competition for territories ensured that breeding birds that had died in the winter were replaced by non-breeders. After this point, the reproductive output of the subpopulation was not sufficient to replace birds dying in the winter. At the same time, winter mortality rates remained high as winter densities were sustained by birds from other subpopulations. These results suggest that conservationists should pay particular attention to monitoring breeding numbers in local subpopulations with below-average rates of reproduction. It is in these subpopulations that the early effects of cumulative winter habitat loss are most likely to be detected. The results also suggest that declines in local breeding subpopulations may actually be attributable to the effects of winter habitat loss and will not necessarily be explained by factors operating in breeding localities themselves. Therefore, not only can responses to habitat loss be influenced by existing population dynamics, but responses to habitat loss may become apparent at some distance away, and at totally different stages in the life cycle of a species.

Understanding of population processes is also vital if individuals or populations are translocated for mitigation purposes. For example, the 'rescue' of great crested newts *Triturus cristatus* and their addition to existing populations might disrupt the structure and dynamics of the receiving population. While a certain limited number of individuals may be 'saved', one distinct population has still been lost and the resulting 'pooled' populations may actually become less stable. Where vulnerable

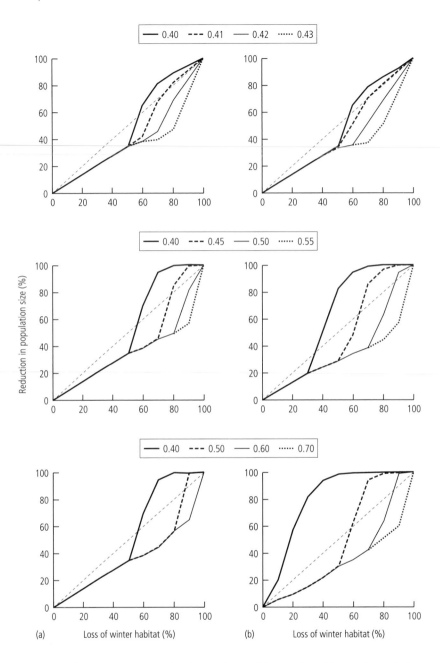

Fig. 4.10 The proportionate reduction in the equilibrium size of four subpopulations to an increasing percentage loss of winter habitat. (a) Breeding area was adjusted so that each subpopulation started at the same size. (b) Breeding areas were the same so that the four subpopulations started at different sizes. $b_T = 0.5$, $b_W = 0.05$ (b_T measures the increasing resistance of territory owners and/or the declining compressibility of territories; b_W is base density-independent winter mortality). The different reproductive rates are given above each graph. The fine dashed line represents $y = x$. (From Durell *et al.* 1997.)

or valued populations are potentially affected by a proposal, key population processes should be identified and included as VECs.

4.6.2 Regulation of population size: density-dependent and density-independent mechanisms

Density-dependent regulation is defined as the regulation of population size by mechanisms that are themselves controlled by the size of that population. The effectiveness of these mechanisms increases as population size increases. Under density-dependent population growth, population growth rate depends on population size, resulting in an 'S-shaped' or sigmoid growth curve. The point of stabilization, at which the growth rate of the population reaches zero, is termed the 'saturation value' or 'carrying capacity' of the environment for that organism.

Density independence occurs when, for example in a new environment, the population density of an organism increases rapidly in an exponential (logarithmic) form but then stops abruptly as environmental resistance or seasonality or some other (external) factor suddenly becomes effective. When density-independent processes are operating, population numbers typically show great fluctuation, giving the characteristic 'boom-and-bust' cycles of some insects. Regulation of population growth is therefore only tied directly to population density in the final stages (Allaby 1994).

Carrying capacity is the maximum population of a given organism that a particular environment can sustain. There are circumstances in which human intervention can enhance the potential of the environment and increase its carrying capacity. Similarly the disruption of the environment for development can have the opposite effect, reducing the quality and amount of available habitat and the overall carrying capacity of the environment for the species it supports. The term 'environmental resistance' is sometimes used to refer the combined environmental limiting factors, both biotic and abiotic, which act to prevent the biotic potential of an organism from being realized. Such factors might include the availability of essential resources, predation, disease, the accumulation of toxic metabolic wastes or behavioural stresses caused by overcrowding and may all be influenced by development actions. Development actions causing habitat loss, for example, are likely to increase environmental resistance and reduce carrying capacity, with consequent effects on the population dynamics of associated species.

While it is generally acknowledged that the conservation and management of populations requires insight into mechanisms of population regulation, unambiguous identification of density dependence can be very difficult. In a study by Middleton and Nisbet (1997) of the acorn woodpecker *Melanerpes formicivorus* (an endangered species in the US), data for a semi-isolated subpopulation was tested using a number of models based on different assumptions about population structure and regulation, i.e.:

• a closed population with density-independent vital rates and an upper limit to population size;
• a regulated, closed population;
• a regulated subpopulation with immigration and emigration.

Tests for density dependence of the reproductive rate and juvenile survival were inconclusive for this species, but adult survival does appear to be density dependent. The study showed that very small interchanges with other subpopulations could have a very large influence on population persistence, highlighting the importance of distinguishing the effects of interchange between subpopulations from density dependencies in determining 'vital rates'.

4.6.3 Species-centred environmental analysis

'Species-centred environmental analysis' (SCEA) is 'a procedure for diagnosing species-specific environmental factors that limit the size of a population' (James et al. 1997). It is used to identify known biotic and abiotic limiting factors, to evaluate their relative importance and to search for new ones. SCEA is a structured process that can be used to frame hypotheses about species requirements in a context that spans population-, community- and ecosystem-level processes while maintaining the focus on those factors which directly or indirectly affect the size of a focal population. James et al. (1997) used SCEA to study the factors limiting population size in the red-cockaded woodpecker *Picoides borealis*, which is found in mature pine forests in the southeastern US. Four types of environmental factors were already known to limit its numbers:

1 insufficient habitat due to hardwood encroachment in the mid-storey forest layer;
2 a shortage of suitable 'cavity' trees in which to nest;
3 loss and fragmentation of habitat;
4 demographic isolation.

Using the SCEA approach to identify other potentially limiting factors in the Apalachicola National Forest of northern Florida, James et al. (1997) developed regression models for predicting within-population variation in the size, density and productivity of social units of woodpeckers from data on habitat variation. It emerged that variation in the social units was 'significantly related to ground cover composition and the extent of natural pine regeneration, both of which are indirect indicators of local fire history'. Additional support for the conclusion that processes driven by fire also limit red-cockaded woodpecker populations came from the fact that females of the species in the Apalachicola Ranger District tend to lay larger clutches of eggs in the first breeding season after their territories have been burned. An SCEA-type approach can therefore be valuable in identifying and comparing the importance of factors limiting population size. It is applicable, not only to determine appropriate management to conserve endangered species like the red-cockaded woodpecker, but also to understand the factors affecting population processes in species potentially affected by proposed development actions.

4.7 Assessment endpoints

The selection of ecologically relevant VECs for further study is the essential first step in the identification of 'assessment endpoints', which will reflect management goals and effectively link decision-making criteria with the specific ecological measures made in the EcIA. Another important step is to establish 'endpoint' criteria or standards against which it will be possible to measure and evaluate ecological responses. In effect, each VEC chosen to reflect assessment endpoints also requires a 'measurement endpoint'.

Definition of endpoints has three prerequisites:

1 scientific understanding of the impacts likely to be generated and the system(s) at risk;
2 the technical ability to provide relevant measurements of system parameters and establish measurement endpoints; and
3 clearly defined societal values.

Evaluation methods are considered in more depth in section 6.3, but assessment endpoints will have to be taken into account in study design and planning. The US Environmental Protection Agency (1992) defines 'assessment endpoints' as 'explicit expressions of the actual environmental value that is to be protected' and they must include or be linked to measurable attributes. In other words, not all ecologically relevant VECs may be suitable as assessment endpoints. In addition to societal or management goals, a key consideration is 'susceptibility to known or potential stressors' (US Environmental Protection Agency 1996). The selection of suitable assessment endpoints for EcIA therefore depends on some form of 'impact screening'. The principles of impact screening were discussed earlier in section 3.5. Once a pool of VECs have been identified, an additional screening exercise may be necessary to take account of the function and 'behaviour' of VECs and the likelihood that they will actually be exposed and vulnerable to the stressors associated with a proposal.

4.8 Screening VECs

When screening VECs, we need to combine exposure assessment with assessment of sensitivity. This enables us to evaluate VECs that will actually be susceptible to defined impacts (which are likely to be exposed and to show a measurable response). It may also be necessary to consider other aspects of species biology that might influence their ability to recover *if* exposed.

For example, a Dutch study of ecological mitigation measures for main roads indicates how information on susceptibility to habitat fragmentation could be used together with information on species rarity, to select species for detailed study in EcIA (van der Fluit *et al.* 1990) (Table 4.5) Clearly relative rarity of species differs between countries, but the approach would be applicable anywhere. In the example shown, the otter, pine marten, badger, black grouse, barn owl, nightjar, corn bunting, ortolan bunting and any amphibians or reptiles present might be considered to merit detailed study on the grounds of both national rarity and susceptibility to habitat fragmentation.

Table 4.5 Species rarity and susceptibility to fragmentation caused by roads in the Netherlands. (After van der Fluit *et al.* 1990.)

Susceptibility	Rarity (limited)			Rarity (great)

(Limited)	Rabbit
	Most mice (some
	  rare species)
		Beech marten
		Hare
		Marsh tit
			Most bats (varies
			  between species)	

		Bittern Goshawk Sparrowhawk Nuthatch Tawny owl	Ruff Roe deer Corncrake	Red deer

		Red squirrel
		Weasel
		Polecat
		Hedgehog
			Wild boar
			Stoat
			Purple heron
			Great reed
			  warbler
			Nightjar
			Great grey shrike
				Many reptiles
				Many
				  amphibians
				Night heron
				Little bittern
				Spoonbill
				Lesser spotted
				  woodpecker

			Corn bunting Ortolan bunting	
(Great)			Badger Many reptiles Many amphibians Hawfinch	Otter Pine marten Black grouse Barn owl

4.8.1   Selecting measurement endpoints

When appropriate VECs have been selected, measurable endpoints or 'goals' must be identified to provide thresholds against which the significance of impacts can be assessed. 'Without goals against which to assess impacts, critical decision points might be missed, resulting in degradation beyond acceptable limits' (Kalff 1995; referring to the use of VECs in cumulative effects assessment in Canadian national parks). To establish appropriate measurement endpoints for VECs in EcIA, it is necessary to know both about current status and about 'limits of acceptable change'. It may not be possible to measure assessment endpoints directly, in which case, some sort of surrogate measurements may suffice. For a freshwater habitat, for example, the assessment endpoint might be 'characteristic composition and diversity of aquatic life', for which a good surrogate measurement endpoint in the UK might be aquatic levels of orthophosphate ($mg\,l^{-1}$), as used by English Nature in setting its pilot water quality objectives (English Nature 1997). The evaluation criteria listed in Chapter 6 are examples of properties that can be used to provide assessment endpoints, but not all of these currently lend themselves to direct measurement. Criteria such as 'extinction risk' and 'minimum viable population' can clearly be expressed numerically in theory, but have not been calibrated for many of the species encountered in EcIA. Others, such as 'replaceability' are more complex to measure and may involve elements of subjectivity.

It is useful to establish general trends and patterns in the abundance, areal extent or concentration of VECs. Also to obtain any available information on rates of change, as this is often correlated with vulnerability: a VEC with rapidly declining status is generally more likely to be at risk than one that is more stable.

When assessing impacts on VECs, proximity to thresholds of viability should be considered: does the proposed activity bring a VEC over its viability threshold or bring it dangerously close, or is there a big safety margin?

4.9 Recommended reading

VEC selection for cumulative effects assessment

Kalff, S.A. (1995) *A proposed framework to assess cumulative environmental effects in Canadian national parks.* Parks Canada Technical Report in Ecosystem Science, No. 1. Parks Canada, Halifax.

5 Identifying and predicting impacts

5.1 Introduction

Ecological impact assessment (EcIA) is all about identifying and quantifying the impacts of defined actions on specific ecosystem components or parameters and evaluating their consequences. If baseline conditions are known, an effort can then be made to predict the extent to which these will alter as a direct or indirect result of a proposed activity (or set of activities). In other words, ecological impacts are measurable, significant departures from baseline. Impact screening (discussed in section 3.5) is used simply to identify the possibility that an impact may occur. The next step is to quantify the likely magnitude, severity, duration and extent of impacts and to estimate their probability of occurrence.

Although prediction lies at the heart of EcIA, it is often the weakest part of the process. Reviews of environmental statements (ESs) show that ecological predictions carried out for environmental impact assessment (EIA) are more likely to consist of vague verbal forecasts than numerical estimates and that rational scientific approaches to prediction for EcIA are the exception rather than the rule. Prediction in the context of EcIA demands more than the straightforward identification of possible impacts ('habitat loss may occur'). Rather, it should provide an estimate of the likelihood or probability that events or consequences of a certain magnitude, duration and extent will follow if defined development actions take place; for example, the probability of a certain magnitude of habitat loss, which species would be affected and where, when the loss would take place and how long it could be expected to last. 'Probability' is the mathematical expression of chance (for instance, 0.20, equivalent to a 20% or one in five chance). It may be applied to the occurrence of a particular event in a given period of time or to the likelihood of occurrence of one among a number of possible events (Department of the Environment 1995). Techniques for estimating the probability of ecological events have been much better developed in the field of ecological risk assessment, where 'risk' is defined as the combination of the probability, or frequency of occurrence of a defined hazard (or event) and the magnitude of its consequences or effects.

> **Box 5.1** Barriers to hypothesis testing using monitoring data in EcIA.
>
> - Failure to involve ecologists sufficiently early in EIA and project design
> - Shortage of monitoring and baseline data
> - Lack of access to control or reference sites
> - Lack of understanding of ecosystem function or response
> - High levels of uncertainty
> - Shortage of consistent evaluation criteria
> - Shortage of field experience in ecological risk assessment
> - Lack of follow-up experience or monitoring

Ideally, ecological predictions for EcIA should be regarded as hypotheses that can be tested using monitoring data (Buckley 1991). Preferably, predictions should be stated in quantitative terms, presented in the form of time series covering the projected duration of project activities and accompanied by estimates of the probability of impacts occurring (Treweek 1995). There are a number of reasons why it is difficult to design impact assessment studies according to this ideal model. These include the factors listed in Box 5.1.

One of the main problems is that opportunities for testing the predictions made in EcIA have been so limited. There has also been a general failure to capitalize on experience gained through EcIA by studying results from previous projects of a similar nature. In many countries, the potential benefits of accumulating a body of relevant experience and knowledge have been weakened by the factors listed in Box 5.2.

Because of the complex and multidimensional nature of ecosystems, it is very difficult to account for natural variation so that any changes observed can actually be attributed to superimposed actions. Multivariate data are likely to be required to distinguish between natural patterns and those attributable to a particular action. To predict an ecosystem's response to a perturbation, it is first necessary to recognize

> **Box 5.2** Barriers to building a knowledge base in EcIA.
>
> - Lack of coordinated monitoring programmes
> - Inconsistent methodologies
> - Inconsistent reporting of methodologies and results
> - Restricted access to impact statements or reports
> - High cost of impact statements or reports
> - Confidentiality agreements

and characterize its multidimensional nature. One of the most difficult analytical challenges in ecology is to identify or predict patterns of change in large datasets. It is relatively common for ecological data to be non-linear, not to conform to parametric assumptions, to have incommensurable units (e.g. length, concentration, frequency) and to be incomplete ('due to both sampling loss and sampling design, whereby different parameters are collected at different frequencies') (Landis *et al.* 1994).

On the other hand, suitable multivariate data for EcIA are rarely available and the data available for prediction in EcIA are often applicable only within narrow spatial and temporal limits. The complexity of ecosystems, combined with a shortage of relevant information means that levels of uncertainty are generally very high (Box 5.3).

Perhaps most importantly, there are two fundamental weaknesses built into the EcIA process: firstly, predictions cannot generally be tested before decisions have to be reached, and secondly, selection of mitigation procedures to minimize or eliminate predicted effects has to be done without knowing what the actual effects will be (Epp 1995). EcIA is therefore required to support defensible and rational decision-making in the face of considerable uncertainty and we cannot know how much confidence to place on an ecological prediction unless these levels of uncertainty are quantified in some way. In other words, we need to know about predictive precision. 'The precision of a scientific theory is a measure of the uncertainty associated with prediction. More precise theories have narrower confidence bands and indicate that a larger range of the possible combinations of predictor and response variables is improbable' (Peters 1991). They provide a sounder basis for future action. It is important that the precision of ecological predictions should be clearly stated. In an ideal world, we would seek to make predictions (or develop theories) that are both precise and of general applicability (Peters 1991). Numerical approaches to dealing with uncertainty are emerging, in particular to strengthen human health and ecological

---

Box 5.3 Sources of uncertainty in EcIA.

- Inherent complexity of ecosystems
- Incomplete knowledge of ecosystem processes (results in predictive models based on over-simplified representation of reality)
- Lack of opportunity for experimental testing (results in shortage of baseline and comparative data)
- Lack of scope for comparative analysis
- Lack of long-term datasets
- Lack of opportunities for testing accuracy of predictions

risk analyses. Some of these are discussed later in this chapter. At the very least, however, primary sources of uncertainty should always be identified. Reasonable decisions can only be reached if the main sources of uncertainty and variability have been identified and their likely effect on predicted outcomes can be judged. As emphasized by Beanlands and Duinker (1983) 'predictions may legitimately be based on any combination of speculation, professional judgement, experience, experimental evidence, quantitative modelling and other methods' so long as the basis on which predictions have been made is explicit.

Finally, predictions are the raw material for evaluation and should be measured and presented in a form consistent with the evaluation criteria that will be used in assessing the significance or importance of impacts. For example, if the importance of different habitats will be evaluated in terms of size and species diversity, these attributes should clearly be measured, and the scale and units used should be consistent between different habitat types. Assessment and measurement endpoints should be chosen to reflect evaluation criteria so that predictions are relevant as well as precise.

5.2 Baseline assessment

Nearly all EIA legislation requires characterization of the 'baseline condition' that would pertain in the absence of any superimposed activity or set of activities. This is important, not only for predicting impacts, but also with respect to consideration of 'zero development options'. Considerable confusion has arisen over what is meant by 'baseline' in this context. It is quite common for UK ESs, for example, to include descriptions of 'baseline' derived from one-off surveys, which really only provide snapshots of current conditions. Obviously ecosystems are dynamic and it is essential to take account of their inherent spatial and temporal variation if any attempts are to be made to attribute subsequent changes in ecosystem parameters to specific stresses or actions. Many systems have 'moving baselines' and if only short runs of data are available, it becomes difficult to distinguish underlying trends from superimposed impacts. Figure 5.1 emphasizes how 'one-off' surveys carried out on isolated occasions could result in a range of markedly different impact predictions for the attribute in question. Without establishing the 'average baseline' and to make meaningful predictions.

Preliminary studies carried out simply to establish the need for further, more detailed surveys and to rationalize subsequent studies rarely go far enough to enable reliable characterization of baseline conditions. A distinction should therefore be drawn between such 'focusing studies' and true baseline studies, which are necessarily based on comprehensive

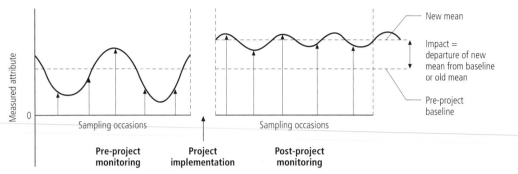

Fig. 5.1 Repeated sampling pre- and post-project implementation is needed to measure difference between new and old means. (Modified from Beanlands & Duinker 1983.)

(and often long-term) monitoring (see Chapter 8). EIA practitioners and development proponents must recognize that it is much harder to make reliable ecological predictions than it is to estimate levels of noise, for example, and that reaching an understanding of the changes likely to occur in the absence of a proposed action will generally require studies to be carried out over a longer period than is customary for other specialist areas of EIA. Few natural systems are understood in sufficient detail for it to be possible to characterize them on the basis of existing data, so longer 'lead times' should be allowed for.

'Quantitative predictions cannot normally be made, nor hypotheses tested, without a firm foundation in measurement' (Beanlands & Duinker 1983). It has been recognized for some time that it is essential for numerical data collection programmes to be focused 'around a statistical definition of natural variation in space and time' (Beanlands & Duinker 1983). However, the high degree of natural variability that characterizes most biological phenomena makes it very difficult to establish baseline conditions with confidence.

There are cases where, by using a consistent approach and establishing clear assessment endpoints, it has been possible to construct suitable baselines for impact assessment. The US 'habitat evaluation procedure' (HEP) (US Fish and Wildlife Service 1981) is a notable example (see section 5.4.4).

5.3 Types of ecological impact

Despite all the problems associated with impact prediction, it is possible to draw some general conclusions about the types of ecological impact that may be encountered in EcIA.

There has been considerable debate over the definition and classification of ecological impact types. 'Direct', 'indirect', 'secondary',

Table 5.1 Broad categories of ecological impact.

Ecological impact category	Examples
Direct	Habitat loss or destruction (e.g. through construction work)
	Altered abiotic/site factors (e.g. through soil removal, compaction or erosion)
	Mortality of individuals (e.g. through road collision, destruction of vegetation)
	Loss of individuals through emigration (e.g. following disturbance or loss of habitat)
	Habitat fragmentation (selective habitat removal and/or introduction of barriers like roads)
	Disturbance (e.g. due to construction noise, traffic, presence of people)
Indirect (including delayed impacts)	Mortality of individuals or populations through reductions in habitat area or quality (overall carrying capacity reduced)
	Reduced population viability due to reductions in habitat area or quality
	Altered population dynamics due to changes in resource availability or distribution
	Increased competition for reduced resources resulting in altered species composition or age structures
	Altered species composition due to changes in abiotic conditions
	Altered species or habitat composition due to increased edge effects (a consequence of habitat fragmentation, for example)
	Reduced gene flow possibly leading to increased vulnerability to stochastic events (a consequence of habitat fragmentation caused by linear infrastructure, for example)
	Habitat isolation caused by a variety of development types, resulting in increased edge effects and sometimes loss of diversity
	Reduced breeding success (e.g. due to disturbance, habitat loss, fragmentation, pollution) possibly resulting in reduced population viability
	Delayed effects (e.g. altered predator–prey relationships due to loss of a keystone species)
Associated impacts (direct and indirect)	Ecological impacts attributable to 'linked' or 'associated' actions or activities (e.g. due to the combined effects of aggregate mining and road construction)
Cumulative impacts (time- and space-crowded effects)	Habitat 'nibbling' (progressive loss and fragmentation throughout an area)
	Reduced habitat diversity, e.g. at the landscape level (associated with reduced biological diversity at other levels in organizational hierarchy)
	Ongoing habitat loss or fragmentation over time, resulting in progressive isolation and reduced gene flow. Reduced genetic diversity can result in loss of resilience to environmental change and increased risk of extinction
	Irreversible loss of biological diversity (e.g. through destruction of unique population units)
	Exceedance of viability thresholds (e.g. falling below regional carrying capacity due to progressive habitat loss)
Synergistic impacts	Toxic effects attributable to 'cocktails' of pollutants which are within individual tolerance limits. May reduce viability of individuals or affect breeding success of whole populations

'cumulative', etc. impacts are all referred to in the EIA literature. So long as the full range of potential impacts is identified and their importance evaluated, how they are categorized seems relatively unimportant.

Table 5.2 Examples of species that have become extinct in the UK in the last 100 years largely due to human activity. (From HMSO 1996. Crown copyright is reproduced with the permission of the Controller of Her Majesty's Stationery Office.)

Species	Main suspected reason for loss 'in the wild'	Date of last record
VERTEBRATES		
Burbot *Lota lota*	This fish is assumed to have been lost as a result of river pollution	1972
Mouse-eared bat *Myotis myotis*	Destruction and disturbance of nursery sites	1990
Sea eagle *Haliaeetus albricilla*	Persecution through poisoning, shooting and egg collection	1916 (subsequently re-introduced)
INVERTEBRATES		
Large blue butterfly *Maculinea arion*	Destruction of grassland habitat. Altered management of remaining habitat (lack of appropriate grazing)	1979 (subsequently re-introduced)
Essex emerald moth *Thetidia smaragdaria*	Coastal defence works reduced population size resulting in pool too limited for successful reproduction	1991
Viper's bugloss moth *Hadena irregularis*	Loss of Breckland heath habitat to agriculture and infrastructure development	1979
Blair's wainscot *Sedina buettnei*	Sole remaining site, a coastal grazing marsh, destroyed by draining and burning	1952
Norfolk damselfly *Coenagrion armatum*	Degradation of the small marshy pools it inhabited through encroachment of reed, willow or alder or through desiccation	1957
Orange-spotted emerald dragonfly *Oxygastra curtisii*	Sewage pollution	1951
Exploding bombardier beetle *Brachynus scolopetus*	Lack of suitable management of chalk grassland habitat	1928
Horned dung beetle *Copris lunaris*	Ploughing of pastures on chalk or sandy soils	1955
A click beetle (*Melanotus punctolineatus*)	Last known site destroyed for golf course construction	1986
Aspen leaf beetle *Chrysomela tremula*	Decline of woodland coppicing	1958
PLANTS		
Thorow-wax *Bupleurum rotundiflorum*	Annual cornfield weed lost due to improved cleaning methods for seed corn	1960s
Lamb's succory *Arnoseris minima*	Arable weed lost due to agricultural intensification (exact cause unknown)	1970
Hairy spurge *Euphorbia villosa*	Cessation of woodland coppicing at only remaining site	1924
Summer lady's tresses *Spiranthes aestivalis*	Drainage of bog habitat	1959

However, the tendency to neglect the more complex ecological impacts does make it important to ask why this is the case and to search for solutions to the problem of over-simplification in EcIA. Some of the broad categories of ecological impact are summarized in Table 5.1.

Whether an impact is considered to be 'direct' or 'indirect', very much depends on how the receiving environment or ecosystem is perceived. Direct ecological impacts are generally said to occur wherever defined activities affect a specific environmental receptor without being mediated in any way through interaction with other ecosystem components. Mortality of an animal due to collision with a road vehicle is an obvious example of a direct impact! The effects of losing that individual on the social structure and dynamics of its own and other interacting populations is less directly associated with the presence of road traffic, but is nevertheless still attributable to it. Few ecological impacts can be assessed effectively in isolation from one another. It is therefore probably less important to define degrees of 'direct causality' than it is to ensure that all the ecological impacts associated with a proposal are identified and taken into account, together with the mechanisms for their interaction.

The main threats to ecological 'value' differ between countries and over time, as the nature and extent of exploitation of natural resources varies. However, it can be salutary to examine the main current and historical reasons for loss or decline of species or habitats. This can help to pinpoint those which bear the brunt of certain development types or to help avert similar extinctions in the future. In the UK, examples of species that are believed to have become extinct primarily as a result of human activity are listed in Table 5.2. Clearly, many of these species became extinct long before the introduction of EIA legislation, but it is also worth noting that many disappeared as a result of changes in agricultural practice and yet there are relatively few agricultural activities currently subject to formal EIA or EcIA in the UK or elsewhere.

5.3.1 Mechanisms of ecological impact expression

Ecological impacts will vary considerably between different impact-generating activities and affected systems, but there are some common mechanisms underlying impact expression.

Project activities may add, remove or redistribute physical, chemical or biotic components or energy. This results, directly or indirectly (depending on fate and functional relationships) in a net loss or gain of valued ecosystem components or functions. Project activities are a potential source of perturbation to ecosystems and prediction of their effects depends on a clear understanding of ecological linkages. Most impacts manifest themselves first through changes in physical or chemical parameters and then through changes in biotic variables, but either can be measured or estimated to predict impacts. Similarly, biological systems can be viewed as nested hierarchies, consisting of individuals, species, populations, communities and ecosystems. All biotic impacts result from collective effects on individuals, but the effects of perturbations may also be expressed at any of the other levels in the hierarchy and responses can be studied at any of these levels. In some cases, for example, it may prove easier to study the responses of lower levels in the hierarchy; for example, individuals or communities rather than whole populations (Beanlands & Duinker 1983). As a general rule, ecological assessments (EcAs) fail to consider how impacts will actually be expressed in ecosystems. The following sections consider some of the techniques that can be used to predict ecological effects. The US Fish and Wildlife Service HEP (see sections 5.4.4 and 5.4.5) is described in some detail because it is a rigorous method that not only requires measurement of habitat loss, but analysis of its consequences in terms of actual carrying capacity for associated species.

5.3.2 Wildlife mortality

Some developments and activities can cause wildlife mortality directly. Removal of existing vegetation precedes most kinds of construction activity. Not only does this result in destruction of plants, it can also kill associated invertebrates and animals using vegetation as cover. Bird mortality may be particularly high during the nesting season. In-filling of water bodies and the diversion or canalization of water courses can also destroy individuals directly. Organisms most at risk from direct mortality are those which are highly habitat specific, or which are relatively immobile and unable to escape and colonize alternative habitat.

Road traffic causes a considerable number of animal casualties in the UK. Precise numbers of animals killed each year are difficult to obtain, but Table 5.3 lists some references from the literature. Additional examples can be found in a report by English Nature (1993).

Note that for developments like roads, impacts on populations due to direct mortality may continue for decades or longer. Figure 5.2 shows a wallaby killed by collision with a vehicle on a Tasmanian road. In such relatively under-populated countries, where roads are sparse and unfenced, additional mortality is less likely to ensue from habitat loss or fragmentation, but this is an area where reliable information is severely lacking.

There seem to be few records of wildlife mortality caused by other categories of development. The extent of wildlife mortality caused by development overall is therefore not known, but may be considerable. Because mortality is rarely measured, its consequences for wildlife

Table 5.3 Selected references to animal casualties on roads.

Species	Numbers	Country	Source of data
Amphibians	20–40% of a breeding population	UK	Langton (1989)
Toads	15–50% of migrating population	The Netherlands	Van Leeuwen (1982)
Birds	5269 per 349 miles (1960–1961), = approximately 2.5 million per year	UK	Hodson and Snow (1965)
	10 million per year		Hill and Hockin (1992)
	30–70 million	UK	TEST (1991)
	653 000	The Netherlands	Van der Zande *et al.* (1980)
Barn owls (a protected species)	5000	UK	TEST (1991)
Mammals	159 000	The Netherlands	Van der Zande *et al.* (1980)
Badgers	47 000	UK	Harris (1989)
Deer	200 000	US	Schafer and Penland (1985)

Fig. 5.2 Road casualty: a wallaby killed by collision with a vehicle in Tasmania. (Courtesy of Owen Mountford.)

population viability are, likewise, largely unknown. The situation is even worse in relation to mortality resulting indirectly from loss of habitat. This is especially difficult to measure as there may be considerable time lags between the destruction of habitat and the death of displaced individuals.

There are many examples where mortality of individuals may lag considerably behind the original causal event, particularly for organisms like plants, which are sometimes able to persist for long periods in suboptimal conditions before they finally succumb.

5.3.3   Habitat loss

Destruction of habitat (associated with increased 'resource patchiness') is a major threat to mammals and birds. Seventy-five per cent and 60% of all threatened mammal and bird species, respectively, are threatened by destruction of their habitat (Mason & Sadoff 1994). In some cases, destruction of habitat is directly linked to mortality. In others, it displaces associated species, in which case their survival depends crucially on the ability to locate alternative habitat. The effects of habitat loss may have local and more global implications, depending on the overall availability and organization of habitat and the behaviour and population dynamics of associated species. Delayed effects of habitat loss are probably very common, but are rarely detected or analysed in EcIA studies.

Habitat loss is a consequence of the majority of developments and is also more often measured (quantified) than other categories of impact. Figure 5.3 shows land-take for development of access to the Channel Tunnel in the UK. Some of this land-take will be indefinite; some was occupied during construction only. Although it is usually relatively straightforward to quantify the area of land-take that will occur as a result of a proposal, however, it is not always as easy as it might seem to measure habitat loss. To do this, we need to know which species an area of land supports, its quality in terms of those species' needs and the relationship between the area under threat and overall carrying capacity. Measuring habitat loss therefore also requires some method for quantifying the ability of land to support different species, such as that developed by the US Fish and Wildlife Service for their HEP. It is also necessary to know how different organisms use their environment before it is possible to define the effective boundary of their habitat. If habitat quality can be defined and the area of habitat needed to support a given individual, population or species is known, it is then possible to predict the likely effects of losing an area of habitat. The approach taken for HEP is described in more detail in section 5.4.4. It is pointless to measure habitat loss for purposes of EcIA without considering the

Fig. 5.3 Construction corridor for development of access to the Channel Tunnel in the UK.

consequences of that loss for associated species of concern. This means that corridor assessments, in which species present are recorded within a corridor, or buffer of set width, will rarely permit an effective analysis of the impacts of habitat loss.

Species–area relationships have been the subject of much ecological research, but for most species, it is difficult to specify the minimum areas of habitat required (see sections 3.6.1 and 6.3.7). As well as affecting individual species, reductions in the area of habitat patches have implications for species richness and diversity and for the range of species able to persist. For example, woodlands less than 2 or 3 ha in size in the Netherlands and the UK (East Anglia) have been found not to support certain 'core' or characteristic woodland bird species. In the Netherlands, 50 ha is regarded as the threshold size for supporting viable populations of core woodland species (English Nature 1993). Studies have shown that, on average, a habitat will lose between 30% and 50% of its species for each 90% reduction in area: in other words, a 100-ha 'habitat island' could be expected to retain only about 70% of the species that would have survived in 1000 ha (Diamond 1975).

Species are threatened not simply by reductions in the overall 'stock' of habitat, but also by spatial reorganization of that habitat which remains. In countries where rates of habitat loss are highest, remaining habitat is also increasingly fragmented, resulting in the restriction of wildlife to isolated areas of suitable habitat, often in an 'inhospitable matrix'.

5.3.4 **Habitat fragmentation**

Fragmentation, quite simply, is the breaking down of habitat units into smaller numbers of units. It is associated with quantitative and qualitative changes. These might include reductions in the size of habitat patches, increased edge effects and changes in species composition. For some species, a degree of fragmentation of habitat can be beneficial. Different life stages of the same species may also respond differently to habitat fragmentation. Collins and Barrett (1997), for example, describe how female meadow voles *Microtus pennsylvanicus* were found to be more numerous in fragmented habitat, particularly during the breeding season. In this case, occupation of smaller, relatively isolated patches of the same habitat type made it easier to defend breeding territories, resulting in a reduced energy requirement for territory defence compared with occupation of territories in unfragmented habitat. Fragmentation is more usually a detrimental process, and overall degrees of habitat fragmentation have been increasing inexorably throughout most of the world.

Habitat fragmentation is something that can be quantified: numerical indices of fragmentation can be derived and this has been greatly facilitated by geographical information systems (GIS). In some contexts it may be relevant to measure the number and size of fragments or patches before and after implementation of a proposal, particularly in studies at the regional level, where cumulative habitat loss and fragmentation may be an important issue. Treweek and Veitch (1996) did this to assess the relative impacts of alternative hypothetical road routes in a study area covering 145 200 ha in the county of Dorset in the UK. They found that the first route would result in fragmentation of an area of lowland heathland to generate two new habitat 'patches'. The second appeared to reduce the overall degree of fragmentation, but in fact it did this by eliminating a number of small patches from the area. For many species, the distribution of habitat patches is also important, so it may be helpful to measure the distances between patches and to determine whether or not they exceed the dispersal distances of associated species.

Some of the consequences of habitat fragmentation include (English Nature 1993):
- decline in species number as habitat patches reduce in size;
- loss of 'core' or characteristic species and concomitant invasion by 'edge' and more catholic species (found for plants and invertebrates in heathland habitats and for birds and mammals in woodlands);
- changes in community composition;
- altered parasitic, symbiotic and predator–prey relationships, sometimes resulting in secondary extinctions;

- altered interspecific relationships;
- altered population dynamics, particularly where species that exist as metapopulations are affected.

For example, fragmentation has been shown to result in reduced reproductive success, particularly for species that are more vulnerable to predation when the ratio of edge/habitat interior increases. A well-known example is the ovenbird *Seirus aurocapillus*, which is believed to be 'edge intolerant' (Porneluzi *et al.* 1993). A study of edge effects in fragmented habitats by Angold (1997a) demonstrated how increasing fragmentation reduces the amount of 'core' habitat and increases the amount exposed to 'edge' effects (Fig. 5.4). In this study, fragmentation was often caused by roads, and typical edge effects included significant changes in plant species composition, plant performance and soil nutrient levels.

Many studies of the effects of habitat fragmentation seemingly fail to take account of how associated species perceive or define their own 'habitat'. The development of GIS has made it straightforward to measure the sizes of vegetation patches that have boundaries clearly visible to us (woodland patches, for example), but how do the organisms using the environment perceive these boundaries? Indices of habitat fragmentation are meaningless without clear understanding of what actually constitutes 'habitat' for a study species or information on the resistance of the landscape's matrix to the movement of organisms through it. For instance, it is relatively straightforward to measure distances between habitat 'patches', but we also need to know how these distances relate to the mobility of species and their ability to disperse between patches. Information on the relative resistance or inhospitability of the intervening areas is therefore just as important as knowing how far a species must travel to find suitable habitat. In terms of effects on species, it may therefore be more appropriate to measure 'insularization',

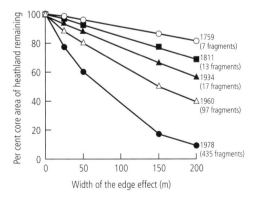

Fig. 5.4 The percentage core area of heathland remaining in the Poole Basin, UK, at various dates, with increasing fragmentation and allowing for edge effects of up to 200 m. (From Angold 1997a.)

as proposed by Spellerberg (1992), as this concept places more emphasis on outcomes for species and less on mere description of habitat geometry.

5.3.5 Habitat insularization

Spellerberg (1992) uses the term 'insularization' to refer to the combined effects of habitat reduction, fragmentation and isolation. Insularization occurs as habitat is lost (often a direct effect) and the distance between areas of similar habitat increases relative to the mobility and dispersal ability of the species associated with it. Effective genetic isolation of species occurs as areas of habitat become so isolated that their associated species cease to interact with populations elsewhere. Losses of individual genetic variation through population isolation and inbreeding may ultimately be disastrous for many species (Ralls *et al.* 1986). In some situations overall genetic diversity in an area may be maintained among isolated groups of populations in a fragmented habitat, but this is likely to be lost if any of these populations are wiped out as a result of continuing habitat destruction or the cumulative effects of inbreeding (Chesser *et al.* 1980, in press). This, and the resultant effects on the long-term viability of the species are indirect effects that can be very difficult to predict. They are also important factors in evaluating ecological effects (see Chapter 6).

In general, the consequences of insularization are likely to be as follows (after Spellerberg 1992):
• losses in keystone species or key species (species on which the ecology of other species depend);
• reductions in populations and extinctions of species at newly formed edges, increased vulnerability to external influences such as disturbance, increased likelihood of invasion by uncharacteristic species;
• inbreeding;
• loss of characteristic species;
• increased vulnerability to stochastic events.
Actually measuring these consequences is not always straightforward, but their potential significance means they should not be ignored. The US Fish and Wildlife Service has therefore built an 'interspersion' factor into its HEP to account for the role of habitat distribution in determining carrying capacity (see section 5.4.4.1). Species most likely to be adversely affected are sedentary species with highly specialized habitat requirements. Often these are species associated with habitats that were once much more widespread (English Nature 1993). Species with more 'catholic tastes', which live in open populations and disperse effectively over long distances, are much less affected by habitat fragmentation or insularization.

5.3.6 Barrier effects

There are a number of linear developments (like roads, railways or sometimes pipelines) that form barriers to wildlife movement through the landscape. Figure 5.5, for example, shows how movements of the ground beetle *Abax ater* between woodland patches separated by roads were less frequent than movements within them. Barrier effects compound those of habitat fragmentation and isolation: not only do animals and plants have to travel longer distances between suitable patches of habitat, but their dispersal routes or corridors may be disrupted or blocked. This is a particular problem for migratory species or species with habitual routes through the landscape. Many amphibians, for example, are found in different habitats during the summer, the winter and in the breeding period. In the European spring, toads, frogs and newts may travel large distances from over-wintering to breeding sites. They follow fixed routes and tend to keep to the same route for many years running.

Barriers may work in two ways: a species may not be able to survive passage across them, or behavioural factors might deter an animal from crossing barriers despite a theoretical ability to do so. For roads, for example, barrier effects may be caused by traffic casualty, or they may simply be caused by a reluctance to venture onto alien surfaces. For some of the smaller birds and invertebrates, turbulence caused by traffic may also be an issue, simply making successful crossings difficult without necessarily causing mortality. In some cases roads appear to be leaky barriers (some animals do manage to cross without being killed), but they may nevertheless restrict the landscape-scale dynamics of species significantly.

A study of small mammals in the UK (Richardson *et al.* 1997), for example, showed that major roads act as a barrier to some degree regardless of traffic levels. Small mammals did not frequently cross paved roads wider than 20–25 m even though they would spontaneously travel distances greater than this along road verges. However, it is not known whether the severity of barrier effect increases gradually with road width, or whether there is a critical width above which small mammals are reluctant or unable to cross. This study also showed that roads were permeable or leaky barriers: when animals were translocated across roads, some were able to return successfully to their territories. However, traffic density played a part in determining cross-over ability. Therefore, while wide roads may constitute barriers in themselves, the degree to which they are permeable barriers depends on the amount of traffic they carry: major roads with constantly high traffic flows, like motorways, are likely to be less permeable than those with low traffic volumes, and narrow roads with high traffic densities may be greater

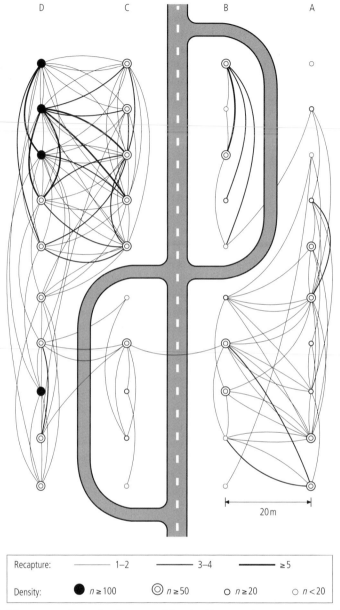

Fig. 5.5 Movements shown by the ground beetle *Abax ater* between patches of woodland isolated by a road and two parking loops. A–D = trap rows; circles = live traps; curved lines represent the movement of a marked animal between capture and recapture. (Reprinted from Mader, H.-J. (1984) Animal habitat isolation by roads and agricultural fields. *Biological Conservation* **29**, 81–96, with permission from Elsevier Science.)

barriers than their width might suggest. Because roads are not complete barriers to small mammals, they do not cause complete genetic isolation, although this may not remain the case if road size and traffic density increase. Not enough is known at present to be able to quantify the normal rate of cross-over or to quantify the relationships between road size, traffic density and degree of genetic isolation.

More research is required to quantify barrier effects for different species, measuring cross-overs in relation to road width, traffic density, periodicity of traffic flow, etc. It may then be possible to go on and investigate the degree of cross-over required to maintain gene flows in the landscape: something we can only speculate on at present.

To some extent, barrier effects can be reduced or mitigated using 'wildlife crossings'. Important wildlife routes for large animals are often used for many years and tend to be clearly visible. If wildlife crossings are carefully tied in with key runs, they are more likely to be successful, but some animals are very distrustful and take some time to adapt to crossings, particularly if they are in the vicinity of potential disturbances caused by footpaths, for example. A wildlife bridge under construction in the Canadian Rockies is shown in Fig. 5.6. This bridge is intended to maintain wildlife movements over a busy highway, and will eventually be vegetated to encourage use. Large wildlife tunnels or underpasses may also be built (see section 7.5). Similarly, to reduce barrier effects for amphibians, it is important to know where to place crossings (tunnels, for example) with respect to habitual routes and how to encourage

Fig. 5.6 Wildlife bridge under construction in the Rocky Mountains, Canada. (Courtesy of Miles Tindall.)

animals to use them. Tunnels under roads work best when they are placed in line with spring migration routes and as close as possible to breeding areas (DWW 1995).

It is important to note that barrier effects can be caused simply by the generation of an inhospitable habitat 'matrix' between areas of suitable habitat. In cases where animals must move between habitat patches to fulfil their daily or seasonal requirements, habitat 'corridors' can play a part in maintaining habitat connectivity. A useful review of the scientific evidence for habitat corridors as conduits for animals and plants in fragmented landscapes has been produced by English Nature (1994b).

5.3.7 Disturbance

Many developments cause disturbance, whether during construction phases or when operational. Wildlife may be affected by visual disturbances, by noise or by the presence of people, dogs, road traffic, etc. Relatively few studies have been carried out to quantify disturbance effects. Notable exceptions are Dutch studies on the effects of road traffic on the density of breeding birds (see section 5.4.1), which demonstrated that disturbance can have a measurable effect on both the quality of habitat and the behaviour of associated species. A review of bird disturbance effects can be found in Hill *et al.* (1997). This review concluded that there had not been enough scientific research to provide a sound scientific basis for impact assessment or the management of disturbance effects for birds. Meanwhile, potential sources of disturbance are increasing all the time. Pritchard *et al.* (1992) concluded that 62 out of 127 Important Bird Areas were affected to some extent by disturbance in Scotland, 56 out of 74 sites in England and 10 out of 14 sites in Wales. At the European scale, Tucker and Heath (1994) reported that 35 of 129 species of conservation concern were either threatened or affected by disturbance. The effects of disturbance may be expressed in a variety of ways, not all of which will be immediately apparent. For example, disturbance of birds can affect the time they are able to spend feeding in a particular area in the short term, or their willingness to use a particular area for feeding or breeding in the longer term. Population-level effects are likely if disturbance affects energy balances and breeding success.

Disturbance effects are complex and difficult to measure, because many species are able to habituate to disturbances and show variable responses. Figure 5.7 shows how waders and wildfowl are likely to respond across a disturbance gradient of increasing severity (Hockin *et al.* 1992). If disturbances are passive, low-level and continuous, waders and wildfowl are able to habituate to them, but if they are active, high-level and continuous only very tolerant species remain. The effects

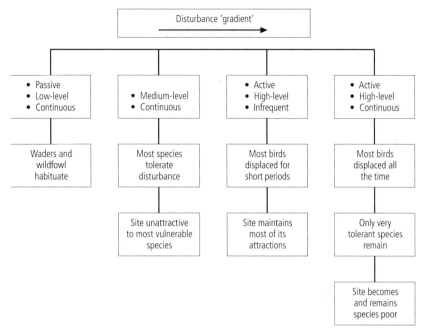

Fig. 5.7 Responses of waders and wildfowl to disturbance of increasing severity. (After Hockin *et al.* 1992.)

of disturbance may also vary depending on life-cycle stage or season. Disturbance tends to be a trigger for displacement, which can have a variety of indirect or associated effects. For example, displacement may result in the following (Hill *et al.* 1997):
• higher densities on some receptor sites;
• more individuals being forced to use suboptimal feeding or breeding habitat;
• direct mortality if no alternative habitat can be found.

In general, larger bird species, those higher up in the food chain or those which tend to feed in flocks in the open, are more vulnerable to disturbance than small species living in structurally complex habitats such as woodland. Hill *et al.* (1997) list the following factors which should be taken into account when estimating the severity and likely impact of disturbance to birds:
• intensity of disturbance;
• duration and frequency (continuous, infrequent, regular, variable);
• proximity of source;
• seasonal variation in sensitivity of affected species;
• the tendency of birds to return following displacement;
• whether regional numbers are affected;
• whether there are alternative habitats available nearby;
• whether rare, scarce or especially 'shy' species are affected.

Understanding of disturbance effects for species other than birds are limited, although there is plenty of anecdotal evidence that many mammals (deer, for example) are prone to disturbance.

Vulnerability to disturbance impacts should also be a factor in the design of mitigation measures: for species that are elusive, shy and prone to disturbance, replacement habitat must be sited where levels of disturbance will be acceptable. Likewise tunnels or bridges constructed to overcome barrier effects should be sited in relatively undisturbed locations.

5.3.8 Effects on species composition

Many developments affect the species or community composition of systems that remain following development. For terrestrial animals, these may follow on from the displacement of keystone species through disturbance. For plants, they may be associated with altered management, pollution or altered hydrological regimes. Angold (1997b), for example, studied the impact of roads on adjacent heathland vegetation in Hampshire in the UK. She found that there was an increase in the abundance of grasses in heathland vegetation near roads and a decrease in the relative abundance and health of lichens. The extent of the edge effect was closely correlated with the amount of road traffic, with a maximum edge effect of 200 m adjacent to a dual carriageway. In the case of roads, changes in vegetation composition can be caused by a variety of pollutants in traffic exhaust. Run-off from modern road surfaces has been shown to contain a number of potential pollutants of neighbouring habitats, including tyre particles, fuel oil, petrol, salt, calcium chloride, hexavalent chromium, phenols, picric acid, fertilizers, organic wastes and bacteria (Detwyler 1971). Vegetation changes can also result from disruptions to surface and subsurface hydrological processes. Changes in species composition can have multiple causes and are often delayed, only becoming apparent some time after they are initiated (see section 5.3.11).

5.3.9 Cumulative effects

Ecosystems are exposed to a wide range of 'stressors', an increasing number of which are attributable to human actions (although 'natural' phenomena can also generate stresses). Much of the pressure to address these cumulative effects has come from conservation agencies who are struggling to protect ecosystems from internal and external threats to their integrity. Although they are established to counter the incremental loss of wilderness areas and wildlife, protected areas are still vulnerable to the effects of land-use activities beyond their boundaries and some are

Box 5.4 Sources of stress in Canadian national parks.

- Development of tourism infrastructure
- Visitor disturbance
- Liquid and solid waste
- Introduction of exotic species of flora and fauna
- Development of service corridors
- Logging activities
- Mining activities
- Commercial fishing
- Recreational fishing
- Reduction of ozone layer
- Climate change
- Visitor traffic (e.g. causing collisions with wildlife)
- Use of pesticides
- Inappropriate management methods

even being degraded as a result of internal development. Savoie and Woodley (1993), for example, listed 22 internal and external sources of stress in Canadian national parks, 43.5% of which were known to be increasing at the time. These include those listed in Box 5.4.

Failure to regulate the collective, or overall impacts of development actions can mean that ecosystems become stressed to the point where their ability to absorb impacts or recover from them is exceeded. In assessing the combined effects of all development pressures on an area or ecosystem, it is therefore important to understand the ability of that area or ecosystem to absorb the various pressures to which it is exposed. Combined modelling of both development patterns and natural systems' responses is required for it to be possible to take account of cumulative effects (Cowart 1986; Sonntag *et al.* 1987; Contant & Wiggins 1991).

Project-level EIA has a number of shortcomings with respect to addressing cumulative effects. Project-EIA is invariably constrained by boundaries in time and space which are inappropriate to the assessment of delayed impacts, those which occur beyond the immediate project site or those expressed through indirect ecological linkages. Despite legislative requirements to take account of cumulative effects, project-EIA often fails in this respect. Another important area of neglect in project-EIA is incremental damage due to environmental changes that are individually minor but become significant when considered collectively (Cocklin *et al.* 1992). In fact, some of the stages in project-EIA are designed specifically to screen out those impacts which, individually, fail to reach significance. McCold (1991) pointed out the risks associated with this phenomenon in the US. In circumstances where it is unclear whether an action will have a significant impact, agencies prepare an EIA. If it is then concluded that the action is unlikely to have a signifi-cant effect, the agency prepares a 'finding of no significant impact' (FONSI). Because there is no central accounting of EIAs prepared by all federal agencies, the exact number of EIA/FONSIs published each year is unknown, although estimates in the region of 30 000 a year have been published (Herson 1986). There is therefore a considerable, but unknown risk of cumulative impact from projects that, when considered on an individual basis, are considered not to generate significant eco-logical effects.

Individual projects often form part of a larger programme of similar or related activities. The US Council on Environmental Quality (CEQ) (1978) regulations on EIA under the National Environmental Policy Act (NEPA) therefore require proponents to consider the incremental impact of a proposed action 'when added to other past, present and reasonably foreseeable future actions' and to group cumulative actions in the same analysis. However, such requirements have been neglected in many other countries.

There are even cases where large projects have been segmented deliberately for assessment. Thus, the potentially important impact of the whole project are down-played and the focus is on the relatively minor, separate impacts of the project's subunits, which are assessed on an individual basis. This phenomenon has been noted for road-building projects in the UK, for example.

Cumulative effects expressed at a regional scale can only be controlled through planning processes directing development at that scale (Sonntag et al. 1987). Effective control of cumulative impacts is therefore likely to require regional planning and cooperation. Failure to deal effectively with cumulative ecological effects is one of the main arguments for strategic application of EIA.

5.3.10 Sources of cumulative impact

Cumulative effects may be due to combined impacts from many, varied sources or to repeated impacts from a single source.

Cumulative actions may be:

- additive, incremental;
- aggregated;
- associated or connected.

5.3.10.1 Incremental or additive actions

Cumulative effects often derive from actions that are, by themselves, insignificant, but that, collectively, result in significant environmental change. For example, the daily decision of millions of people to drive to work is partly responsible for global warming (Kalff 1995). Odum (1982) referred to this phenomenon as 'the tyranny of small decisions'. Additive actions are repeated similar activities, which are concentrated in time or space, for example series of small dams constructed to generate hydro-electric energy (Ortolano & Shepherd 1995). Other examples include the concentration of waste incinerators in a region or the inexorable development of national road systems.

5.3.10.2 *Aggregated actions*

Aggregated actions are dissimilar actions that are concentrated in time or space, such as might occur in a 'development zone'. These can impose significant space-crowding effects on ecological resources. In these circumstances, great care must be taken to ensure that locally distinct ecosystems are safeguarded. In fact, aggregated actions are not uncommon in areas of high ecological sensitivity. Coastal areas popular for tourism are a classic example. The marine environments surrounding Hong Kong are another. Here, rapid economic development has brought many problems for wildlife in its wake. Liu and Hills (1997) describe threats to the Chinese white dolphin *Sousa chinensis* due to Hong Kong's new airport. This species is already threatened by other major development projects, marine pollution, fishing and shipping activity, and if current rates of decline continue it is possible that it will become extinct by the beginning of the next century. EIAs for a variety of development projects have mentioned the dolphins, but have failed to take account of the combined cumulative impacts on their survival. Liu and Hills (1997) stress that, in situations where a particular species may be endangered due to aggregated actions, there is a case for a species-based environmental assessment (analogous to strategic environmental assessment (SEA) for a particular economic sector or activity such as transport), which would seek an integrated and coordinated solution.

5.3.10.3 *Associated or connected actions*

Actions that lead predictably to other, related actions can be referred to as 'associated' or 'connected'. The former term is generally used to describe looser associations: transport routes (roads and railways) invariably result in associated 'ribbon' development, for example. Where access has been a limiting factor on infrastructure development, development of transport routes can 'open up' whole new areas. In-filling is also common between roads and existing urban areas. Connected actions are mutually dependent. These might be gas compressor stations associated with transmission pipelines, for example. There are many developments, currently assessed on individual basis, which have obvious connections with other development actions. These generate impacts that should be considered in their totality. Examples are:

- development of roads in limited sections which are clearly intended to be linked;
- power stations dependent on pipeline development for fuel supply;
- on-shore oil terminals dependent on pipeline development to export oil;
- timber harvest and roadway construction required for extraction of timber;
- mining of aggregates for road construction.

The construction and repair of roads in the UK uses considerable amounts of primary and secondary aggregates: about 90 million tonnes a year (RCEP 1994). These aggregates are often mined in areas of high landscape value which are some distance from the road for which they are to be used. It is therefore inappropriate to consider the impacts of road construction only in local terms. Because the mining and transport of aggregates is an absolute prerequisite for road construction, it should be included in any EcIA carried out with respect to a new road proposal. Not only is it often treated as an entirely separate issue but, in the UK, there is additional concern because it is theoretically possible for new quarries to be opened on environmentally sensitive sites without the need for planning permission under the terms of interim development orders issued in 1947 to help post-war reconstruction. There is therefore the potential for significant ecological impacts to slip through the legislative net completely.

Growth-inducing actions also take place, which act as triggers for specific follow-on development. An example would be the growth of housing following development of a sewage system in a previously unserviced area (Kalff 1995).

5.3.11 Responses of ecosystems

Natural systems rarely react in a simple, direct or straightforward way to external pressures. There are many examples where incremental accumulation (Baskerville 1986; Clark 1986) or delayed responses (Baskerville 1986) have led to discontinuous impacts, exceeding of thresholds or the crossing of stability boundaries which cannot be attributed to any single action (Sonntag *et al.* 1987; Preston & Bedford 1988). There are also many recorded examples of non-linear functional relationships (e.g. Bedford & Preston 1988) and synergistic or interactive effects.

5.3.11.1 *Incremental effects*
Incremental effects are often characterized by a slow rate of change or expression. Effects may be 'readily observable, i.e. measurable, but not of sufficient magnitude to warrant immediate corrective action' (Beanlands

1992). There is historical evidence of the potential damage that can result from slow, cumulative effects; for example, one theory for the decline of Mesopotamian civilization is that it was due to the slow accumulation of salts in agricultural soils following centuries of irrigation.

5.3.11.2 *Interactive effects*

Cumulative effects may also be antagonistic or synergistic: impacts may interact such that their combined effects are smaller or greater (respectively) than would be expected if they were summed. Possible interactions between individual impacts should therefore be taken into account when attempting to measure overall or 'total' impact. In other words, 'cumulative impacts' should be measured to include 'the totality of interactive impacts over time: the sum of incremental or synergistic effects caused by all current and reasonably foreseeable actions over time and space' (Vlachos 1982). Synergistic effects are commonly observed in ecotoxicology, in situations where a number of chemical pollutants have been released into a single environmental medium (e.g. gaseous emissions to the atmosphere, or leaching of contaminants to groundwater). These may interact to form 'cocktails', which are more damaging than would be expected from their individual toxicity profiles.

5.3.11.3 *Time- and space-crowded effects*

Human activities generate environmental effects that overlap in time and space. Cumulative effects occur when the time required for an ecosystem to remove or dissipate the effects of an impact is greater than the time between impacts (Kalff 1995), for example addition of radioactive substances to water at a rate exceeding their natural rate of radioactive decay (Peterson *et al.* 1987). Time-crowding effects are more likely when there are frequent and repetitive impacts on a single environmental medium, ecosystem or area (Sonntag *et al.* 1987). Similarly, cumulative effects occur when impacts are crowded in space at such a density that they exceed an area's ability to absorb their effects. Progressive fragmentation of wildlife habitat is a good example of space-crowding. These impacts may come from a variety of sources.

5.3.11.4 *Thresholds*

There appear to be thresholds, or points at which added system perturbations, no matter how small, can result in the sudden deterioration or collapse of ecosystems. Cumulative effects tend to become evident when such thresholds are exceeded.

5.3.12 Measuring cumulative impacts and effects

The ability to measure or predict cumulative ecological effects depends on knowledge of combined impact sources and understanding of eco-system responses. It therefore demands:

• understanding of natural variation (monitoring data);
• knowledge of ecosystem resilience;
• understanding of recovery mechanisms;
• removal of time and space constraints on EcIA process and administration;
• institutional frameworks for integrated assessment and planning.

Clark *et al.* (1993) described the problem of assessing the impacts of pesticides on coastal wetlands in the US. This example demonstrates how difficult it can be to quantify potential sources of impact or to identify the mechanisms by which impacts are expressed in the receiving environment. The coastal wetlands that formed the focus of Clark *et al.*'s (1993) study received pesticides indirectly from agricultural and forest control of insects in upland drainage areas, indirectly or directly from insect control for development of adjacent lands for agricultural, recreational and residential uses and directly from control activities practised within the wetlands themselves for protection of public health and pest control. The wetlands were therefore exposed to the cumulative effects of pesticides from a variety of sources. However, the magnitude and importance of this was unclear, because the actual input of pesticides was unknown. This would have required consideration, not only the amounts and types of pesticide applied, but also the biodegradability of the pesticides in relation to the proximity of their application sites to the wetland. The rate and extent of localized flushing, mixing and stratification within the wetland could also have greatly affected exposure concentrations and durations for the wetland biota but these functions were difficult to measure directly. This is a case where there have been repeated exposures to multiple chemicals at sublethal concentrations, resulting in as yet unquantified cumulative effects. Not enough was known about the 'normal' functioning of the wetlands for it to be possible to identify pesticide effects with certainty. Regulation of individual pesticide applications would not help, unless an analytical framework was in place that would permit assessment of the combined effects of those releases on potential target organisms.

This emphasizes the fact that, although cumulative impacts can be considered as just another category of impact, they may nevertheless show certain properties which mean that their assessment requires new approaches to legislation and policy. Although there are technical barriers to the measurement of cumulative ecological effects, often their assessment is hampered simply by lack of access to the necessary study sites. A more tricky problem is that of recognizing or detecting adverse impacts when their expression in individual receptors is close to detection limits. Very sensitive indicators may be needed in such cases.

5.4 Techniques for impact prediction

5.4.1 Description and correlation

Straightforward techniques of 'description' and 'correlation' can play an important part in ecological prediction for EcIA. Observed correlations between the distributions and abundance of species and physical factors like temperature, water regime, soil pH and mineral composition, for example, can be used as a basis for predicting the likely composition of vegetation on a site where future physical conditions can be specified. Sanderson *et al.* (1995), for example, describe a method for predicting plant species distributions using the British National Vegetation Classification (NVC) as a framework and taking account of information on physical site factors and management. The NVC is based on a series of multivariate classifications of data derived from plant quadrat surveys throughout the UK. The NVC is published as a series of manuals that include maps of the distribution of sampled quadrats of each vegetation community, the community species composition and summary environmental and management regimes considered responsible for its development (e.g. Rodwell 1991a,b, 1992, 1994). The NVC could therefore be used to provide a consistent framework 'within which to characterize the effects of both environmental conditions and management regime on seminatural vegetation communities' and to develop models for predicting the occurrence of plant species under specified management scenarios. In this example, the accuracy of prediction varied with the scale of resolution: the model was much more successful at predicting habitat suitability for widespread communities and species than it was for more restricted or localized plant communities. With respect to EcIA, one of the limitations of such models is the fact that they are 'steady-state' models, which essentially predict the vegetation that would occur given sufficient time under a given set of environmental and management conditions, i.e. they predict the likely 'plagioclimax' community. In EcIA, it is also necessary to know how quickly changes can be expected to occur following a specific perturbation in environmental conditions or management, something which Sanderson *et al.* (1995) acknowledged in their study, would be far from straightforward.

Similarly, accurate description of vegetational successions in the past might enable us to make reasonably accurate predictions about the nature of future successions. However, if the causal factors that explain observed patterns and trends on one site are not known, it is very difficult to make generalizations that permit predictions to be made for ecologically dissimilar areas.

One example where careful observations of the responses of species to a particular impact have proved useful in developing predictive techniques, is the method developed by Reijnen *et al.* (1995b) at the Road and Hydraulic Engineering Division of the Dutch Institute for Forestry and Nature Research, for predicting the effects of motorway traffic on breeding bird populations.

The method was based on extensive research in woodland and grass-land landscapes throughout the Netherlands between 1984 and 1991 (Reijnen & Foppen 1991a,b, 1994, 1995; Foppen & Reijnen 1994; Reijnen *et al.* 1995a). The effects of roads used by high-speed traffic on the breeding density of birds were studied by comparing the density of breeding birds in the vicinity of roads with their density at distance from roads. The study was based on 69 locations throughout the Netherlands. This research showed that the density of breeding birds is reduced along roads mainly used by high-speed traffic. There appeared to be a clear relationship between noise loads due to traffic and the quality of breeding habitat (Fig. 5.8).

Typical 'dose–effect' relationships for a number of woodland and grassland species are shown in Fig. 5.9. The transition in the curves indicates the threshold, or noise load at which density decreases. The distance between a road and the locations at which noise thresholds are reached (the 'effect distance') increases with increasing traffic density and speed. Effect distances are reduced by the presence of tall standing vegetation (especially trees), which block noise. For example, in an area with 75% woodland, adjacent to a motorway used by 75 000 vehicles

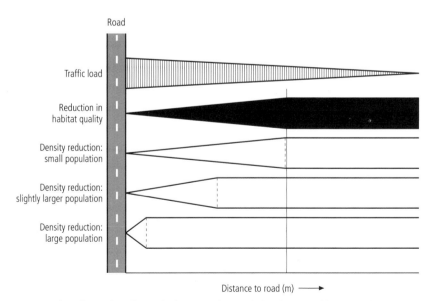

Fig. 5.8 The effect of traffic on habitat quality and the density of breeding birds at various population sizes. Dashed lines indicate the distances up to which densities are reduced. (From Reijnen *et al.* 1995b.)

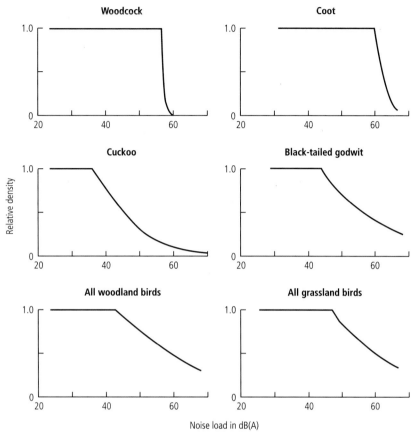

Fig. 5.9 Dose–effect curves for the relationship between breeding bird density and noise load on woodland and open grassland. (From Reijnen *et al.* 1995b.)

per day, the smallest effect distance was 81 m (for woodcock) and the largest was 990 m (for the cuckoo).

5.4.2   Experimental systems

Experimental systems have a potential part to play in quantifying and validating ecosystem responses. However, they can be costly and difficult to set up and will not always yield useful results within the timeframes generally dictated by EIA procedures. Results can generally be expected over periods of years, rather than months. Experimental systems therefore probably have most to offer if used as part of large, ongoing programmes set up to tackle fundamental issues relevant to EcIA in specific situations.

For example, experimental mesocosms and microcosms have been used for registering pesticides in the US. These systems are large, constructed ponds (up to 1000 m³ in volume) and of sufficient depth to support fish, such as bluegill *Lepomis* sp. The purpose of these experiments is to validate exposure models and determine the potential hazard of exposure to a variety of aqueous and sediment-bound chemicals. Such pond mesocosms are valuable for determining the exotoxicity of pesticides in ponds, lakes and riverine backwater habitats, but extrapolation of results to other surface-water situations is less straightforward (La Point 1995). One of the main problems associated with the use of experimental systems is the validity of extrapolating from highly controlled and artificial systems to 'field' situations where responses may be complicated by a range of factors not operating under controlled conditions. In particular, there is a tendency for a limited range of species to be included in experimental programmes. As a general rule, results obtained from experimental systems involving a limited number of species are more reliable than those obtained from more complex systems, where it is much harder to identify clear relationships.

There are many problems associated with small, artificial experimental systems. Large-scale processes in particular are very difficult to mimic. There is also a tendency to over-simplify complex systems in an attempt to isolate effects from 'noise'. For many environmental management problems, direct experimental manipulations of whole ecosystems may have more relevance. Such manipulations may be carried out over considerable areas, such as catchments (watersheds) or the entire home range of a large vertebrate predator (Carpenter *et al.* 1995). Experimentation at the ecosystem level is a field science, and involves experimental conditions that are less easily controlled. It is therefore a challenge to determine whether observed changes are the result of historical trends or inherent patterns rather than being attributable to an imposed manipulation. For this reason, the use of unmanipulated reference ecosystems is beneficial. Ideally, long-term pre- and post-manipulation data are gathered for several reference ecosystems and experimental systems are monitored for long enough to assess trends before any experimental manipulation occurs. However, replication is often difficult, whether because of cost, or the uniqueness of the ecosystem being studied. Ecosystem experiments are therefore less likely to be based on testing null hypotheses and their statistical significance than laboratory experiments under controlled conditions. Extrapolating from the results of single ecosystem experiments to other systems and situations depends on knowledge of mechanisms, or the ability to draw on comparisons of many ecosystems. Where large-scale ecosystem experiments have been established by numerous research groups to tackle a clearly identified ecological process, there has been considerable progress in understanding. For example, Carpenter *et al.* (1995) refer to the numerous experiments on lake eutrophication and acid rain, which have been carried out in several countries. Comparing such experiments makes it possible to determine which conclusions can be considered to be of general applicability and which depend on local conditions. The effects of acid rain on surface water have been studied very effectively through ecosystem experiments based on paired catchments. In Norway, for example, the RAIN (Reversing Acidification In Norway) study used two experimental catchments to compare the effects of rain exclusion using a roof with addition of acidified rain to another 'pristine' catchment. The study demonstrated that acidification of surface waters could be attributed to acid deposition and also that the effects of acidification could be reversed if this deposition was reduced. Follow-on experiments were set up to evaluate potential mitigation measures.

The ability to establish large-scale ecosystem experiments for EcIA is generally restricted by the tight timing of decision-making processes. Their potential value is only likely to be realized when the integration of assessment with ecological monitoring is more widely accepted (see Chapter 8). In some countries, particularly where highly sensitive ecosystems are potentially at risk, the establishment of ecological monitoring programmes with realistic lead times has become more common. This provides a procedural framework within which design of ecosystem experiments to test impacts becomes a realistic prospect. For example, biological monitoring has become a routine part of assessing mining impacts in the Kakadu National Park of northern Australia (Humphrey *et al.* 1995). On the other hand, there are particular problems associated with testing impacts experimentally in sensitive areas. One study proposed to investigate the effects of anchors on pristine parts of the Great Barrier Reef had to be abandoned, although previously protected reefs have been opened up to scientists to study the impacts of line fishing (Anderson 1997). To ensure that the controversy surrounding experiments in sensitive areas can be handled objectively, an Australian working group has proposed some guidelines or ground rules that will govern research in protected and environmentally sensitive areas. These may become enshrined in law and there is scope for their implementation in other countries. EcIA is often carried out in sensitive areas. This not only means that scope for impact testing in the field may be limited, but there may be important ethical aspects to consider when designing field trials.

5.4.3 Models

Modelling involves constructing symbolic representations of the functioning of ecosystems and modelling in some form is an essential part of EcIA, even if it stretches to little more than a crude prediction about the consequences of an action. One demand is for models that can be used to predict 'normal' states of ecosystems. Predicted 'normal' states can then be compared against observed states, whether these are current or future, 'polluted' or 'altered' states. Modelling can also play an important part in predicting the likely distribution and magnitude of potential stressors so that ecological responses and outcomes can be estimated. For example, it might be necessary to use atmospheric deposition modelling to identify areas likely to be exposed to deposition of atmospheric pollutants and to estimate their likely concentrations at different locations under different weather or output scenarios. Generally, modelling in EcIA is used to make predictions about the responses of ecosystems, or components of ecosystems to specific stresses or disturbances.

Walters (1993) identifies four possible approaches to modelling for the environmental manager.

1 *Dogmatic*: beliefs and actions are based on past precedents and 'accepted' practice, without attention to their empirical basis.

2 *Empiricist*: predictions are based on examination of past results in similar situations.

3 *Reductionist*: attempts are made to decompose the problem into a set of component relationships (submodels) that are linked by some deductive or computational engine (model).

4 *Experimentalist*: experimental comparisons are designed that will reveal the best option.

In more simple terms, models used in EcIA are likely to be either conceptual or numerical. Stochastic models use mathematical representations of ecosystems, which take account of probability, so that a given input may yield a number of possible results. Deterministic models, on the other hand, rest on mathematical descriptions of ecosystems in which relationships are fixed, so that any given input invariably yields the same result (Allaby 1994).

From a purely scientific point of view, ecological modelling originated in attempts to explain why fluctuating populations behave the way they do. The usefulness of models for forecasting the future behaviour of ecosystems has been the subject of a considerable amount of research (Pielou 1981). Whereas mechanistic models are intended to explain observed events in natural ecosystems, heuristic models are intended to forecast future events. In doing so, they do not necessarily set out to explain them, but heuristic modelling may nevertheless contribute to our understanding of ecosystem function and is obviously an important tool for predicting ecological impacts due to development. Deterministic linear population models are most commonly used to predict population level effects. These models have been used in forestry, fisheries and wildlife management, and ecological risk assessment (for example see Emlen 1989). Models are usually age, growth-stage or size specific, tracking the abundance of different age, size or ontogenetic stages separately. A 'bookkeeping' approach is usually adopted, birth, death and growth rates being applied to age, stage or size class abundances to predict abundances at a subsequent time interval. The most common interval is 1 year, to reflect annual cycles, but the interval can be modified depending on life-cycle length (Environment Canada 1994).

All modelling begins with some assumptions about the 'true state of things', and these assumptions are invariably restrictive to some degree (Haimes *et al.* 1994). Models are abstractions of reality and their effectiveness and applicability depends crucially on the inclusion of all relevant information. 'Models retain their validity under sufferance of what they have left out' (Barlow *et al.* 1992). In practical terms, the failure of models to predict actual outcomes is often due to omission of particular, key interactions. In this respect, 'adaptive environmental assessment and management' (AEAM) has also proved beneficial in developing models for EcIA. The AEAM approach emerged in the 1970s (Holling 1978) with the intention of bringing scientists and environmental managers together to build predictive models that have a sound scientific basis but also address fundamental management issues. In essence, it is a pragmatic approach. It is also iterative and adaptive and often involves use of different methods to model each part of a complex problem; for example, review of the literature might be used together with ecosystem experimentation to derive input data. In practice, it is rarely possible to identify all key variables and inter-actions and then write equations to quantify how they behave (Walters 1993).

Walters (1993) describes how AEAM modelling was used to assist in water management in the Florida Everglades, a relatively well-studied ecosystem. A major ecological change in the Everglades has been a drastic decline in wading bird populations and a shift from successful breeding along estuarine mangrove zone to unsuccessful breeding further 'upstream' in freshwater marshes. This is believed to be linked with widespread drainage of the marsh ecosystem, and re-supplying water to the estuarine zone is one possible solution to reversing the decline in wading bird populations. Collaboration between hydrologists, biologists, engineers and wildlife managers was necessary for model development. It proved relatively easy to model the basic hydrology of the system and the model appeared quite precise, based on ability to hindcast historical water depths in the area given known historical water patterns. The team was able to use the hydrological model to re-construct a 'target' pre-drainage hydrological regime and also to develop submodels that produced realistic results concerning changes in the density of aquatic prey for wading birds. However, it proved impossible to link such simple trophic predictions to submodels for larger-scale bird distributions or feeding and reproductive success. An experimental pro-gramme was therefore developed to tackle two key questions: the feasibility of large-scale changes in water distribution in the system (to be tested through hydrological manipulations) and the responses of wading birds to broad shifts in the distribution and timing of water delivery to the system. This particular example illustrates the difficulties of modelling complex ecosystem responses and indicates the level of investment in modelling and research that may be required to tackle important issues of ecological management. The high cost of ecological research is often used as an argument for bypassing important stages in the EcIA process, but the consequences of failing to predict ecological responses can be even more costly. In this case, considerable investment in hydrological engineering could be completely wasted if the dynamics of breeding bird populations are not understood. On the other hand, the consequences of inaction due to uncertainty or incomplete under-standing can be equally serious. In other words, use reliable information to construct models, but do not go overboard!

Different models have different strengths and limitations. There is no rigorous (mathematically defensible) way to decide which model should be used under which circumstances, but choice of model can have a significant effect on predictions. It is important to match the data and assumptions on which models are based to the decision that is to be made. Guidelines for community-level habitat models were presented in a report by Schroeder and Haire for the USDI Fish and Wildlife Service in 1993. They emphasized the need for:
- clearly defined, testable output;
- testing with empirical data;
- documentation of sources of information;
- clearly stated assumptions and limitations;
- clearly defined variables;
- adequate verification of model performance.

In addition it is important to select models that are consistent with decision-making criteria. Decisions about which models are to be used should be made as early as possible in the impact assessment process so that the input parameters required by the chosen models can be identified and data collection programmes designed accordingly. Ideally, sensitivity analysis should be carried out to identify the critical para-meters that influence model output, and goodness-of-fit tests should also be used. This may demand identification of distributions for input parameters, as they may have a range of values. If distributions for input parameters are not known, it may be necessary to estimate them. Most importantly, it is important to establish whether the model and its associated input parameters produce results that can discriminate among different decision alternatives. If not, it will be necessary to go back to the beginning (in other words to iterate the process) (Haimes *et al.* 1994).

5.4.4   Habitat evaluation procedure

The HEP developed by the US Fish and Wildlife Service (1980) demonstrates the utility of habitat-based approaches to EcIA. HEP is routinely used for assessing impacts on wildlife in the US. It combines theoretical knowledge of species' habitat needs with field survey to document the quantity and quality of habitat available and to compare it with 'ideal' or 'optimum' conditions. It can be used to compare the relative value of different areas for a selected wildlife species at a given point in time, or to compare the relative value of the same area at different times. In EcIA these types of comparison can be used to carry out baseline assessments and impact assessments, respectively. They can also play an important part in comparing options for compensation or mitigation. However, it is important to note that HEP only addresses the issue of habitat availability and carrying capacity for selected evaluation species. In itself, it does not address other categories of ecological impact, such as changes in overall species composition.

HEP is based on the assumption that certain habitat variables can be measured (e.g. vegetation composition) which are strongly correlated with the ability of an area to support given wildlife species (its carrying capacity) and can be used to derive numerical habitat suitability indices (HSIs). These numerical indices (from 0.0 to 1.0) can be multiplied by the area of available habitat to obtain the habitat units (HUs), which form the basis of impact assessment. The reliability of the HEP and the significance of HUs are directly dependent on the ability of the user to assign a well-defined and accurate HSI to the selected evaluation species and, more specifically, to identify clear relationships between 'carrying capacity' and the modification of specific environmental variables. The development of HSIs is considered in the following section (5.4.4.1).

HEP as a whole involves certain steps or stages as summarized in Box 5.5.

The selection of the evaluation species that will form the focus of the HEP requires some sort of screening, as it is never possible to include all species in an assessment. Under HEP, evaluation species are selected as representatives of terrestrial 'feeding' and 'breeding' guilds (see section 4.2.7). By necessity, as in EcIA as a whole, the derivation of study limits for HEP must often be an iterative process. Study area boundaries may have to be revised after vegetation cover types have been delineated and evaluation species have been selected. Unless the general habitat requirements of evaluation species are relatively well known at the outset, it may not be clear, for example, what home-range sizes should

Box 5.5 Stages in HEP.

- Selection of key wildlife indicator species
- Review of habitat requirements for selected wildlife indicator species
- Definition of study limits
- Identification of plant community types and associated landscape units needed to estimate habitat suitability
- Collection and field measurement of a range of significant habitat variables by plant community
- Development of habitat suitability index (HSI) models
- Determination of HSIs (inadequate to optimum) for wildlife evaluation species by plant community type
- Determination of habitat supply (habitat units, HU): a product of HSI values and habitat availability (land area of plant communities or landscape units)
- Description of baseline habitat conditions in terms of HUs
- Projection of future habitat conditions
- Evaluation of difference

be assumed for them. In fact, home-range size is influenced by habitat quality, which is clearly an unknown quantity at the beginning of the HEP. Study limits should therefore always be flexible enough to take account of changed knowledge about site-specific species behaviour.

5.4.4.1 *Developing HSI models*

HSI is derived by comparing habitat conditions in the study area with optimum habitat conditions for the same evaluation species. It can therefore be defined as study area habitat conditions/optimum habitat conditions.

The HSI has a minimum value of 0.0 (totally unsuitable habitat) and a maximum value of 1.0 (optimum habitat). To obtain an HSI, models based on life-history characteristics and knowledge of habitat requirements are used. For some evaluation species, there may already be existing models. Otherwise it will be necessary to produce one. Models produce a 0.0–1.0 index on the assumption that there is a linear relationship between the HSI value and carrying capacity (Fig. 5.10). The ability of an area to 'cater for' the needs of evaluation species is fundamental to HEP, which essentially provides a consistent and objective method for calculating the combined effects of habitat loss, or the alteration of habitat variables on the carrying capacity of the study area. The method cannot work unless the relationships between habitat variables and carrying capacity are well understood. The method can only be as good as the information on which it is based, so it remains as vital as ever for practitioners to establish what the key variables are and what data are required to derive reliable functional relationships. There

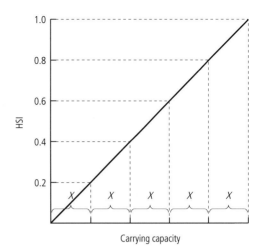

Fig. 5.10 Assumed linear relationship between habitat suitability index (HSI) and carrying capacity. (From US Fish and Wildlife Service 1981.)

are a number of theoretical problems associated with the use of indices. The guidance issued with respect to HEP emphasizes that 'word models' may be just as good as numerical models, which sometimes imply a level of precision not merited by the data on which they are based. Note that there may be cases where it is unreasonable to assume a direct linear relationship between habitat suitability and carrying capacity, for example there may be abrupt changes when thresholds are reached for some variables, or more extreme responses as habitat suitability decreases towards zero.

For US species there are a large number of existing models. In other countries, where the HEP approach to EcIA has not been widely adopted, there would be some way to go to ensure that adequate species–habitat models could be generated. For example, HEP has been used with some success as a tool for ecological rehabilitation of wetlands in the Netherlands by developing specific HSI models as required, but to make HEP a standard method, it would be necessary to develop habitat models for a full range of evaluation species (Duel *et al.* 1995).

Models may be numerical or descriptive. A simple ranking approach can be used to convert descriptive models into a numerical ranking, as shown in Table 5.4. The HSI in this case is equivalent to the output rank for the area of interest divided by the highest rank the model could provide, i.e. for 'excellent' it would be $4/4 = 1.0$.

Similarly, for models with defined output units, a denominator corresponding to the maximum regional value for the predicted measure is used to derive the index, for example the highest long-term density observed for the evaluation species in a region might be 60 individuals per square mile, and this would be used together with the current population density estimate for the evaluation species in the study area to derive the HSI.

Table 5.4 Method used in habitat evaluation procedure to convert descriptive measures into numerical ranks. (After US Fish and Wildlife Service 1981.)

Output	Numerical rank	HSI value
Excellent	4	1.00
Good	3	0.75
Average	2	0.50
Below average	1	0.25

HSI, habitat suitability index.

An HSI model is needed for each evaluation species included in the HEP analysis. Setting model objectives involves:
- defining the ideal and acceptable model outputs;
- defining the geographical area to which the model is applicable;
- defining the season of year for which the model is applicable.

The ideal model outputs is a 0–1.0 rating that has a direct linear relationship to carrying capacity (e.g. defined as units of biomass per unit area or units of biomass production per unit area). If such ideal output is unobtainable, it is necessary to define limits of reliability considered acceptable in each case, given the amount of time, information and funding available. In other words, to define a level of reliability at which the model can be considered ready for application. Criteria used to determine acceptable levels of reliability might be as follows:
- model outputs based on sample data appear reasonable to evaluation team;
- model outputs appear reasonable to species expert;
- model outputs rank study sites in a manner similar to independent expert rankings;
- model outputs predict carrying capacity as measured by population estimates;
- model outputs predict carrying capacity as measured by populations, within 10% with a confidence limit of 90%.

Every model should be applicable to a defined geographical area within which it can be expected to yield consistently reliable HSI values. The geographical area of model applicability should therefore be clearly defined for each species. It may include the entire range of the species, but if a species displays significant differences in habitat preference for different geographical areas, regional models may be appropriate for each area. The area of model applicability should therefore be referenced to some standard unit such as a watershed or catchment, an administrative boundary or an ecoregion. Models should also take account of residency status and differing requirements throughout year. It is common for separate models to be developed for different seasons.

Components   Variables

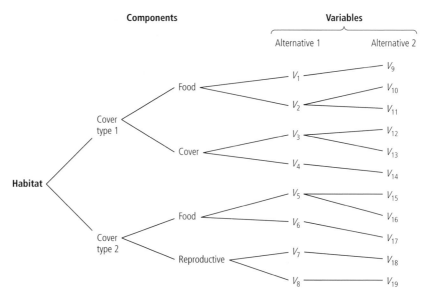

Fig. 5.11 Identification of model variables through definition of habitat components. (From US Fish and Wildlife Service 1981.)

Identifying suitable model variables is vital. This entails identifying those environmental variables which, if altered or modified, would be expected to affect the capacity of the habitat to support a particular evaluation species (e.g. see Fig. 5.11). Potential variables for a typical habitat assessment include measurable physical, chemical or biological characteristics of habitat. Species population variables are rarely included in habitat models because they are costly to measure and difficult to predict.

In identifying model variables it is always advisable to check the following conditions:
• the variable is related to capacity of habitat to support species;
• there is at least a basic understanding of the relationship between the variable and habitat (e.g. best and worst conditions for the variable and how it interacts with other variables are known);
• the variable is practical to measure within the constraints of model application.

The level to which a species' use of habitat is separated into components depends considerably on the quantity and quality of life-history information available. Ideally, division should continue to the point at which each component can be related to measurable variables that can be quantitatively described with some degree of replicability using standard field sampling and mapping techniques.

There may be alternative variables for the same component. For example, a measure of food availability for a species might be insect abundance through the summer. While insect abundance can be measured directly, it may be more straightforward or cost-effective to use some indirect or surrogate measure instead. For example, measures of vegetation structure might provide an indirect measure of insect abundance.

Habitat variables for an evaluation species tend to be grouped according to the following components:
• seasonal habitat (e.g. winter range, hibernating or breeding season habitat);
• life requisites (such as food, cover, water or other 'special' resources);
• life stages (e.g. larval, adult);
• cover types (areas of land with similar physical, chemical and biological characteristics).

Division of variables on the basis of cover type permits rationalization of field survey. For example, as in Fig. 5.12 not all variables need necessarily be measured in every cover type. Delineation of cover types is also useful when defining habitat suitability in terms of spatial relationships ('interspersion') of food, cover and reproductive resources.

Having identified measurable variables for life requisites (e.g. see Fig. 5.13), it is then possible to consider overall habitat suitability. Many evaluation species will use a variety of cover types to satisfy their needs. If measurable variables have been identified for more than one cover type, then overall habitat quality is likely to be influenced by spatial variables as well as the overall amount of each type available.

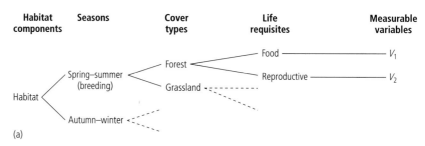

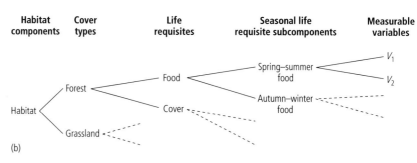

Fig. 5.12 Tree diagrams used to derive measurable variables for terrestrial habitat. (From US Fish and Wildlife Service 1981.)

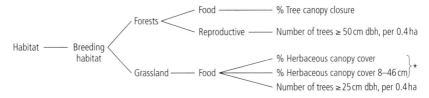

Fig. 5.13 Cover types, 'life requisite components' and measurable variables for breeding season habitat for the red-tailed hawk. dbh, Diameter at breast height. *Food depends on small-mammal populations and their availability/accessibility. The assumption is made that these can be inferred from the vegetation structure. (From US Fish and Wildlife Service 1981.)

Spatial aspects of habitat provision are less straightforward to model. Spatial factors vary considerably between species and for many species have not yet been defined. It is generally necessary to search for relevant information in the literature to find likely upper and lower limits, for example with respect to maximum acceptable distances between feeding and roosting sites. A stylized illustration of the general relationship between overall habitat suitability and the proximity and composition of life requisite variables is given in Fig. 5.14.

Often species use a variety of cover types to satisfy their requirements. The value of an area to a species will be influenced by factors such as the relative contribution of one cover type to the overall availability of all useful cover types. For example, 'optimum red-tailed hawk habitat is composed of at least 70% optimum food-producing areas (i.e. grasslands and forest lands) and at least 15% of cover and reproductive

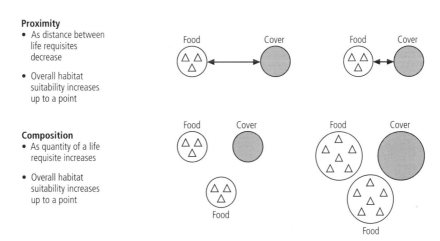

Fig. 5.14 Stylized relationship between overall habitat suitability and the proximity and composition of life requisite variables. (From US Fish and Wildlife Service 1981.)

habitat (forest lands) which are located within 1.2 km of each other. Food and reproductive areas separated by more than 3.6 km are not suitable habitat' (US Fish and Wildlife Service 1981).

When applying an HSI model in HEP, you will need to establish the following.

• Does the model produce suitability indices for available habitat from different individual cover type indices?

• If cover type suitability indices are calculated, does the available habitat for the species consist of more than one cover type?

• If available habitat consists of more than one cover type, is interspersion between cover types important for the species? If yes, it will be necessary to produce a combined index that takes interspersion into account. If no, and all habitat requirements can be met by any cover type, a different approach can be used to derive a simple weighted mean of the suitability indices for the cover types weighted by the area of each cover type.

Each variable identified must be combined with other model variables to actually produce the HSI. This demands definition of relationships between variables. Relationships may be expressed in the form of graphical display, written statements or mathematical equations. Likewise, a variety of modelling approaches may be used. Often, it is necessary to produce 'word models' to express habitat suitability.

These generally express 'rules of thumb' about species requirements. For example: 'The suitability of red-tailed hawk nesting habitat increases with tree diameter'. This might be refined further, for example by adding a threshold value: 'The best red-tailed hawk nesting locations are in trees > 50 cm dbh (diameter at breast height)'. Application of a threshold in this way makes it possible to deduce that trees less than 50 cm are not suitable (index of 0.0) and that trees greater than 50 cm dbh are (index of 1.0). Often there is a range of suitability classes, for example 'red-tailed hawk nests are only found in trees greater than 20 cm but the best locations are in trees greater than 50 cm' (US Fish and Wildlife Service 1981).

For overall habitat suitability, an example 'word model' for the red-tailed hawk is given in Box 5.6.

'Word' models can be converted to simple mechanistic models by determining a suitability index for each measured variable and aggregating these to form a component (or combined) suitability index. A suitability index graph can be constructed using a variety of approaches, as follows for example.

• Plot specific habitat variable measurements against an observed measure of abundance such as standing crop. The maximum observed abundance measure merits a suitability score of 1.0. Other measures of the variable are assigned a suitability score equal to the observed

Box 5.6 An example 'word model' of habitat suitability for the red-tailed hawk. (After US Fish and Wildlife Service 1981.)

Red-tailed hawk breeding habitat is composed of a mixture of feeding areas (grassland and forest) and reproductive areas (forest) within a certain, specified distance of one another
• Optimum habitat consists of 70% optimum feeding areas and 15% optimum reproductive areas on average within 1.3 km of each other
• Medium suitability habitat occurs when feeding areas occur over 35% of the area, nesting sites occur over 7% of the area and feeding and reproductive habitat is on average separated by 2.4 km
• Marginal suitability habitat occurs when food or cover–reproductive suitability is marginal or when food and cover–reproductive areas are separated on average by more than 3.6 km

abundance at that measured value of the variable divided by the maximum abundance. (Note that problems may arise because it is often necessary to base such graphs on evidence from studies carried out in very different circumstances, or to rely on a certain amount of subjective judgement.)
• Construct suitability graphs based on general statements from the literature (e.g. 'the species prefers to nest in trees above 15 m tall').
• Consult experts and use their opinions to rank suitability.

Suitability index scores are obtained by comparing existing or predicted conditions in the study area with the relationship depicted by the suitability curve, for example see Fig. 5.15.

Note that the implied precision of the relationships from which suitability indices are derived is often greater than that of the biological data from which the graph was actually constructed.

HEP has established certain rules concerning the aggregation of index relationships for different variables. There are a number of factors that need to be taken into account about the specific variables involved in

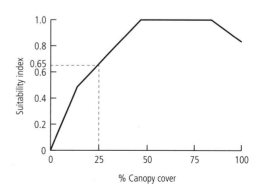

Fig. 5.15 Deriving a suitability index for a variable. (From US Fish and Wildlife Service 1981.)

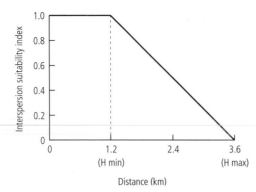

Fig. 5.16 Interspersion suitability index graph for the red-tailed hawk. (From US Fish and Wildlife Service 1981.)

any one case. For example, is there any one factor that can be considered to be limiting or that overrides others? Are there a number of different ways in which overall habitat requirements might be met satisfactorily or are there compensatory relationships where low suitability of one variable is compensated for by high suitability of others?

For example, in situations where all measured variables are important, combined scores should be zero if any of these variables has scored zero. In other words, if an essential life requisite has not been met, then overall habitat requirements cannot be met from the study area. On the other hand, compensatory relationships occur when a threshold level exists that can be met by any one of several variables or a combination of variables. For example, an optimum density of potential nest sites for a bird species might be provided by trees, cliffs or both. In this case, the combined index score cannot exceed 1.0.

An important point to note is that averaging functions tend to become insensitive to extremely low or high values in situations where four or more variables are used.

Just as for individual variables, when combining indices for different cover types, it is necessary to adjust the suitability of 'cover type life requisites' downwards as the distance between cover types increases. In HEP, 'interspersion' is used to adjust cover type life requisite values (Fig. 5.16). Final determination of habitat suitability is made using composition parameters, comparing calculated composition (taking account of interspersion) with the optimum. The closer the study area approaches optimum composition figures, the higher its suitability.

5.4.5 Using HEP for impact assessment

HEP analysis is structured around the calculation of HUs for each evaluation species in the study area. The number of HUs is defined as the product of the HSI (quality) and the total area of available habitat (quantity). The total area of available habitat for an evaluation species includes all areas that can be expected to provide some support to that species.

Habitat models used in HEP must be in an 'index' form, comparing values of interest with a standard of comparison. The development of HSI models has been considered in the previous section. The main point to stress is that models must be structured around measurable variables that are directly related to carrying capacity in a linear relationship (or in a relationship that can be transformed to linear), as impact assessment using HEP is based on the assumption that a change in HSI from 0.1 to 0.2 will be of the same magnitude as a change from 0.8 to 0.9.

Baseline assessments are used to describe existing ecological conditions and to provide a reference point from which it will be possible to compare existing conditions in two or more locations and to predict the changes that can be expected to occur in the absence of a proposed action (in other words under a 'do nothing' option). Doing a baseline assessment properly can be a time-consuming and intensive exercise, but it is well worth investing in because it makes it possible to identify wildlife resource capabilities and ensure that future actions are directed away from particularly valuable or vulnerable areas at an early stage in the design and planning process. Baseline assessments under HEP depend on the selection of suitable evaluation species for study (see section 4.2.7), the definition of study limits (including definition of the study area), delineation of vegetation cover types, the acquisition (or in some cases development) of appropriate habitat models as outlined in the previous section (5.4.4.1) and the characterization of the study area in terms of HUs.

For baseline assessments, the main objective is to calculate the number of habitat units provided for each evaluation species, by multiplying the area of available habitat by the HSI. Results for different areas at the same point in time can be compared directly, for example to determine which areas are more valuable. However, baseline data can also be used to provide a reference point for impact assessments, in which case HEP requires additional calculations to be carried out. By comparing predicted future conditions with and without a proposed action against 'baseline year' provision of habitat units for a species, it is possible to quantify the net impact of a proposal using consistently measured units (US Fish and Wildlife Service 1980).

Impact assessments are performed by quantifying habitat conditions at several points in time throughout a defined period of analysis. Assessment of land-use impacts is facilitated by dividing the study area into impact segments, or areas in which the nature and intensity of future land use can be considered homogeneous, such as the flood-pool area in a reservoir project, a recreational area, or the area devoted to a particular agricultural practice or land use.

Habitat units must be calculated for evaluation species at each of the future points in time for 'future-with' and 'future-without' project conditions. This process includes predicting total available habitat and HSIs for each evaluation species, using the same HSI models that were used for the baseline year. For each proposed action the area of available habitat must be estimated for future years. Some cover types will increase in total area, while others may decrease. In some cases, new cover types will be created or existing ones totally lost under projected

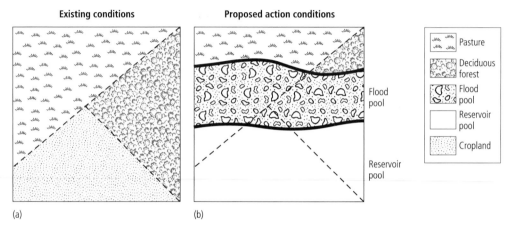

Fig. 5.17 Examples of cover type maps illustrating (a) existing and (b) predicted future conditions. (From US Fish and Wildlife Service 1980.)

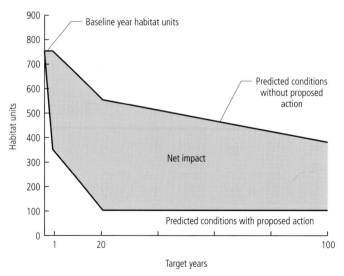

Fig. 5.18 Relationship between baseline conditions, conditions with a proposed action, conditions without the proposed action and net impact. (From US Fish and Wildlife Service 1980.)

future conditions. One recommended method for determining the future area of cover types is to overlay 'impact segment' boundaries on a base-line cover map in GIS. Each proposed action requires its own series of overlays in order to determine changes in area of available habitat between selected years (see Fig. 5.17).

HU gains or losses are annualized by summing HUs across all years in the period of analysis and dividing the total by the number of years in the life of the project to obtain average annual HUs (AAHUs). See the manual published by the US Fish and Wildlife Service (1980) for full details about the calculations required. Figure 5.18 shows the relationship between baseline conditions, conditions with and without a proposed action, and the net impact that will result.

5.5 Recommended reading

Impact characteristics

Spellerberg, I.F. (1992) *Evaluation and Assessment for Conservation: ecological guidelines for determining priorities for nature conservation*. Chapman and Hall, London.

Ecological basis for prediction

Peters, R.H. (1991) *A Critique for Ecology*. Cambridge University Press, Cambridge.

Impact assessment techniques

US Fish and Wildlife Service (1980) *Habitat Evaluation Procedures (HEP)*. ESM 102. Division of Ecological Services, Department of the Interior, Washington, DC.

US Fish and Wildlife Service (1981) *Standards for the Development of Habitat Suitability Index Models for Use in the Habitat Evaluation Procedures (HEP)*. Division of Ecological Services, Department of the Interior, Washington, DC.

6 Evaluation

6.1 Introduction

As emphasized by Smith (1993), resource management decision-making before the 1950s was generally based on the following questions.

- Is the proposal technically feasible?
- Is it financially viable?
- Is it legally permissible?

Environmental impact assessment (EIA) changed things by requiring systematic analysis of environmental acceptability: 'Is it environmentally acceptable?'. Ecological impact assessment (EcIA) provides a formal procedure for taking ecological information into account when making decisions about the exploitation of natural resources, land use, or the acceptability of a proposed change in land use. EcIA therefore demands the evaluation of ecological consequences to determine their acceptability in a particular societal, as well as an ecological context. At the end of the day, it will be necessary to decide what relative emphasis should be given to ecological factors compared with others included in the decision-making process. This demands the translation of ecological findings into a form that permits comparison with other categories of impact (social impacts, for example). The need to find a uniform 'currency' for comparing different categories of environmental impact in decision-making has been one of the main incentives for developing tools of ecological economics (see section 6.5).

By definition, 'evaluation' demands measurement. Ecological evaluation for EcIA generally demands two forms of measurement: measurement on an internal scale and then comparison against some external standard. Impact prediction should generate estimates of impact magnitude, range, probability and frequency. It is then necessary to determine their ecological 'significance', for example to establish whether a predicted population reduction is significant when measured against control, unaffected or pre-impact conditions, reference standards or thresholds such as 'effective population size' (measurement against an internal scale). It does not always follow that statistical significance will be reflected in biological or ecological significance, so both aspects should be considered together to ensure an objective and scientifically defensible approach. A distinction can also be drawn between criteria used to estimate the ecological significance of an impact and those used to consider impact magnitude and significance in the context of external standards to estimate impact 'importance': a measure that must, invariably, take account of societal values and standards.

Evaluation for EcIA might therefore involve the following stages.

• Estimate ecological significance with respect to numerical thresholds on an internal scale, for example evaluate habitat loss using habitat evaluation procedure (HEP) or reduction in population size in the context of minimum viable populations.

• Evaluate ecological outcome in the context of external standards or objectives, for example does a predicted reduction in population number constitute an unacceptable loss in terms of national conservation status objectives (such as agreed under the Convention on Biodiversity) or will important ecological 'services' be maintained? (This can involve subjective judgement in cases where suitable reference data are limited.)

• Consider economic implications (economic evaluation).

• Compare with other factors in decision (any common comparative standard?).

There is little point in making precise ecological predictions if these are then evaluated using loosely defined, inconsistent and purely subjective criteria. From the very start of the EcIA process, information gathering and impact prediction should therefore be structured within a sound evaluative and decision-making framework, and informed by the questions that will have to be answered to make 'good' decisions. Some examples of the types of question for which answers might be sought in EcIA are listed in Box 6.1. Absence of reliable numerical predictions and evaluation criteria has resulted in a tendency to fall back on subjective judgement, but it is best if the limits of objective analysis are reached before subjectivity comes into play. It is equally important not to obscure subjective judgements behind a 'front' of apparent objectivity by using arbitrarily derived numerical weights and indices. Ecological evaluation should be a transparent process, open to challenge as the knowledge base improves.

There has been much debate over the efficiency and validity of different techniques for ecological evaluation (van der Ploeg & Vlijm 1978; Gotmark *et al.* 1986; Usher 1986; Anselin *et al.* 1989). In particular, difficulties have arisen over the combination of quantitative and qualitative factors. Several authors have proposed the use of multiple evaluation criteria, which are then given weights or priorities using scoring system (Yapp 1972; Wright 1977; Usher 1986). Particularly where objective scoring systems are to be combined with more subjective weighting, it is important that there should be absolute clarity about:

Box 6.1 Examples of ecological questions likely to be encountered.

- Will loss or redistribution of habitat affect the long-term viability of associated species?
- Will carrying capacity, stress thresholds or assimilative capacity be exceeded?
- If this habitat is destroyed, will associated species find alternative habitat?
- If this habitat is destroyed, will remaining habitat be adequate to support associated species?
- If this habitat is destroyed, can it be replaced using current technology and within a reasonable timeframe?
- Will ecosystem resilience or stability break down?
- Will predicted population reductions for a species result in loss of long-term population viability?
- Will significant, irreversible loss of biodiversity occur?
- Will reduced genetic diversity result in reduced ability to withstand environmental change in future?
- Will loss of one habitat type be more damaging than loss of another?
- Will the post-development state of an ecosystem be significantly different from its pre-impacted condition?
- Should any losses of ecosystem components or functions be mitigated or compensated for and if so, which ones?
- Will proposed mitigation measures guarantee the maintenance of natural resources within acceptable limits, i.e. will the residual condition of ecosystems (post-impact and taking account of mitigation) be acceptable?

- what is being evaluated;
- which criteria are to be used;
- how they are to be measured.

Evaluation may be achieved by applying certain criteria to measured ecological parameters. No standard set of criteria has emerged and a multitude have been used in practice. Attributes that have often been tested in the UK are described in some detail by Ratcliffe (1977) and Usher (1986) and are used routinely in the selection of Sites of Special Scientific Interest (SSSI). Ratcliffe's (1977) criteria for evaluation of nature conservation importance have been widely used in EcIA in the UK (Box 6.2), but there are some problems with their application in this context, notably the lack of criteria relating to the recoverability or replaceability of natural resources. These criteria are also difficult to apply consistently because they cannot all be measured using defensible, consistent, objective techniques. For example, while it is possible to derive objective, numerical estimates of 'diversity' and 'rarity', it is much less straightforward to measure attributes like 'intrinsic appeal'.

These criteria can be used to evaluate sites, biotopes, habitats, habitat assemblages or the species supported by them. It is not always easy to

Box 6.2 Criteria for evaluating nature conservation importance. (After Ratcliffe 1977.)

Primary criteria
• *Size*. Including both area of vegetation types and population sizes for individual species
• *Diversity*. Applied either as simple species richness, or by giving different weightings to species according to their 'interest'
• *Rarity*. Applied either to habitats or to species. The latter most commonly tested by comparison with national or county population size or distribution by 10-km squares
• *Naturalness*. Habitats that are least intensively modified by humans are generally more highly regarded
• *Typicalness*. A measure of how well the study area represents habitats or vegetation types on a wider scale
• *Fragility*. Some habitats or species are especially vulnerable or sensitive to anthropogenic change. Those with restricted area or ranges are generally held to be more vulnerable

Secondary criteria
• *Recorded history*. Can be useful in confirming that a site has been 'important' for some time. Sites with a long history of study may contribute significantly to our understanding of ecological processes
• *Potential value*. Relates to the likelihood that appropriate management could restore or enhance an area's ecological interest
• *Position in geographical or ecological unit*. Some areas of fairly low 'intrinsic value' may be more important because they form successional stages between more important areas. Also, nationally common habitats or species might be very rare locally
• *Intrinsic appeal*. Habitats or species with public appeal promote the cause of nature conservation and can 'pull' funds. This criterion can also be interpreted to include estimates of public use, access, amenity value, etc.

apply them consistently. For example, with respect to 'size', both small and large sites can be equally important, albeit in different ways. A large site is more likely to support diverse assemblages of characteristic species and less likely to be subject to 'edge effects' than a small site, but for less mobile species with small home ranges, the preservation of a few remaining small sites might be crucial. Similarly, high diversity of species on a site is not invariably a desirable attribute. Some areas have innately low diversity but support many species that occur nowhere else. In such areas increased diversity is often a consequence of invasion by uncharacteristic or invasive species, which may be of less conservation value. It is also important to note that diversity within one taxonomic group will not always be reflected in others. For example, *Calluna* (heather) heathlands in North Yorkshire in the UK are plagioclimax communities with low plant diversity, but often support very rich invertebrate assemblages (Usher 1992) and a complex and unique bird fauna (as a result of periodic management by fire and grazing, intended to prevent woodland succession and sustain grouse populations).

6.2 Important sites or areas

In most countries, EIAs for proposed developments are required to identify potential impacts on protected species or any sites that are designated for nature conservation. It can be 'taken as read' that such resources are valued by society and that the consequences of their potential destruction or damage should at least be given careful consideration. However, the methods used to designate sites or confer protection on species are not always rigorous or consistent. Not only does the process of legislation to protect natural resources frequently lag far behind their rates of decline, but it is not always clear to what extent protected area networks are designed to protect a country's 'critical natural capital' in its entirety. In other words, can society afford to lose a proportion of its protected sites every year, as happens in the UK, and still maintain all its characteristic native wildlife resources? Before the consequences of ecological impacts can be weighed up against the economic benefits of development or compared against other categories of impact, therefore, evaluation techniques are required to determine the *ecological* significance or importance of predicted impacts.

Much of the motivation for research on ecological evaluation has come from the desire to select nature reserve networks that are fully representative of biological variation. The effectiveness of nature reserves in protecting wildlife depends on adequate knowledge of the distribution and status of natural features and well-founded methods for locating reserves in relation to those features and other, potentially damaging, land uses. Evaluation in terms of representation of habitats and species requires explicit targets for the number or extent of each feature that is required for adequate safeguard. These may need to be adjusted for local, regional, national or international contexts. They may also need to be spatially adjusted to take account of 'design criteria' for wildlife areas, for example to ensure the maintenance of wildlife areas that are sufficiently large to support a range of species, that are well connected to other wildlife areas and that are compact enough for edge effects to be kept down to an acceptable level. In areas where the degradation of natural areas outside reserves is rapid, the failure to locate reserves sensibly with regard to distributions of species and their habitats places them at risk of at least local extinction. Exactly the same logic can be applied to EcIA. Our ability to divert development or exploitation away from critical natural resources is severely hampered without good knowledge of their distribution and status. How can we evaluate the importance of a local extinctions and losses if we do not know any-thing about the overall status of a species, the availability of alternative habitats and its ability to colonize them?

Bedward *et al.* (1992) outline an iterative approach for nature reserve selection in which the potential contributions of unselected sites to full representation are recalculated each time a site is added to the reserve network as a whole. In this context, the highest 'efficiency' in selection is associated with the ability to represent all features in the smallest number of sites. Even so, the proportion of the total available land area needed to represent all features in an area can be considerable. Estimates vary from about 8% (Pressey & Nichols 1989) to 45% (Margules *et al.* 1988) depending on the scale of definition of the features and the size of the areas being examined. There are obvious applications for this approach to EcIA, where the damage or destruction of sites through development influences the extent to which remaining wildlife sites permit full representation of biological diversity in an area. In evaluation for EcIA, instead of determining how many nature reserve areas are needed to represent certain features, we might ask the question: 'how many ancient woodlands can we afford to lose from an area before the species associated with them will become non-viable in that area?'. Answering this type of question not only demands clarity about the factors contributing to viability, the links between habitat distributions and species distributions and the consequences of spatial reorganization, it also demands clear standards for deciding how much the conservation of biological diversity matters. In other words, we need to decide exactly how much we care if a species becomes 'non-viable'.

6.3 Ecological evaluation criteria

If EcIA is to play any part in safeguarding the long-term viability of natural resources, there is an obvious need for regional, ecosystem-based approaches to evaluation. It is not enough to know that a species or habitat is vulnerable and threatened. We also need to know how much, why, where and exactly what the thresholds of viability are. There are a number of concepts and ecosystem properties that have potential appli-cations to ecological evaluation for EcIA and that can be used to derive suitable assessment and measurement endpoints (Box 6.3). Some are con-siderably more easily measured than others. For example, it is possible to determine characteristic levels of diversity and rarity for habitats or ecosystems, to quantify 'minimum viable habitat' for valued ecosystem component (VEC) species or to define thresholds for population viability. It is much less straightforward, however, to quantify properties like 'endangerment' 'vulnerability', 'replaceability' or 'ecosystem health', which are complex and based on many interacting factors.

Box 6.3 Ecological concepts and ecosystem properties applicable to EcIA evaluation.

- Extinction risk
- Population viability (the 'minimum viable population')
- Metapopulation dynamics
- Genetic diversity
- Effective population size
- Minimum viable habitat
- Home-range size
- Rarity
- Diversity/complexity
- Fragility
- Stability
- Resilience
- Replaceability
- Endangerment

As a general rule, a number of criteria will be used to evaluate impacts, such that damage to populations of species that are rare and also threatened throughout their range might be taken more seriously than damage to populations of species that are rare, but showing no general signs of decline.

6.3.1 Extinction risk

Extinction is a 'natural' process, but while we do not necessarily miss all the species that have disappeared, we should be very wary of allowing our actions to cause species extinctions unchecked. As mentioned at the beginning of the book, study of the fossil record suggests that current rates of species extinction are unprecedented and a great many extinctions are directly attributable to human action. Some estimates have put levels of extinction in the order of several species a day (Prance 1991), but we are not in a position to make reliable estimates because our knowledge is so limited. The fact that species are disappearing before we have even named them, ascertained their ecological roles or elucidated their potential importance to us, suggests that it is wise to take a precautionary approach. If we 'take it as read' that it is undesirable for human actions to cause extinction of species, then we must develop methods for detecting the warning signs of impending extinction at an early enough stage for avoiding action to be taken.

In the context of EcIA, avoidance of species extinction is a logical benchmark for evaluation, where we might ask the question: 'will the predicted impacts result in a significant increase in extinction risk for the species affected?'. However, 'extinction risk' is not a simple property to measure. There are many interacting factors that determine the chances of extinction, the events that might trigger it, the speed at which it will happen, etc. While rare species are generally more vulnerable than common ones, for example, it is also necessary to take account of factors that make a particular species vulnerable to specific threats. Undoubtedly there are a number of species that have become increasingly rare as a result of human activities, but some species are inherently localized in distribution simply by virtue of specialization to extreme or unusual circumstances. Similarly, in any one country or locality, a species may be rare simply because it is at the geographical limit of a much more extensive natural range, rather than by virtue of anthropogenically induced decline. Because rarity can be quantified, it is a useful property to use in ecological valuation (see section 6.3.10). However, measures of rarity alone cannot be used as a reliable indicator of extinction risk. Often it is necessary to consider a whole range of factors that may be affecting the viability of populations. The emerging discipline of population viability analysis (PVA) is considered in the following section.

6.3.2 Population viability

PVA involves the 'structured, systematic, and comprehensive examination of the interacting factors that place a population or species at risk' (Shaffer 1990). PVA has an obvious role in EcIA, for any case in which endangered species may be affected by a proposed action. It also has more general applicability in an age when species that are generally considered to be ubiquitous and common can slide suddenly towards extinction. The decline of British songbirds in recent years is a notable example. Not only is it necessary to estimate the likely magnitude of a reduction in population number, it is also necessary to estimate risks posed by population reductions with respect to the long-term viability of a species.

If applied in this way, it is necessary to consider what timeframes are appropriate for PVA and what probabilities of persistence can be considered sufficient to be confident about species or population viability. In other words, what thresholds of persistence should be used if PVA is used in the evaluation phase of EcIA? Importantly, the use of PVA has emphasized the need for an objective forward-look when evaluating impacts on endangered species. It is not enough just to use current conservation status for evaluation: it is necessary to know whether species or their populations are declining, stable or recovering.

Assessments of population viability can be considered as a form of risk analysis applied to issues of species conservation, in that they attempt to assess the likelihood of extinction by some specified time in the future under various scenarios of management. PVA is a relatively new technique, christened by Gilpin and Soule in 1986 but it has already proved useful to conservation biologists as an aid in making management decisions. For example, it has been used in the US to aid management decisions for the grizzly bear *Ursus horribilis* (Shaffer 1981), the spotted owl *Strix occidentalis* (Marcot & Holthausen 1987) and the red-cockaded woodpecker *Picoides borealis* (US Fish and Wildlife Service 1985). Although earlier applications of PVA tended to be for relatively large, wide-ranging, long-lived and slowly reproducing species in critically low numbers, the concepts and techniques of PVA are equally applicable to small, short-lived species with rapid reproductive potential that are restricted to one or a few relatively small habitat areas but may be present at high numbers within them (Shaffer 1990).

6.3.3 Minimum viable populations

The concept of a 'minimum viable population' is based on the premise that species have a minimum population size below which extinction is likely to occur through ecological or genetic factors. In 1981, Shaffer defined the 'minimum viable population' as 'the smallest isolated population having a 99 per cent chance of remaining extant for 100 years despite the foreseeable effects of demographic, environmental and genetic stochasticity, and natural catastrophes'. It remains an elusive number for most species, but various attempts have been made to define threshold sizes below which populations will effectively cease to be viable and some studies have come up with size ranges that can be used as rules of thumb for management purposes. For example, a study by Berger (1990) on the persistence of different-sized populations of bighorn sheep *Ovis canadensis* in south-western North America showed that native populations of bighorn sheep below a threshold size of 50 were unable to resist rapid extinction: 100% of populations with fewer than 50 individuals went extinct within 50 years. Similar PVAs are needed for other species to derive minimum viable population numbers for use in ecological evaluation as 'measurement endpoints'. It is then possible to evaluate predicted population reductions in terms of whether or not thresholds of viability will be exceeded.

6.3.4 Metapopulation approaches: their role in PVA

When conducting PVAs, it is not always easy to be precise about exactly what constitutes 'a population'. Particularly in fragmented landscapes, it may be necessary to decide whether or not metapopulation dynamics are operating.

Murphy *et al.* (1990) carried out a PVA for the threatened bay checkerspot butterfly *Euphydryas editha bayensis*, which demonstrated the need to take what they called an 'environment-metapopulation approach'. The bay checkerspot butterfly is a threatened species confined to two areas in California. It is restricted to discrete patches of native grassland associated with outcrops of serpentine rock. The population dynamics of this insect are largely driven by variation in thermal conditions and rainfall (macroclimate) and local topography (topoclimate). Both larvae and host plants are highly sensitive to thermal differences between slope exposures and to annual variation in rainfall, and extinction of local butterfly populations is common. The sensitivity of bay checkerspot butterfly populations to year-to-year differences in rainfall and insolation makes local populations of the butterfly particularly susceptible to extinction. The ability of the species to recover from known local extinctions implies that it may have been persisting as a metapopulation, or a collection of interdependent populations affected by recurrent extinctions and linked by recolonizations. Indeed studies of the butterfly at Morgan Hill (Harrison *et al.* 1988) suggested that the occupancy of habitat patches depended not only on their quality in terms of size, topographic diversity and resource abundance, but also on proximity to a large 'habitat island', which may have been acting as a reservoir for recolonization. Patches more than 7 km away from this reservoir were less likely to be occupied, even if they appeared suitable for the butterfly. The regional distribution of this species at any given time could only be explained with an understanding of metapopulation dynamics. Likewise, it would only be possible to evaluate the significance of losing individual habitat patches by studying the overall distribution of the butterfly and the availability of alternative habitat with respect to sources of recolonization. Ideally, understanding of the population dynamics of a species makes it possible to determine whether a proposed action or development proposal is likely to cause a degree of population reduction or isolation that will increase extinction risk significantly.

Metapopulation dynamics may have an important part to play in maintaining a favourable conservation status for many species. Another example that has been studied is the acorn woodpecker *Melanerpes formicivorus*. This species naturally occurs in small, isolated populations throughout much of the American south-west, but it is neither rare nor endangered. Data gathered over 10 years showed that the species survives through immigration from other populations (Stacey & Taper 1992). Such a 'rescue effect' can only be relied upon if the various populations within the metapopulation are exposed to relatively independent environmental stresses as, if all the populations declined together, the entire metapopulation would be at risk of extinction due to environmental stochasticity. Early theories that metapopulations consisted of 'source' and 'sink' populations (*sensu* Pulliam 1988) need to be modified to take account of the fact that each population may shift between source and sink whenever environmental stochasticity is great. In other words, metapopulations may be highly dynamic for some species.

This has considerable implications for EcIA where, in deciding whether a predicted loss of habitat or associated reduction in population size will translate into a significant loss of viability, it is necessary to have a detailed understanding of all the factors that maintain populations or, conversely, lead to their extinction. Not all habitat patches may be critical to the survival of a species in a region. However, where metapopulation dynamics are operating, even populations that are apparently stable may depend on frequent, regular immigration for their survival. Progressive loss, fragmentation and isolation of habitat reduces opportunities for interchange and can have serious implications for the ability of species to recover from local extinctions caused by climatic factors.

For many invertebrate species, population extinctions tend to result from the combined effects of habitat loss, habitat deterioration and extreme weather events. Proposed actions that reduce habitat area, size and quality may not cause extinction of a population directly, but may increase vulnerability to extreme weather or other events that do cause extinction. For species like butterflies, which are so influenced by environmental variation, demographic stochasticity may only come into play when populations have declined to really small sizes. Demographic stochasticity is thought to have played a significant part in the extinction of the large blue *Maculinia arion*, for example, but only *after* environmental factors reduced the population to just five individuals (Thomas 1984).

6.3.5   Reductions in genetic diversity: implications for viability

Reduced genetic diversity is one potentially damaging feature of small populations which should be taken into account when establishing minimum viable population sizes. Spellerberg (1991) gives examples of species with populations so small that their genetic variation has become severely depleted, including the rhinoceroses of Africa and India (Merenlander *et al.* 1989; Ashley *et al.* 1990). Between 1970 and 1990, numbers of black rhinoceros *Diceros bicornis* in Africa declined from 65 000 to 3800. The remaining animals were distributed between 75 populations, only 10 of which had more than 50 animals. The situation was even worse for the northern white rhino *Ceratotherium simum cottoni*, with only 22 individuals known to exist in the wild by 1987 (Western 1987). The southern white rhino *Ceratotherium simum simum* fared a little bit better during the same period, recovering from a bottle-neck at the beginning of the century to about 3000 individuals (Penny 1988). There was a proposal to supplement the northern race with members of the southern population, but genetic comparisons between the two suggested that they may have persisted as subspecies isolated from one another, possibly for 2 million years. Paradoxically, in this case, the proposed 'rescue' technique had to be discounted because of obvious genetic differences between the two races. For black rhino, on the other hand, the pooling of small, isolated populations to form larger combined populations (which might be easier to safeguard in designated rhino reserves), may be appropriate because the remaining populations show a high degree of genetic similarity. Clearly, genetic diversity has considerable implications for the conservation management of endangered species.

Reduced genetic diversity is not just a feature of small population size, it is also a feature of the isolation associated with fragmentation of wildlife habitat. As the distances between remaining areas of habitat and the hostility of the intervening 'matrix' increase, it becomes ever more difficult for species populations to interact. One of the consequences of such isolation can be reduced genetic exchange and variability.

In general, smaller and more isolated populations are more likely to be subjected to genetic drift and inbreeding and are less likely to be able to adapt to changing environmental conditions as a result. The implications of reduced genetic diversity for persistence of populations in the longer term are not well understood, but there is some evidence to suggest that it increases the chances of extinction for some species. Berger's (1990) study of bighorn sheep, for example, suggested that genetic factors might have played a significant part in the observed, rapid extinctions of smaller populations. In captivity, inbred bighorn sheep suffer higher juvenile mortality than those from less inbred lines (Sausman 1984) and it is possible that this also occurs in wild populations.

With respect to the deleterious effects of reduced genetic diversity, it has not proved possible yet to specify minimum viable population sizes for many species. Studies of British butterflies, for example, suggest that if populations become small enough to manifest the effects of inbreeding, extinction is likely to occur for other reasons (Murphy et al. 1990).

Variation within and between organisms or groups of organisms can be assessed at a variety of levels of biological organization. If the biodiversity of a whole ecosystem is to be assessed, all these levels of organization should be taken into account. This might include, for example, the following (after Rhodes & Chesser 1994):
• communities distributed across a landscape;

- species guilds within communities;
- single species within guilds;
- populations within species;
- individuals within populations;
- genes within individuals.

One way to take account of variation at a number of different organizational levels is to use common measures of genetic diversity or similarity. This can make it possible, for example, to derive estimates of the uniqueness of different populations, species or guilds: something that could have useful applications in EcIA. In many cases, estimates of species diversity alone will not be enough (Smith & Rhodes 1992). It is also important to bear in mind that 'biological diversity' is not a static property. It may be more appropriate to think in terms of 'biological diversity processes' rather than in terms of a 'state that is attained and maintained' (Solbrig 1991 in Rhodes & Chesser 1994). Rhodes and Chesser (1994) refer to 'functional biodiversity' in which processes such as mutation, genetic recombination and natural selection can all alter the characteristics of individuals within populations, while immigration, emigration and gene flow affect the distributions of characteristics among populations and communities. Processes like succession and competition can also influence biodiversity in addition to random or stochastic events. It is important to understand the rates and directions of changes attributable to such processes if 'biodiversity' is to be managed in any coherent way as a property of ecosystems. There is ongoing debate about the degree of genetic complexity required to promote ecosystem stability, but it is generally acknowledged that for most ecosystems, a threshold exists below which present complex systems will lose their stability and become subject to radical alteration (Solbrig 1991). 'Estimates of how gene diversity is maintained and of the rate at which it is lost, are crucial to the maintenance of a biologically diverse and sustainable landscape' (Rhodes & Chesser 1994). One of the most useful measures for obtaining estimates of the loss of gene diversity is 'effective population size'.

6.3.6 Effective population size

Genetic diversity has been studied most intensively at the level of populations and individuals, the level at which the majority of management and conservation efforts are focused. The concept of 'effective population size' was introduced by Wright (1931, 1938). The number of organisms within any given population which actually transmit their genes successfully to the next generation rarely corresponds to the total number of organisms in the population. 'Effective population size' represents the number of individuals in an ideal population that would undergo the same amount of genetic change, via random union of gametes, as the actual population under consideration. Two types of effective population size that are often used are 'inbreeding effective size' and 'variance effective size'. Inbreeding effective size is based on the number of gene correlations within individuals in a population and can be used to determine the rate of loss of heterozygosity within populations. Variance effective size is based on gene frequency drift and can be used to determine the rate of loss of overall polymorphism across populations.

There has been little research done to examine the relationships between development activities and genetic diversity. For example, the genetic implications of barriers caused by roads are not well understood. This makes it very difficult to predict the longer-term ecological consequences of isolation. Most of the research on 'effective population size' has focused on single population models, which do not address the behaviour of groups of organisms with complex social structures and patterns of gene flow (Rhodes & Chesser 1994). Rates of loss of genetic diversity within individuals or among populations vary considerably from species to species depending on their biology and the magnitude of damage to their environment. Unless there is more investment in research to understand the relationships between gene diversity at different levels of biological organization (including the wider landscape level), habitat destruction and fragmentation and the survival of species, it will remain very difficult to prevent the cumulative pressures that are so often responsible for 'slides towards extinction'. Concepts like 'effective population size' offer scope for management at the population level which might ensure the safeguard of gene diversity at other levels within ecosystems.

6.3.7   Minimum habitat area

It is possible to specify a minimum area needed for viable populations of a species or characteristic levels of diversity to be sustained. The 'Minimum Critical Size Ecosystem Projects' of Lovejoy et al. (1986) began in 1979 and continue as the 'Biological Dynamics of Forest Fragment Project' (Spellerberg 1991). They involved study of newly created fragments of tropical forest ranging in size from 1 to 10 000 ha. Diminishing numbers of species and increased extinctions of plants and animals were recorded as fragment size decreased (see also section 5.3.3).

In more simple terms, if the area of habitat available to an individual or a population falls below that required to satisfy its needs, the individual or population is likely to disappear. In many cases, 'minimum habitat area' and 'home range' will be comparable, although it is possible that species will opt for home ranges slightly larger than needed to satisfy their minimum requirements if they can. If minimum habitat areas are known, it becomes possible to interpret land-take in a meaningful way, and to evaluate the likely consequences of habitat area reduction for associated species. However, it can be incredibly difficult to determine, particularly for migratory species that have breeding and wintering habitat separated by huge distances.

Pickett and Thompson (1978) suggested that, to conserve some species, it might be necessary to identify a 'minimum dynamic area'. Species might occupy habitat patches within a larger 'biotope' at different times. Within each patch, the species might go extinct, but re-establish later from other patches (see also section 6.3.4). The 'minimum dynamic area' is the area of biotope patch necessary to retain enough such patches to prevent overall extinction. This is a useful concept in theory, but the 'minimum dynamic area' has not yet been defined for most species.

6.3.8 'Home range'

Most species have characteristic ranges, or mobilities within the landscape (see section 3.6.1). The use of information on 'home-range size' has proved most useful, or popular, for vertebrates. It has relevance for EcIA with respect to evaluating the impacts of habitat loss or, conversely, quantifying habitat requirements or suitability. Home ranges may vary seasonally and may also vary according to the quality or type of available habitat. A review of home range sizes for the herpetofauna of Scotland (Langton & Beckett 1995) records three seasonal 'phases' for the adder *Vipera berus* (lying out and mating, feeding and returning to hibernation), but none for the slow-worm *Anguis fragilis*. Home ranges may also differ for males and females of the same species and for different life stages. For the sand lizard *Lacerta agilis*, for example, home-range sizes are in the region of $1365 \, m^2$ for males and $398 \, m^2$ for females (Stribosch 1978). For this species there is also some evidence that females may move out of 'home range' to lay their eggs (Nicholson 1980). For species such as the sand lizard, which are protected under the Wildlife and Countryside Act (1981) in the UK, it is illegal not only to kill, injure or take the species without a licence but also to destroy breeding and resting sites. Surveys of these species to predict habitat loss should therefore take account of habitat requirements and home ranges throughout the year, not just on isolated occasions. Knowledge of species requirements is essential to avoid the damage or destruction of vital habitat for protected species, but many of them are rare, relatively secretive and therefore difficult to survey. The sand lizard, for example, is secretive in comparison with many birds and mammals, being small, nocturnal and dormant in cold and dry weather. It is also difficult to mark and re-identify. For species that are hard to find and monitor, information on home-range size is both particularly useful and also particularly difficult to collect in the first place. In some cases it may make sense to choose VECs likely to have larger home ranges than other potentially affected species. Conservation of the home range for such species ought to ensure that those having smaller home ranges within the same area will also be conserved.

Analysing the loss of a part of a species' home range is also difficult without including some measure of the importance or quality of the habitat that is to be lost or damaged. Habitat suitability indices as used in the US Fish and Wildlife Service's HEP (see section 5.4.4) should therefore be incorporated where possible to help evaluate the consequences of reduced access to home range.

6.3.9 Capability and suitability

Estimates of capability and suitability are particularly useful in EcIA because they provide benchmarks relating to the ability of an area to support associated populations, species and communities. Habitat capability, for example, has been defined as the ability of a habitat to provide the life requisites of a species, under optimal natural (seral) conditions, irrespective of its current habitat conditions. In other words it has been used as a measure of potential, while habitat suitability is generally used to refer to the ability of a habitat to provide the life requisites of a species in its current condition (Demarchi *et al.* 1996). Using the concepts of capability and suitability to plan for wildlife management or to evaluate ecological impacts requires a consistent ecological definition of land units ('ecoregions'), habitats and species requirements. Reductions in habitat suitability due to specific impacts are measurable through comparison with defined habitat 'capability' and can therefore be used to evaluate the consequences of habitat loss or degradation.

Information about potential distributions is particularly important in EcIA where impacts should be assessed and evaluated against a true baseline: in other words, not only conditions as they are found now, but as they might or could be in the future. For endangered species and declining habitats, the potential for them to exist in an area in future is an important consideration, particularly where metapopulation dynamics may be operating.

6.3.10 Rarity

The rarity of a site, species, habitat or system is relevant to EcIA evaluation because it affects risk of irreversible loss. Rare entities are more vulnerable to extinction and are less 'buffered' against the impacts of development. The extinction record for British species is peppered with instances where the last known site for a rare species has succumbed to an individual development proposal. Degrees of rarity should therefore be considered when evaluating ecological impacts. It is also important to consider why something is rare, however, as this has a bearing on its likely resilience and replaceability.

6.3.11 Resilience

Holling (1973) used the term 'resilience' to refer to the ability of ecosystems to absorb changes and still persist. Measures of resilience can also be used to quantify the amount of disturbance (in terms of kind, rate and intensity) a system can absorb 'before it shifts to a fundamentally different behaviour' (Holling & Clark 1975) or structure. A 'resilient' system can absorb and perhaps even benefit from stress or disturbance (Okey 1996). Systems that lack resilience are more likely to be adversely affected by disturbances and are less likely to recover easily from some types of stress. Clearly, measures of ecosystem resilience can be useful in EcIA evaluation in estimating the likelihood of recovery following impact. Also, to sustain human activities, we need to ensure that the ecological systems on which our economies depend are resilient with respect to the stresses we impose on them.

In economic terms, loss of ecosystem resilience may be important for three reasons:
1 discontinuous change in ecosystem functions as a system flips from one equilibrium to another could be associated with a sudden loss of biological productivity and use-value;
2 loss of resilience may imply irreversible change, for example loss of biodiversity or of ecosystem 'supply functions' such as provision of reliable groundwater supplies;
3 discontinuous and irreversible changes from familiar to unfamiliar states increase the uncertainties associated with the environmental effects of economic activities.

Because all economic activity ultimately depends on a finite environmental resource base, we should act in a 'precautionary way' to maintain both the diversity and resilience of ecosystems (Arrow *et al.* 1995). However, ecological resilience is difficult to measure and varies both between systems and kinds of disturbance. We have only rudimentary understanding of the dynamics of single populations and much less understanding of the behaviour of complex systems (Ludwig 1996), making it difficult to identify suitable indicators or early warning signs. Ultimately, the resilience of ecosystems may only be tested by intelligently perturbing systems and observing responses. In the meantime, much could be gained from classifying and ranking ecosystems and habitats in terms of their apparent or average resilience to certain types of stress and disturbance.

6.3.12 Fragility

'Fragility' has been defined as the 'inverse of ecosystem stability' (Nilsson & Grelsson 1995). The identification of species, communities or ecosystems likely to be strongly damaged by human activity is an essential part of EcIA. Fragility may be regarded as an inherent property of an ecosystem, such that an ecosystem has a certain fragility whether or not it is ever actually exposed to the disturbances that would cause that fragility to be expressed. To be of value as an assessment endpoint in EcIA, it would be necessary to find expressions or properties of fragility that could be measured in response to various categories of impact. For example, both 'fragility' and 'stability' can be measured in terms of the degree of change in species abundance and composition following disturbance (whether natural or man-induced). Fragile ecosystems are characterized by high rates of species turnover or population fluctuations while stable ecosystems are able to maintain their equilibrium.

6.3.13 Stability

A general definition of 'stability' is the ability of an ecosystem to maintain some form of equilibrium in the presence of perturbations (Allen & Starr 1982; Holling 1986). Stability and resilience are closely related properties, but Holling (1973) draws a distinction between 'resilience', which refers to the behaviour of a system far from equilibrium, and 'stability', which refers to its behaviour very near to equilibrium points. In other words, resilience tends to be used with respect to ecosystem behaviour following a shift away from equilibrium, whereas stability pertains more to the ability of an ecosystem to resist such shifts in the first place (Okey 1996).

'Stability' is not a straightforward concept and there is much confusion about the terminology used to define its various aspects. A useful review can be found in an article by Connell and Sousa (1983). For example, the ability of a system to remain at equilibrium when potentially affected by a stressor has been variously referred to as 'inertia', 'persistence' and 'resistance'. The tendency of a system to retain constant numbers (e.g. of organisms in a population) has been referred to as 'constancy', 'persistence', 'conservatism' and 'endurance'. Another aspect of the concept of 'stability' is the behaviour of a system following a perturbation: the speed and readiness with which it is likely to return to equilibrium and the extent of departure from equilibrium from which it is able to recover. The speed of return has been termed 'elasticity' or 'resiliency' while the distance or degree of departure from which a system is able to return to equilibrium can be termed 'amplitude'. Connell and Sousa (1983) conclude that the terms 'amplitude' and 'elasticity' be used to refer to the two main components of adjustment following perturbation or disturbance. Stable systems therefore resist change, a property that may be expressed through conservatism or constancy in population or species number. They are also relatively 'elastic' (recovering equilibrium quickly) and can withstand perturbations that cause departures from equilibrium states of relatively great 'amplitude'. These are all aspects of 'stability' that demand measurement of ecosystem state and, in particular, definition of 'equilibrium states'. However, the term 'stability' is also used to refer to the ability or tendency to exist at all, in other words, to evaluate the extent to which systems or populations are 'buffered against extinction in adverse periods' (Whittaker 1974).

For EcIA evaluation, it would be useful to rank ecosystems (or populations of different species) in terms of their characteristic stability, based on measurements of their elasticity of response to imposed stresses and the amplitude of departure from equilibrium they have been demonstrated to withstand. Relevant measurement endpoints might be numerical abundance of populations, or relative abundance of species in guilds and communities.

Another important aspect is to consider the minimum area in which a population or community is stable and/or persistent (Connell & Sousa 1983), i.e. the smallest area that provides adequate conditions for continued viability (see also section 6.3.7). It is possible that many experimental studies have been carried out at a scale that is too small for stability or persistence to be recognized. For EcIA evaluation, it would be very useful to define the conditions that are required for stability and, likewise, the circumstances in which stability is likely to be jeopardized.

6.3.14 Diversity/complexity

'Diversity' can be defined in terms of genes or species, where 'diversity' is measured as the number of genes or species present (Allen & Starr 1982). It can also be defined for other levels of ecological organization. At the landscape scale, habitat diversity is an important factor in the maintenance of species distributions. Within individual habitats, structural diversity can be a major factor in the range and quality of habitat available to associated species. When considering diversity or complexity, it is therefore always important to ask 'of what?'. Okey (1996) defines diversity and complexity jointly as the 'structural and functional variability of an ecosystem'. However it is defined, diversity is a property that is often affected by human activity and that is closely related to some of the other properties already discussed. There is ongoing debate, for example, about the relationships between diversity and stability in ecological systems. In herbaceous plant communities in particular, species diversity has been shown both to enhance stability and to reduce it (Collins & Benning 1996). Regardless of the implications of diversity or complexity for system stability and resilience, it can be argued that some systems have characteristic levels and types of diversity that should be maintained. The implications of reduced genetic diversity for population viability were considered earlier, in section 6.3.5. Where proposed human activities are likely to affect the diversity or complexity of impacted ecosystems, the implications of this change should be evaluated on the basis of existing knowledge about the likely viability of that system, any species that are particularly affected and their status in the affected area and elsewhere.

6.3.15   Replaceability

The replaceability of species, habitats or ecosystems is a key factor in evaluating the implications of adverse impacts. Systems that are very difficult to reconstruct or reinstate are more likely to be lost 'for good' than those which can be easily replaced and their destruction should therefore be taken more seriously.

There is evidence to suggest that some habitats are more replaceable than others, complexity being a key factor. It would be useful to study different systems for evidence of hysteresis and to produce characteristic curves (where possible) to indicate the extent to which the characteristics of ecosystems in recovery depart from those observed during degradation. Westman (1985) referred to this as 'differences in paths of alteration and recovery' and pointed out that, in cases where there is little evidence of hysteresis, knowledge of succession can be used to predict the path of recovery. In cases where hysteresis does occur, however, changes during recovery may be very hard to predict. Practical experience suggests that replacement habitats are often poor replicas of undisturbed or pristine examples of the same type, but for many habitats the time required for characteristic species assemblages to be attained is unknown.

There are many examples in EcIA for EIA where undertakings have been given to replace species or habitats, but degrees of relative success or failure have rarely been recorded. It is vital that more attention should be given to the replaceability of ecosystems and their component parts, given the techniques currently available.

6.3.16 Endangerment

Ecosystems or their components may have inherent properties that make them vulnerable to decline or extinction. Some of these properties have been considered in preceding sections. As we have seen, some populations are more prone to extinction than others by virtue of their small size and narrow genetic base. In EcIA, we are often dealing with situations where ecosystem decline is just as much a function of external influences. While inherent ecosystem properties are crucial in determining responses to impacts (including recovery from them) we need to take external pressures into account when evaluating the overall significance of adverse impacts. For example, we need to ask whether a species is exposed to similar threats throughout its range, or only in a few isolated cases.

Such questions are addressed explicitly in some relevant legislation. The US Endangered Species Act of 1973, for example, requires status and threats to be taken into account when determining 'jeopardy' or adverse modification of critical habitat for endangered species. In determining jeopardy, it is necessary to look at the present status of an endangered species (improving or deteriorating) and also to take account of the possible direct, indirect, interrelated and independent effects of Federal actions undergoing consultation. Those State and private actions reasonably certain to occur constitute potential sources of cumulative effects, which must also be taken into account. In other words, threat due to any one, specific proposal must be evaluated in the context of all other potential sources of threat. This combined analysis is measured against the formal definition of 'jeopardy', which is 'an action that reasonably would be expected, directly or indirectly, to reduce appreciably the likelihood of both the survival and recovery of a listed species in the wild by reducing the reproduction, numbers or distribution of that species' (US Fish and Wildlife Service Division of Endangered Species 1997 web page). Endangerment is not a straightforward property to measure, but is potentially very useful in EcIA where degrees of endangerment could be used to determine the acceptability of a proposed action.

6.4 Multiple criteria in evaluation

For most EcIAs, a number of criteria will be used to evaluate impact significance and importance, and practitioners must select those evaluation criteria which are most appropriate for each case. For example, it might be concluded that destruction of a habitat type that is rare, non-resilient and difficult to replace should be avoided if at all possible, while loss of a habitat that is more widespread and replaceable might be considered acceptable so long as suitable mitigation measures can be implemented. Similarly, impacts on species that are endangered, declining and difficult to replace might be taken more seriously than impacts on species that are currently endangered, but for which recovery plans have been shown to be effective and are in place. As suggested in preceding sections, it is possible to set thresholds (numerical or qualitative) for many evaluation criteria and to score ecosystems or ecosystem components against them. It is also possible to rank systems using multiple evaluation criteria. For many evaluation criteria, however, we are not yet in a position to set reliable thresholds. In these circumstances it can be constructive to take a pragmatic approach and to use reasonably well-established rules of thumb to compare the impacts of alternative actions or to estimate the overall ecological significance of a proposal.

Box 6.4 lists some general rules of thumb that could be used in combination to prioritize potentially affected ecosystems or their components.

Box 6.4 Criteria for prioritizing potentially affected ecosystems or their components.

High priority	Less priority
Natural or pristine	Altered by human activity
Undisturbed	Disturbed
Large, unfragmented areas	Small, highly fragmented areas (unless last remaining examples of habitat or system very rare elsewhere)
Characteristic (e.g. of a particular tradition of management)	Not associated with any particular management tradition
Unique (e.g. unusual combinations of species or highly localized habitats associated with particular conditions)	Commonly occurring
Ancient	Newly created
Species-rich	Species-poor
Diverse	Less diverse (unless low diversity is characteristic of community or habitat)
Keystone	No key role/position or importance for other components
Rare or restricted in distribution	Common or widely distributed
Declining throughout range	Stable throughout range
Highly specialized	Generalist
Threatened throughout range	Not threatened
Non-resilient (slow to recover)	Resilient (quick to recover)
Irreplaceable (or very slow/expensive/difficult to replace)	Replaceable (quick, cheap and easy to replace)
Poorly understood	Well understood

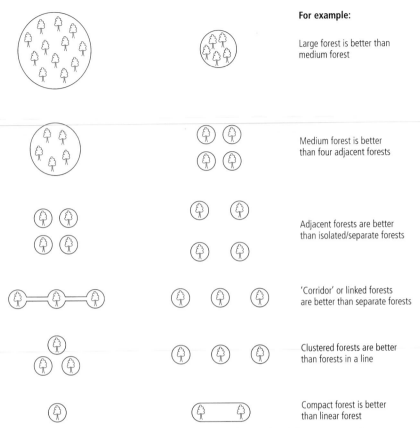

Fig. 6.1 Rules of thumb for habitat design based on 'island theory'. In each case, species extinction rates should be lower for the alternative on the left. (After Diamond, J.M. (1975) The island dilemma: lessons of modern biogeographic studies for the design of natural preserves. *Biological Conservation* 7, 129–146; with permission from Elsevier Science.)

(It does not include explicit socio-economic considerations that may be relevant in any particular case.)

It is also important to consider aspects of spatial organization in the landscape. Figure 6.1 gives some examples of how the organization of forests affects their general quality, based on the theory of island biogeography. All other things being equal, large forests are better than smaller ones, and more connected forests are more valuable than isolated ones. Such rules of thumb have often been developed to enhance the design of wildlife habitat, but they are equally applicable when we wish to compare potentially affected systems or evaluate mitigation proposals.

Trying to compare different ecosystems or their components and to evaluate their relative importance is probably the most difficult part of EcIA. The need to compare across ecosystems in decision-making has been the main driving force behind development of systems for scoring, weighting and the derivation of combined indices.

6.4.1 Scoring and weighting

Systems of scoring and weighting are used frequently in EcIA to measure and adjust criteria and impacts. Indices represent the difference between a set of objects by reference to a scale (Westman 1985). For example, it is possible to develop complex, quasi-mathematical representations in which potential impacts or interactions are transformed into units on a common, notional scale, are weighted in terms of their relative importance and are then manipulated mathematically to derive indices of 'total impact'. Examples include the Environmental Evaluation System (EES), devised by Dee *et al.* (1973) and the method developed by Sondheim (1978). The production of single, numerical indices using combined scoring and weighting systems, however, is not always advisable, as it can generate a false impression of precision and conceal uncertainties in evaluation.

Scoring enables abstract evaluation criteria to be expressed numerically and used more readily in decision-making. Without some form of scoring, measurement endpoints cannot be identified. It also becomes difficult to link assessment and measurement endpoints. Various scoring procedures have been used and reviews are available in Margules and Usher (1981), Smith and Theberge (1986) and Usher (1986). Westman (1985) summarized the four main types of scale and associated permissible mathematical and statistical operations (Table 6.1). For criteria that can be quantified, like 'diversity' and 'rarity', scoring may be based in interval or ratio values, or ordinal values. Thus, 'diversity' might be measured on the basis of the actual number of species present, or given a 'low', 'medium' or 'high' ranking of diversity (Smith & Theberge 1986). In an attempt to standardize 'evaluation terminology', which is not at all consistent, Bedward and Pressey (1991) refer to the former as 'scores' and to the latter as 'score classes'. Use of score classes can facilitate the measurement of criteria, the combination of multiple criteria into overall indices of 'value' and the standardization of criteria for comparison with one another. However, these benefits of convenience are often accompanied by a tendency to follow invalid procedures like the addition and multiplication of score classes for different criteria (Smith & Theberge 1987). No theoretical basis for the use of composite indices has been established, although they are used commonly in EcIA.

Systems of weighting are based on the premise that criteria such as species richness, habitat diversity, stability and naturalness have relative importances that can be quantified. However, different criteria may be assessed using both quantitative and qualitative methods, giving values

Table 6.1 Four types of scales and their permissible mathematical and statistical operations. (From Westman 1985.)

Scale	Nature of scale	Permissible mathematical transformations	Measures of location	Permissible statistical procedures	Examples
Nominal	Classifies objects (e.g. in categories of 'hot', 'warm' or 'cold')	One-to-one substitution	Mode	Information statistics	Classifying species, numbering soil types
Ordinal	Ranks objects (e.g. according to whether they are 'hottest', 'warmest' or 'coldest'). Ordinal scales indicate order, but do not indicate magnitude of relative differences	Equivalence to another monitonically increasing or decreasing function	Median	Non-parametric statistics	Ranking land use plans in order of preference, ranking development alternatives
Interval	Rates objects in units of equal difference (i.e. quantifies the degree of difference between objects)	Linear transformation	Arithmetic mean	Parametric statistics	Time (in hours and minutes), temperature
Ratio	Rates objects in units of equal difference and equal ratio (i.e. quantifies degree of difference relative to an absolute starting or reference point)	Multiplication or division by a constant or other ratio-scale value	Geometric, harmonic or arithmetic mean	Parametric statistics	Height, weight

('scores') on different scales and often expressed in different units. By weighting these values, inconsistencies in scoring can be concealed, together with the ability to take account of differences in the precision with which criteria were originally measured. A method of weighting nature conservation interest frequently used in EcIA is in terms of relative importance within a geographical range. Thus, an area may be important within an international, national, regional or local context.

Weighting generally involves some element of subjectivity and is used to convert different systems of measurement and scoring into a common format for decision-making purposes. The EES, described in many texts on EIA, is a well-known example of a combined scoring and weighting method which attempts to combine scientific measurement of 'value function' for selected indicator variables (e.g. carrying capacity for a certain species or levels of dissolved oxygen in water) with weighting, using values allocated by members of the public chosen to represent relevant sectoral interests. More 'weight' is given to socially valuable characteristics and less to characteristics regarded as insignificant. Clearly this approach is only valid if the public 'panel' has been well informed about the indicator variables under consideration and the likely consequences of any predicted change. Predicted changes in environmental quality units for individual indicator variables can then by multiplied by value weightings and summed to derive an overall total. While the derivation of an overall 'score' can be useful to compare a range of development options, there is always the risk that the concensus reached will fail to take due account of important individual qualitative aspects, as the method dilutes the relative importance of specific indicator variables.

Nevertheless, there is nothing inherently wrong with the use of weighting in EcIA so long as the criteria and systems used are always explicit and transparent.

6.4.2 Biological integrity or ecosystem health

It can be argued that ecological impacts disrupt the 'integrity' or 'health' of ecosystems in a measurable way and that it is possible to measure and evaluate the significance of departures from undisturbed 'reference' states. However, the metaphorical use of the concept of health in relation to ecosystem condition has been highly controversial (Okey 1996). Many argue that ecosystems cannot be compared in any meaningful way with individual organisms (e.g. Suter 1993b), or that there are insufficient conceptual and empirical foundations to define ecosystem health (for example Merriam 1994; Shrader-Frechette & McCoy 1994). Most importantly, 'ecosystem health' is not an 'observable property' and Suter (1993b) argues that it is much better to focus on properties of ecosystems that are both measurable and considered worthy of protection. Measured declines or damage to any one of these properties might provide a basis for remedial action, without the need to produce combined indices of 'disease' or decline from a 'healthy condition'. On the other hand, proponents of the concept argue that evaluating ecosystem health offers a pragmatic approach in the absence of clear ecological principles (Norton 1991) and that ecosystems demonstrate

clear functions or properties that can become impaired in measurable
ways. One of the main problems in applying the concept of ecological
health for purposes of EcIA evaluation is the difficulty of identifying
clear assessment and measurement endpoints for these functions or
properties. As pointed out by Okey (1996), the 'notion of health is
perhaps more applicable to agroecosystems than natural ecosystems
because the former possess definite goals (e.g. food production) which
may be impaired, whereas it is arguable whether or not a natural
ecosystem is organized for any particular purpose'.

Evaluation of biological integrity or ecosystem 'health' generally
involves reference to a variety of factors. Three criteria commonly used
to assess ecosystem health include (after Rapport 1989):
1 systemic indicators of ecosystem functional and structural integrity;
2 ecological sustainability or resilience (relating to the ability of a
system to withstand 'natural' or anthropogenic stresses);
3 absence of detectable symptoms of ecosystem disease or stress.

Ecological 'health' therefore has both positive and negative indicators:
it can be recognized or defined through the presence of certain structural
and functional attributes and the absence of conditions or processes
resulting from known stressors. Clearly the validity of the concept in
practical terms rests on full knowledge of the range of ecosystem
structure and function and the ability to recognize 'good' (or healthy)
and 'bad' (unhealthy) examples.

In terms of EcIA evaluation, Suter's (1993b) reservations should be
borne in mind. The important thing is to retain emphasis on the selection
of suitable indicators or VECs for which meaningful assessment and
measurement endpoints can be established.

The concept of 'integrity' is often used interchangeably with that of
'health', but is in fact used deliberately to avoid medical analogy (Okey
1996), with a greater emphasis on defined 'reference states'. 'Indices of
biotic integrity' have been developed both in the US and the UK for
purposes of ecological evaluation. Measuring the status of impacted
ecosystems relative to comparable undisturbed examples has proved a
particularly useful evaluation technique for aquatic systems. For example,
in the UK, the presence of certain invertebrate groups in freshwater has
been used to derive scores that reflect the quality of the water. The River
Invertebrate Prediction and Classification System (RIVPACS) uses an
environmental quality index with a scale ranging from 10 ('clean' water
with a diverse fauna) to 0 (grossly polluted water with no fauna) (Wright
et al. 1984). Similarly, the index of biotic integrity (IBI) developed by
Karr and colleagues in the US (Karr et al. 1986) is based on the responses
of fish communities to changes in water quality. The IBI for fish uses
three broad groups of measurement to characterize watersheds: species
taxonomic composition and richness, trophic assemblages and total

Table 6.2 Characteristics measured to derive the fish IBI. (After Karr *et al.* 1986.)

Species richness and composition
1 Total number of native fish species
2 Number and identity of darter species (benthic species)
3 Number and identity of sunfish species (water column species)
4 Number and identity of sucker species (long-lived species)
5 Number and identity of intolerant species
6 Percentage of species as tolerant species

Trophic composition
1 Percentage of individuals as omnivores
2 Percentage of individuals as insectivores
3 Percentage of individuals as top carnivores

Fish abundance and condition
1 Number of individuals in sample
2 Percentage of individuals as hybrids or exotics
3 Percentage of individuals with disease, tumours or skeletal abnormalities

abundance and condition (taking account of the incidence of diseases, skeletal malformations, tumours, etc.). Ratings of 5, 3 and 1 are assigned according to whether the values measured for the characteristics listed in Table 6.2 approximate to, deviate somewhat from, or deviate strongly from the value expected at relatively undisturbed sites.

The difficulties of measuring all ecosystem attributes to assess overall ecosystem 'health' have prompted a number of approaches based on the use of selected indicator values. The US Environmental Protection Agency, for example, uses indicators to assess the condition, or health of ecological resources in its Environmental Monitoring and Assessment Program (EMAP) (US Environmental Protection Agency 1994a). Indicators must distinguish between ecosystem conditions anywhere on the spectrum from 'healthy' to 'unhealthy'. They must also link with assessment endpoints. For EMAP, information about assessment endpoints falls into the following categories:
• ecosystem condition;
• ecosystem exposure to potential stressors;
• availability of conditions necessary to support desired ecosystem states.
Four types of indicator were selected to reflect these endpoint conditions (Hunsaker & Carpenter 1990):
1 *response indicators* (to provide evidence of the ecological condition of a resource at the organism, population, community, ecosystem or landscape level of organization);
2 *exposure indicators* (to provide evidence of the occurrence or magnitude of a response indicator's contact with a physical, chemical or biological stressor);

3 *habitat indicators* (to characterize conditions necessary to support an organism, population, community or ecosystem in the absence of pollutants);

4 *stressor indicators* (to quantify natural processes, environmental hazards or management actions that can effect changes in exposure, such as climate fluctuations, releases of pollutants, introductions of species).

Indicators are identified using conceptual models of ecosystems in the first instance. These models may be based either on current understanding of stress impacts or on the structural, functional and recuperative features of ecosystems considered to be in 'good' condition. Indicators then need to be measured in units consistent with the level of ecosystem organization for which they are defined (US Environmental Protection Agency 1994a). Evaluation is based on a suite of response, exposure, habitat and stressor indicators. This can be achieved by constructing complex indices that aggregate indicator measurements to derive a single variable as demonstrated by Karr *et al.* (1986) for the IBI. One of the main advantages of doing this is that it is generally easier to compare indices of ecosystem condition across different areas or regions than it is to compare original measurements directly, but aggregation can obscure important qualitative differences and make it harder to identify key diagnostic functions or attributes. Ecosystems with the same 'score' could in fact be very different in fundamental ways. In most cases it is important to know the specific reasons underlying a departure from a 'healthy' state as well as the overall extent of the departure. EMAP deals with this by examining monitoring data for statistical associations between ecosystem 'condition' and possible causes using information on potential stressors, degrees of exposure to them and habitat indicator data. These analyses are correlative and do not necessarily 'establish causality', but they play an important part in focusing more detailed monitoring and research efforts on those 'geographical areas, stressors and resource classes of greatest concern' (US Environmental Protection Agency 1994a).

6.5 Social and economic evaluation

Decisions about the use of natural resources cannot be made in isolation from social and economic considerations. It is now widely accepted that social impact assessment (SIA) should play some part in policy and project assessment, so that the consequences of proposed actions are considered with respect to the ways in which people 'live, work, play, relate to one another, organize to meet their needs, and generally cope as members of society' (Burdge & Vanclay 1995). There are many possible linkages between social, cultural and ecological factors and these should form an integral part of the decision-making process. Often these linkages have an economic dimension. Some natural resources produce obvious benefits for society and can therefore be considered to have 'economic value'. The ability to conserve or sustain natural resources is heavily dependent on our ability or willingness to forego or constrain their economic exploitation. While there are well-developed techniques for economic appraisal and social assessment, little progress has actually been made in integrating these techniques with those for EcIA in order to reach balanced decisions about the overall acceptability of ecological change. This is a deficiency that urgently needs to be addressed.

It is outside the scope of this book to provide a comprehensive account of social and economic evaluation for EIA. For readers interested in SIA, a useful introduction is given by Burdge and Vanclay (1995). Because of the increasing interest in environmental and ecological economics, a brief introduction is given to some of the approaches and techniques that may be encountered.

6.6 Environmental economic valuation

6.6.1 Values of environmental assets

The overall value of environmental assets can be broken down into a number of components (Brown & Moran 1993). 'Natural environments' produce a variety of benefits for society, some more obvious and easily measurable than others. For example, natural resources are exploited directly to produce food and other marketable commodities. They are also exploited, and appreciated, for amenity and recreation. The non-market values of natural resources can be divided into two broad categories: 'use values, in which the benefits of a non-marketed resource accrue directly to those who use it, and non-use values, in which people who do not use the resource directly nevertheless derive some benefit from it and, consequently, are willing to pay for its preservation' (More *et al.* 1996).

The selection of assessment endpoints in the US Environmental Protection Agency's EMAP, for example, is based on recognition of certain 'environmental values' identified by the seven EMAP 'Resource Groups'. These are summarized in Table 6.3. They include obvious 'use values' such as 'fishability' and the 'productivity' of forests and 'non-use' values like 'aesthetics'. They also include some that cannot clearly be assigned to one camp or the other. 'Biodiversity', for example, can have both 'use' and 'non-use' values. Non-use values are not always easy to pin down. In the context of EcIA, it may be more appropriate to argue, like More *et al.* (1996) that it is actually more helpful to distinguish instead between on-site benefits (those a person receives from being in

Table 6.3 Environmental values identified by EMAP resource groups for selection of assessment endpoints. (After Hunsaker & Carpenter 1990.)

EMAP resource group	Environmental values identified
Estuaries	Ability to support harvestable and contaminant-free fish, maintenance of habitat structure, aesthetics
Surface waters	Fishability, biological integrity, trophic condition
Wetlands	Wetland integrity, habitat diversity, hydrological function, water-quality improvement
Forests	Spatial extent, sustainability, productivity, aesthetics, biodiversity
Arid ecosystems	Sustainability, biodiversity, aesthetics
Agro-ecosystems	Supply of agricultural commodities, quality of natural resources, conservation of biological resources
Great Lakes	Fishability, water quality, trophic conditions

close proximity to a resource) and off-site benefits. In EcIA, adverse impacts are frequently experienced in one locality, while the benefits of natural resource exploitation are exported elsewhere. The impacts of hydro-electric power generation, for example, are predominantly attributable to dam building, local infrastructure development and hydrological alteration, while the benefits of the electricity generated may be experienced many miles away, even in another country.

'Use' and 'non-use' values ('instrumental values') represent the range of direct or indirect services and functions that are important to society. Together they constitute 'total economic value' (Brown & Moran 1993). Non-instrumental values may be intrinsic or inherent values: what Ehrlich and Ehrlich (1992) call ethical values. Both instrumental and non-instrumental values must be included to estimate 'total value' (Fig. 6.2).

However, intrinsic value is particularly difficult to define precisely. In its most extreme, and perhaps most common form, intrinsic value is the idea that objects (things, individuals, species, ecosystems) have an

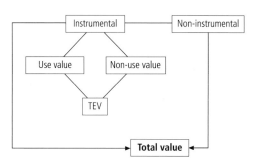

Fig. 6.2 Total value of an environmental asset. TEV, total environmental value. (From Brown & Moran 1993.)

inherent worth that makes them valuable in and of themselves, regardless of any human benefit or cost they may generate. This is close to the idea of 'existence value' (although this gives a stronger emphasis to human satisfaction). Does biological diversity have intrinsic value, or is it a commodity to be 'valued' using traditional economic tools and traded in a market-place? Whatever the answer to this question, demands for valuation techniques to apply to natural resources have been escalating. While it is generally acknowledged that it is wrong for economic growth to take place at any environmental price, it is not always easy to weigh up the costs and benefits. The following section gives brief summaries of some of the economic environmental valuation techniques currently in use. Fuller descriptions of the techniques can be found in Hanley and Spash (1993) and Turner (1993).

There are few methods that have been developed specifically for application to EcIA evaluation. However, there are some circumstances where economic valuation has an obvious potential application in EcIA. In particular:

• to measure the cost-effectiveness of alternative mitigation options where clear ecological limits or quality standards have been set (e.g. alternative strategies intended to maintain minimum viable habitat for an endangered species);
• to make replacement costs (e.g. in shadow projects) explicit;
• to internalize the economic value of unavoidable ecological impacts into economic appraisal to ensure that these play a part in justifying or rejecting a proposal.

As a general rule, economic appraisal in EcIA is most effective in cases where there are clear end-point objectives, targets, thresholds or intentions. Otherwise, it can add to the opacity of situations in which a variety of 'fudge factors' have already been applied to convert different impact types to a common measurement system (see section 6.4.1 for warnings about the use of weighting in EcIA). The main challenge to ecologists is therefore to develop methods of ecological evaluation that will provide defensible management objectives. The cost of achieving these can then be measured using straightforward techniques of economic valuation.

6.6.2 Cost–benefit analysis

Economic valuation of environmental costs and benefits is often applied within EIA to compare alternative mitigation strategies or development options. It can also play an important part in overall project appraisal, to assess the economic viability of a proposal. The main reason that cost–benefit analysis plays such a prominent part in EIA is that money provides a common denominator or yardstick for comparing economic benefits (often the main incentive for a proposal) with adverse environmental impacts. The problem is that not all environmental costs are easily quantified. Cost–benefit analyses can also be complicated by various temporal and spatial asymmetries, for example over what timescale should economic returns from different land use options be compared and at what scale is it appropriate to judge benefits? Wells (1992) noted that the benefits of biodiversity protection through national parks tend to be lower locally than they are nationally or even internationally, while the costs associated with protection are more likely to be incurred locally. On the other hand, intertemporal resource allocation issues have generated considerable debate about discounting (Markandya & Pearce 1991) and the need to collapse the net benefits or costs likely to accrue over the lifetimes of alternatives into the current period to provide a yardstick for efficient resource allocation (Brown & Moran 1993).

The methods considered in the following sections may be used to derive values that can be incorporated into cost–benefit analysis for EIA.

6.6.3 Methods based on market values

Natural resources play a part in production and therefore in determining levels of earnings from production. Some methods of economic valuation are based directly on market prices for environmental goods and services or, if market prices cannot be used as an accurate guide to scarcity, they may be adjusted by shadow pricing (Brown & Moran 1993). They are applicable when a change in environmental quality affects the actual quantity or price of marketed inputs and outputs, for example through effects on production, loss of earnings or the need to prevent or rectify damage that would otherwise affect production or earning capacity. In other words, they apply when there is a clear 'dose–response' relationship. For example, the production value of fish catches foregone as a result of polluting a watercourse or the reduction in timber production attributable to air pollution damage can be estimated directly using actual or projected market prices.

In some cases it is not possible to estimate the benefits of environmental quality or protection because there are no direct market measurements of cost. It may then be necessary to use 'experimental market techniques'. These techniques are based on attempts to elicit preferences, usually using questionnaires.

6.6.4 Contingent valuation method

The contingent valuation method (CVM) is one of the techniques that can be used to estimate economic values for commodities that cannot be traded conventionally in markets. The method was proposed originally by Davis (1963). CVM works by directly soliciting from a sample of 'consumers' their willingness to pay (WTP) for, or accept (WTA) a 'change in the level of environmental service flows in a carefully structured hypothetical market' (Hanley & Spash 1993). Briefly, CVM can be undertaken in six stages as outlined in Box 6.5.

Box 6.5 The six stages of CVM. (After Hanley & Spash 1993.)

Set up a hypothetical market
Specify reasons for payment, 'bid vehicle' (e.g. income tax) and 'provision rule', i.e. how the decision on whether or not to proceed with a proposed project will be made

Obtain 'bids' for environmental goods or services
Carry out survey, usually through face-to-face interview, by telephone or by mailshot. WTP may be derived through bidding (higher and higher amounts are suggested to respondents until their maximum WTP is reached), as a closed-ended referendum in which a 'yes/no' reply is sought for a single recommended payment, as an open-ended question in which respondents are asked to specify their maximum WTP or as a 'payment card' in which respondents are able to choose from a range of values

Calculate average WTP or WTA
As a general rule report both mean and median (to take account of very large bids) and exclude 'protest bids'. Protest bids are zero bids given for reasons other than a zero value actually being placed on the resource in question. For example, some respondents might refuse to accept any amount of compensation for unique environmental resources like the Grand Canyon

Estimate bid curves
Bid curves are used to explore the determinants of WTP/WTA bids. WTP/WTA amounts can be used as the dependent variable and regressed against independent variables such as income, education or age. They are also useful to test the sensitivity of WTP amounts to changes in the amount of environmental quality being bid for

Aggregate data
This is the process whereby mean bids derived from a population sample (the sample mean) are converted to a total value figure for the whole population. For example a sample mean might be multiplied by the number of households in the population. Clearly this only works if the sample can be considered representative of the whole population

Evaluate exercise
CVM exercises should be evaluated to establish how successful the application of CVM has been. For example, results should be considered in the light of whether a high proportion of protest bids was received, whether respondents appeared to understand the hypothetical market and whether they had direct experience of the environmental asset in question

6.6.5 The hedonic pricing method

The hedonic pricing method (HPM) 'attempts to impute a price for an environmental good by examining the effect which its presence has on a relevant market-priced good' (Bateman 1993). For example, land prices might be influenced by land characteristics. Typically, hedonic pricing (HP) is used to find a relationship between levels of environmental services or attributes (e.g. noise levels) and the prices of marketed goods. Most HPM studies have looked at situations where the property market is influenced by surrounding environmental characteristics such as air quality, water quality or noise. HPM is based on a number of assumptions (after Bateman 1993 and Hanley & Spash 1993):
- WTP is an appropriate measure of benefits;
- individuals can perceive environmental quality changes (these changes affect the future net benefit stream of a property and therefore it can be assumed that people are willing to pay for environmental quality changes);
- the entire study area can be treated as one competitive market with freedom of access across the market and perfect information regarding house prices and environmental characteristics;
- this housing market is in equilibrium, in other words, individuals continually re-evaluate their location such that their purchased house constitutes their utility-maximizing choice of property given their income constraint;
- all buyers in the housing market are perfectly informed of environmental quality at every possible housing location.

These assumptions are relatively restrictive. In particular, the degree to which individuals are able to perceive 'environmental quality' (or its absence) is a crucial factor affecting the validity of any HPM study (Bateman 1993). In many cases there may be thresholds below which people are unable to sense changes in environmental quality and the method can run into problems if it is not possible to assume that environmental change is a continuous variable. Individuals may also be more influenced by historic levels of pollution or by future predictions than by current estimates. In situations where there is a risk of environmental disasters occurring (in earthquake zones, for example), individuals' perceptions of possible losses rarely match the scientific probability of loss or damage, resulting in hedonic prices which over- or under-estimate welfare changes (Hanley & Spash 1993). Because of the rigid assumptions on which it is based, HPM is often impractical to use. However, it can be reasonably effective for evaluating impacts like aircraft-noise around airports (O'Byrne *et al.* 1985) and urban air quality (Brookshire *et al.* 1982).

HPM has little obvious application in evaluation for purposes of EcIA, but there are circumstances in which ecological impacts might be closely correlated with changes in environmental quality which do lend themselves to analysis using HPM. Losses in amenity value due to destruction or damage of woodland, for example, might affect housing prices (Willis & Garrod 1991).

6.6.6 The travel cost method

The travel cost method (TCM) seeks to place a value on non-market environmental goods by using consumption behaviour in related markets. The costs of consuming the services of the environmental asset are used as a proxy for price (Hanley & Spash 1993). For example, TCM is typically applied to estimate the recreational value of a site by analysing the travel expenditures (e.g. petrol costs, train fares) incurred by visitors. Bateman (1993) gives a full description of the technique. The method is predominantly used in outdoor recreation modelling, and is widely used by government agencies in the US and the UK. TCM draws on standard empirical models and uses a survey approach. Visitors are provided with questionnaires and requested to provide information about their normal place of residence, their frequency of visits to the site in question and other sites, the length and costs of their trip, etc. Visit costs (and other relevant factors) can be related to visit frequencies to establish a 'demand relationship'. 'In essence, the TCM evaluates the recreational use value for a specific recreation site by relating demand for that site (measured as site visits) to its price (measured as the costs of a visit)' (Bateman 1993).

   Limitations of the method include its general failure to take account of situations in which people actually move in order to live near to a particular recreation site, thereby reducing the cost of travel to it. The method can also run into problems in cases where sites have become congested, in which case the observed demand curve is likely to be an under-estimate of true demand. Estimating true demand is also difficult because visitor surveys do not include 'non-visitors', so TCM cannot be used to estimate non-user values. There are various techniques available for overcoming some of these difficulties, but it remains difficult to quantify those aspects of a site which determine its overall recreational value. In some cases, ecological parameters may play an important part (presence of diverse wildlife, for example) in contributing to the overall 'recreational experience'. However, 'environmental quality factors' are often closely correlated, making it difficult to tease out their relative importance. The size of a wildlife park, for example, is often correlated with the number of access points and routes, which can also have a considerable influence on the distribution of wildlife and the extent to which visitors are able to observe it and benefit directly from its presence. Where at least two of the attributes potentially explaining site environmental quality are highly co-linear, a 'suppressor variable problem' may occur. While there are some techniques available to value the contribution to overall site value made by individual attributes, there is no single definitive solution to the treatment of suppressed variables. When it comes to valuing sites in EcIA, it is clearly important to know which particular attributes contribute to a site's perceived overall recreational value and to what extent. If a site is to be altered, damaged or reduced in size as a result of development, how will this affect its recreational use in the future? In cases where a proposal will affect travel costs or distances directly and there are existing models, it may be possible to model 'before' and 'after' situations, but this will only be effective if the relationship between 'use' and 'attributes' is clear. The TCM is most effective if it is applied 'to the evaluation of well-defined recreation sites or a well-perceived, separable, environmental attribute within such a site' (Bateman 1993).

6.6.7   Replacement costs

In EcIA, it is common for ecosystems that are damaged or destroyed to be replaced. The cost of replacement can be estimated for some ecosystems or habitat types. For others, techniques are still in a developmental or experimental stage and it is difficult to estimate how much it will cost to achieve a desired end-result. It would seem reasonable for the costs of necessary research to be incorporated into 'replacement cost' for a habitat or system, as without it, desired end-results may never be achieved and the resource is effectively lost without due compensation. Replacement habitats or ecosystems are often poor replicas of those for which they are substituted, and this loss of ecological value needs to be accounted for in some way. Inter-generational equity should also be considered (see also Chapter 7).

In the US, the concept of compensating people for loss of natural resources has become a reality, with restoration-based measures of compensation being built into natural resource liability statutes. In the US, the atmosphere, oceans, estuaries, rivers, and plant and animal species are regarded as public trust resources, and public policies emphasize their protection from injury and depletion. Several major environmental statutes passed in the 1970s contain provisions for recovery of damages for injuries to public resources (Jones & Pease 1996). The main federal statutes containing provision establishing liability for injuries to resources in the public trust are the Comprehensive Environmental Response, Compensation and Liability Act (CERCLA, generally referred to as 'Superfund') and the Oil Pollution Act. Natural resource damage claims generally have three basic components. The Oil Pollution Act specifies the measure of damages as:

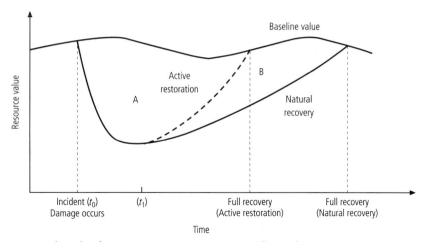

Fig. 6.3 The role of restoration in recovery. Area A allocated to active restoration, areas A + B allocated to natural recovery. (After Jones & Pease 1996.)

• the cost of restoring, rehabilitating, replacing, or acquiring the equivalent of the damaged natural resources ('primary restoration');
• the diminution in value of those natural resources pending recovery of the resource to baseline, but for the injury ('interim lost value');
• the reasonable cost of assessing those damages.

Primary restoration can reduce interim lost value by speeding the rate of recovery to baseline conditions (Fig. 6.3).

Natural resource trustees are authorized to act on behalf of the public to protect natural resources. For example, the National Oceanic and Atmospheric Administration (NOAA) serves as a trustee for coastal and marine resources and determines claims or damages to be filed against parties responsible for damage resulting from discharges of oil, releases of hazardous substances, etc. To determine appropriate compensation, habitat equivalency analysis (HEA) may be used. This is based on the premise that the public can be compensated for past losses of habitat through habitat replacement projects providing additional resources of the same type. The ultimate goal of HEA is to ensure that the value to the public of the increase in services from a replacement project should be equivalent to the value of services lost due to the injury. HEA is typically implemented using a resource-based metric, for example hectares per year, numbers of fish, etc. as a proxy for 'services', but problems occur when services at a compensatory restoration site are not of the same type and quality as those injured, making one-to-one trade-offs difficult (NOAA 1996).

HEA also has potential applications under European conservation legislation. For example, The Conservation (Natural Habitats, etc.) Regulations 1994 (SI no. 2716) demand that if a decision is taken to permit development that will damage sites designated as part of the

Natura 2000 network (due to overriding public interest), 'the Secretary of State shall secure that any necessary compensatory measures are taken to ensure that the overall coherence of Natura 2000 is protected'. HEA or an equivalent procedure would be necessary to judge whether or not 'overall coherence' was maintained. However, in Europe, problems of the kind identified above are highly likely to occur, due to the limited extent of so many important wildlife habitats and the difficulties of obtaining suitable compensatory sites.

It is the problem of deciding what constitutes a reasonable level of 'replacement' of damaged ecosystems that has prompted calls for more integrated, regional approaches to ecological valuation and management. These are often closely tied to the concept of sustainability and to a longer-term perspective on the economic use of natural resources as considered in the following section.

6.6.8 Sustainability objectives

At a more strategic level, it may be useful to determine whether the post-impact state of an ecosystem or area will be consistent with regional or national targets for sustainability. A commitment to sustainability makes it essential to consider the potential redistributional effects of destroying a natural asset. Even if mitigation is proposed to reinstate it, there may be a delay before it is fully restored. This raises intergenerational distribution issues. For example, if a habitat takes a long time to restore, current generations may effectively be deprived of it, even if future generations may be no worse off. There are other circumstances in which the costs of making demands on limited natural resources may, effectively, be paid by future generations.

The ability to recognize thresholds of sustainability (both spatial and temporal) is therefore important. Despite a general acknowledgement of this need, little progress has been made in deciding what is sustainable and what is not. Eckholm (1982) suggested that 'the biosphere seldom presents human society with imperatives; rather we face choices about what sort of world we want to live in. Responses to environmental threats can be formulated only in relation to broader human goals'. In the context of EcIA, it is important that some way should be found to express these 'broader human goals' in ecological terms.

The concept of sustainability originated with the 1980 World Conservation Strategy of the International Union for the Conservation of Nature and Natural Resources (IUCN). The IUCN advanced sustainability as a strategic approach to the integration of natural resource conservation and development consistent with:

- ecosystem maintenance;
- preservation of genetic diversity;

- sustainable utilization of resources (Smith 1993).

There are a number of ecological properties that affect the continuing/long-term viability of species and habitats, a number of which were outlined earlier in this chapter. In compliance with the requirements of the Biodiversity Convention, a number of governments have identified indicators of sustainable development. In the UK, the categories of indicator include (*inter alia*):

- fish resources;
- climate change;
- acid deposition;
- air;
- freshwater quality;
- marine;
- wildlife and habitats;
- land cover and landscape;
- soil.

Many of these have implications for how the concept of sustainability can be expressed in ecological terms and used to provide criteria for ecological evaluation.

English Nature (the statutory nature conservation agency for England) has defined 'sustainable development' as development compatible with the following objectives (English Nature 1992):

- maintenance of characteristic wildlife communities and natural features;
- retention of the diversity and character of different ecological zones;
- establishment and maintenance of populations of wildlife species distributed across their traditional ranges;
- establishment and maintenance of viable populations of rare and vulnerable species and improvement of their status wherever possible;
- fulfilment of international responsibilities and obligations for nature conservation.

Clearly, as discussed earlier, it is impossible to meet such objectives by considering individual sites and impacts on them in isolation. Information on the national distributions of natural features, habitats and species is necessary before it is possible to plan and manage development in such a way that irreversible damage to the 'natural environment' can be avoided. It is essential to know when species, local populations, habitats, etc. have reached a state in which they are unable to withstand any further loss without risk of extinction. It is mainly for this reason that the concept of 'critical natural capital' has proved useful as a means of defining thresholds for acceptable loss. Critical natural capital includes those aspects of biological diversity that cannot be replaced using known techniques. Constant natural assets, on the other hand, are components of biological diversity that are not individually unique or irreplaceable, but for which the overall stock must be maintained. By focusing attention on the overall status of ecological resources and the extent to which they can be substituted (or traded) for each other, the concept of 'environmental capital' can prove useful for deciding what form ecological mitigation should take and in evaluating its likely success.

6.7 Recommended reading

Ecological evaluation criteria

Spellerberg, I.F. (1992) *Evaluation and Assessment for Conservation: ecological guidelines for determining priorities for nature conservation.* Chapman and Hall, London.

Weighting and scoring methods

Westman, W.E. (1985) *Ecology, Impact Assessment and Environmental Planning.* John Wiley and Sons, New York.

Social impact assessment

Burdge, R. & Vanclay, F. (1995) Social impact assessment. In: *Environmental and Social Impact Assessment* (eds F. Vanclay & D.A. Bronstein), pp. 31–67. John Wiley and Sons, Chichester.

Environmental economics

Hanley, N. & Spash, C.L. (1993) *Cost–Benefit Analysis and the Environment.* Edward Elgar, Aldershot.

Sustainable resource use

Smith, L.G. (1993) *Impact Assessment and Sustainable Resource Development.* Longman, Harlow.

7 Ecological mitigation

7.1 Introduction

The actual consequences of a proposal, once implemented, will depend not only on the adverse impacts associated with it, but on the effectiveness of action taken to mitigate their effects. Most environmental impact assessment (EIA) legislation therefore makes some provision for mitigation. The European and UK legislation, for example, as well as emphasizing the need to include the data necessary to assess any significant environmental effects, also requires proponents to recommend suitable mitigation measures. The UK Department of the Environment's 'guide to the procedures' for EIA includes a checklist of matters to be considered for inclusion in an environmental statement (ES) and suggests that 'where significant adverse effects are identified, a description of the measures to be taken to avoid, reduce or remedy those effects' (Department of the Environment 1989) should be included. This includes adverse ecological impacts.

However, a review of proposed ecological mitigation measures in UK ESs suggested that there might be problems under the current legislation in ensuring that ecological impacts are mitigated effectively (Treweek & Thompson 1997). The review suggested that there is considerable confusion over the extent to which ecological mitigation is required for different categories of environmental impact. The need for ecological mitigation is widely acknowledged, but the mitigation measures proposed do not always relate directly to the ecological impacts identified and a number of potentially serious ecological impacts remain unmitigated. There is therefore a high risk of residual adverse effects, which will result in the erosion of natural capital. The review also emphasized the lack of any generally accepted method for evaluating the likely effectiveness of proposed mitigation measures despite the clear recommendation from the Department of the Environment (1989) that an 'assessment of the likely effectiveness' of mitigation measures should also be included in an ES. Similarly, in a review of 32 impact assessment methods, Atkinson (1985) found that only 10 addressed mitigation requirements and recommended appropriate mitigation measures.

7.2 What is mitigation?

Mitigation includes any deliberate action taken to alleviate adverse effects, whether by controlling the sources of impacts, or the exposure of ecological receptors to them.

Measures may be taken to avoid impacts altogether, to reduce their magnitude or severity, to alleviate their effects if they do occur, or to compensate for any damaging effects, for example through the provision of alternative sites for conservation. The definition of mitigation used by the US Council on Environmental Quality (CEQ) (1978) suggests that the following approaches to mitigation should be implemented sequentially, with avoidance measures assuming priority (Canter 1996):
- avoiding the impact altogether, by desisting from a certain action or parts of an action;

Box 7.1 Types of ecological impact mitigation.

Avoidance
- Sensitive design
- Siting based on least damage criteria
- Avoidance of key areas (e.g. protected habitat)
- Avoidance of key periods (e.g. breeding season)
- Desisting from impact-generating actions

Reduction, moderation, minimization
- Emission controls
- Noise barriers
- Screens
- Oil interceptors
- Controlled access during construction/operation
- Wildlife bridges, tunnels, 'ecoducts'
- Wildlife fences

Rescue (relocation, translocation)
- Translocation of plants or animals
- Translocation of habitat
- Removal of turves for reinstatement

Repair, reinstatement, restoration
- Reinstatement of habitat (woodland, wetland, grassland, etc.)
- Re-seeding of grassland
- Restoration of damaged hydrological function (e.g. reinstatement of raised water-level areas)

Compensation
- Donating substitute habitat areas
- Creating new habitat on alternative sites
- Provision of resources for 'creative' management

- minimizing the impact by limiting the degree or magnitude of an action;
- rectifying impacts through repair, reinstatement or restoration of the affected ecosystem;
- compensating for the impact by replacing or providing substitute resources or environments.

Under what circumstances should mitigation come into play and how much should be invested in it? To a great extent, this depends on the value placed on the natural resources that are at risk of damage and the amount of loss that is regarded as acceptable. Mitigation measures can be costly to implement and some are more effective than others. It is important to determine what criteria should be used in deciding which adverse ecological impacts should be mitigated. A requirement to mitigate every single adverse impact would not only be expensive, but would stifle development and might result in investment in mitigation projects of little value with respect to the overall conservation of biodiversity.

Ecological impact mitigation might take any of the forms listed in Box 7.1.

7.3 Avoidance

The best form of mitigation is avoidance through design, so that potential ecological damage can be stemmed at source. Siting and design based on 'least damage criteria' is by far the most effective way of ensuring that the integrity of natural resources is preserved. It can be argued that it is inappropriate to refer to impact avoidance as a form of mitigation. If EIA is introduced early enough into the design and planning process and is managed as an interactive process, it should be possible to build avoidance measures into projects or plans as they evolve. Post-implementation mitigation should be considered as a last resort.

However, the deliberate avoidance of ecological impact through objective consideration of ecological constraints at the design stage appears to be relatively unusual. In many countries, modifications to siting, construction and operation procedures to take account of ecological constraints are more common after the design stage, when major alterations are likely to be too expensive to implement. The failure to take account of ecological constraints during project design and planning is a major cause of delay in the impact assessment process and can also result in unnecessary investment of effort in survey and impact assessment.

It might be expected that deliberate avoidance of impact would dominate in situations where rare or protected species or habitats are at risk. However, in the review of UK statements referred to earlier (Treweek & Thompson 1997), ecological 'importance' did not appear to be the main determining factor with respect to avoidance measures.

Many sites with international or national designations for nature conservation were potentially affected by the proposals reviewed. For example, 112 Sites of Special Scientific Interest (SSSIs) were identified as being subject to potential direct effects. This figure includes 34 SSSIs that were on the route of one proposal to replace an overhead power transmission line. The presence of important, designated areas for nature conservation appeared not to act as a deterrent to development proposals or to trigger deliberate avoidance measures. Deliberate avoidance of SSSIs was only recommended by proponents in eight of the 112 cases.

In other countries, notably the Netherlands, formalized scoping and independent review of scoping studies means that ecological constraints are more often taken into account early in the planning and design of new proposals. Scope for avoidance is then considerably greater.

Even if ecological constraints are given full and careful consideration in the design process, it is possible that the final proposal or design might generate adverse ecological impacts if implemented. These impacts are the focus for targeted mitigation measures, i.e. measures undertaken to reduce the magnitude and/or severity of adverse ecological impacts.

7.4 Rescue

Important or protected habitats and/or species may be 'rescued' through translocation to alternative sites. However, habitat translocation is rarely 100% effective.

It is vital for recipient sites to match the physical conditions and landscape context of donor sites, a requirement that is stipulated in all the available guidelines on translocations and re-introductions (e.g. IUCN 1987; Nature Conservancy Council 1990; IUCN 1995; Sheppard 1995). Case studies on the translocation of grassland indicate that changes in community composition, and the abundances of individual species, can be expected due to factors such as damage to plants during the transplant process (Fig. 7.1), severing of roots by shallow turf stripping, or differences between the donor and receptor sites in terms of environmental conditions or management (Bullock *et al.* 1995). In general, wetter communities are harder to transplant than drier types due to differences in the hydrological regimes between sites and discontinuities between the turf and substrate.

Where mitigation involves relocation of habitat or the relocation or re-introduction of specific organisms, detailed ecological knowledge of species and habitat requirements is essential. There are well known examples of where translocations have failed due to a lack of ecological understanding. Morton (1982), for example, describes failure to transplant the brown bog rush *Schoenus ferrugineus* from its site to nearby sites that appeared superficially similar.

Fig. 7.1 (a) Lifting turves of heathland vegetation for reinstatement elsewhere. (b) Removal of heathland turves. (Courtesy of R. Pywell.)

If the intention is to use mitigation as a measure to help safeguard natural resources and to ensure that development is compatible with conservation of biodiversity, it is important that resident organisms on recipient sites should not suffer as a consequence of translocation attempts, for example as a result of genetic alteration or the transmission

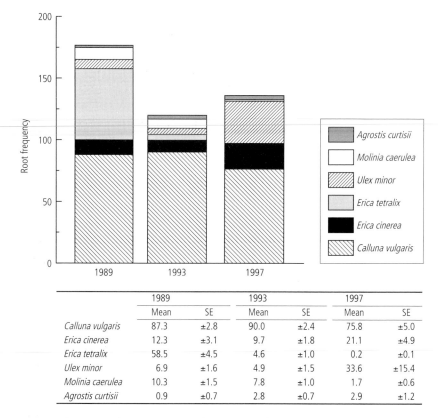

Fig. 7.2 Changes in the composition of translocated heathland turves. (After Pywell *et al.* 1995.)

	1989		1993		1997	
	Mean	SE	Mean	SE	Mean	SE
Calluna vulgaris	87.3	±2.8	90.0	±2.4	75.8	±5.0
Erica cinerea	12.3	±3.1	9.7	±1.8	21.1	±4.9
Erica tetralix	58.5	±4.5	4.6	±1.0	0.2	±0.1
Ulex minor	6.9	±1.6	4.9	±1.5	33.6	±15.4
Molinia caerulea	10.3	±1.5	7.8	±1.0	1.7	±0.6
Agrostis curtisii	0.9	±0.7	2.8	±0.7	2.9	±1.2

of novel pathogens or parasites from translocated individuals to members of any original populations (Nature Conservancy Council 1990).

In practical terms, translocated systems are invariably altered from their original state and evolve differently, as shown in Fig. 7.2. In this example, while the dominance of *Calluna vulgaris* has continued, there are significant changes in the frequency of other components of the vegetation. *Erica tetralix* had almost disappeared from the vegetation by 1997. *Molinia caerulea* has also undergone a marked decline, while the frequency of *Ulex minor* has increased. Not only may there be changes in existing components of the vegetation post-translocation, but new species may be incorporated (see also section 4.5.2).

For the following reasons, translocation should not be used to compensate for the loss of 'high-value' sites (after English Nature 1993):
• vegetation changes are likely to occur following translocation;
• only a small proportion of associated animals are likely to be translocated, and not all will be able to recolonize;
• if part of a habitat area is removed, both donor and recipient sites

may then be too small to support viable populations: reduced diversity and increased edge effects are likely;

- soils may be irretrievably altered;
- hydrological regimes are difficult to mimic;
- management continuity may be difficult to maintain;
- landscape-scale dynamics of species are likely to be disrupted;
- knowledge of many systems and processes is inadequate.

In situations where more common habitats and species are affected, however, translocation can be more effective than attempts to recreate habitat from scratch.

7.5 Management of receptors

One way of reducing the overall impact on organisms is to regulate their exposure. This is particularly common for development types that generate operating impacts of long duration. To avoid wildlife mortality on roads, for example, it is easier to restrict the access of wildlife than to regulate the density of traffic. In the UK, most major roads are fenced off. For example, deer fencing is used on the M40 motorway in areas where densities of deer are known to be high, and badger tunnels are commonly used to maintain badgers' access to habitual routes that have been bisected by new roads. Wildlife tunnels and bridges are similarly used to maintain migration routes across busy highways in Canada. The extent to which different wildlife species are able to adapt to the use of bridges and tunnels is not really known, but the anecdotal evidence suggests that larger mammals are able to adapt their behaviour reasonably well. There are also examples of fencing to keep amphibians off roads, notably toads.

Figure 7.3 shows tunnels designed to overcome barrier effects caused by roads for amphibians and mammals, respectively.

Siting of tunnels with respect to habitual migration or travel routes is important (see section 5.3.6), and for some species it may be necessary to provide visual cues or associated habitat arrangements that lead animals towards the tunnel and encourage them to use it. For larger animals, the construction of tunnels and bridges can be very expensive and they tend to be used only when important habitat areas or migration routes have been severed.

Ecoducts are wildlife bridges vegetated as naturally as possible to maintain habitat continuity. Sometimes several habitats may be reconstructed (Fig. 7.4).

7.6 Repair, reinstatement, restoration

Temporary damage may be caused during construction, which can later

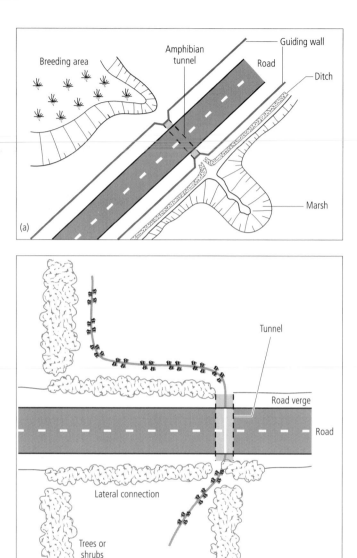

Fig. 7.3 (a) Connecting over-wintering and breeding areas for amphibians with 'attractive' tunnel. (b) Tunnel for large mammals (e.g. badger) with connecting landscape elements. (After DWW 1995.)

be repaired. For example, laying of underground pipelines entails removal of surface vegetation. The various options for its restoration will depend both on characteristics of the affected vegetation and on site factors as listed in Box 7.2.

Reinstatement of vegetation is difficult unless soil structure is maintained. Retention of topsoil is therefore advisable if at all possible.

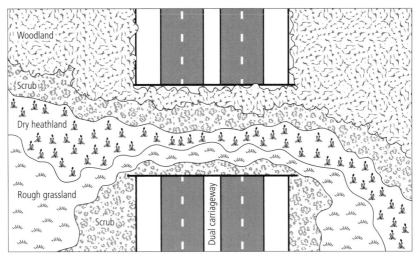

Fig. 7.4 Ecoduct with several vegetation types. (After DWW 1995.)

Box 7.2 Factors affecting reinstatement of damaged vegetation.

Vegetation characteristics, including:
- age/successional stage of vegetation
- complexity
- reproductive strategies of component species
- availability of propagules

Site characteristics, including:
- soil type
- extent of disturbance to soil structure
- extent of disruption to site hydrology

7.7 Compensation

In cases where resources are lost, or sites are damaged to the extent that remediation is impossible, compensation measures may be recommended. This might entail the purchase and donation of an alternative, existing example of wildlife habitat by the proponent, or the creation of new examples of the habitat or system that has been damaged or lost. One of the problems with compensatory mitigation is that it tends to occur without explicit analysis of losses and gains in ecosystem function and value (Brinson & Rheinhardt 1996). Not only is there a frequent failure to reinstate important ecosystem functions, but one type of habitat may be substituted for another. In other words, there may not be full compensation 'in kind'.

Effective compensatory mitigation depends strongly on an understanding of ecosystem function, particularly where restoration or creation projects are to be undertaken. The Road and Hydraulic Engineering Division of the Ministry of Transport, Public Works and Water Management in the Netherlands, for example, advises that new breeding ponds for amphibians should be sited in the known migration zone and within a reasonable distance from over-wintering habitat (DWW 1995). Artificial wetlands are often created in attempts to replace lost wetland ecosystems in the US. Such attempts should be made with an understanding of the complexities of wetland hydrology, but this is often lacking. This makes it very difficult to predict the long-term development of newly created compensatory wetlands of their eventual 'equivalence' with those lost or damaged. Simenstad and Thom (1996), for example, analysed 16 ecosystem functional attributes of a wetland restored to tidal inundation in 1986 (the Gog-Le-Hi-Te Wetland in the Puyallup River Estuary, Puget Sound, Washington, US). After 7 years, only a few of the 16 ecosystem attributes showed any sign of equivalency with natural wetlands in the area and many appeared to demonstrate dysfunction relative to reference wetlands.

Restored or newly created wetlands may fail to reach functional equivalence with 'natural' wetlands at all, or very long timescales may be required. Variability among reference sites can also make it difficult to predict developmental trajectories, as Simenstad and Thom (1996) found in their study. Since it is often necessary to evaluate compensatory mitigation success in the short term, as dictated by regulatory timeframes, it is important that evaluation should be based on attributes that predict longer-term trends in the development of restored or created systems.

The extent to which newly created habitats or ecosystems really compensate for lost resources varies considerably. Complexity is a key factor. For some wetland types, it may well be possible to create a fully functioning ecosystem that is comparable with 'natural examples'. For example, a paper by Dobberteen and Nickerson (1991) reported promising early results from the creation of cattail (*Typha*) marshes following destruction of wetlands at a municipal landfill site in Danvers, Massachusetts. The artificial, new wetlands had very similar structure and growth patterns to remaining 'natural' cattail marshes after a couple of years. However, further studies would be needed to evaluate function and viability in the longer term and there is growing evidence to suggest that newly created systems are often poor replicas of those which are lost.

In the UK, there has been so little follow-up of mitigation projects that it is difficult to judge the extent to which compensatory mitigation has replaced lost ecosystem function. However, it is clear that here, too, no explicit analysis occurs to determine which functions should be replaced through compensatory mitigation. In countries such as the UK, where remaining wildlife habitats have become degraded and fragmented following centuries of urban, agricultural and industrial development, the availability of equivalent land is an important issue. Where habitat restoration or creation is proposed for compensatory mitigation, it will not always be possible to locate suitable alternative sites.

In the US, 'habitat equivalency analysis' (HEA) has evolved to determine compensation for 'resource injuries' (see section 6.4), and to ensure that the public can be compensated for past losses of habitat resources through a habitat-replacement project providing additional resources of the same type (NOAA 1996).

7.7.1 Mitigation banking

In the US there has been particular focus on damage to limited wetland habitats and the development of strategies for mitigation and compensation. Despite wetlands protection legislation, considerable areas of wetland are lost each year to development. 'Mitigation banking' has emerged as a pragmatic solution to the problems of obtaining and managing suitable compensation sites for wetlands damaged or lost through development.

Section 404 of The Federal Water Pollution Control Act Amendments of 1972 requires proponents of developments in the US to obtain a permit from the Army Corps of Engineers ('the Corps') before discharging dredge or fill material into water. In 1989, the Corps, the Environmental Protection Agency, the Soil Conservation Service and the US Fish and Wildlife Service developed a method for identifying and classifying 'jurisdictional' wetlands and specified preferred approaches to 'mitigation sequencing'.

1 Identify alternatives that could avoid damage.
2 Minimize the damage.
3 Compensate for damage with:
   • wetland restoration to the 'natural' state;
   • wetland creation;
   • wetland enhancement;
   • wetland preservation.

Compensating for damage is often hampered by the difficulties of identifying and obtaining suitable sites for wetland restoration, creation, enhancement or preservation. Furthermore, individual development proponents may not be well equipped to undertake wetland restoration, creation or management. The concept of 'mitigation banking' emerged out of the need for access to compensation wetlands and for specialized wetland management skills. Mitigation banks have been created over the last 15 years and are now fully integrated with mitigation procedures in some States. They have been particularly successful in California and Florida. The banks purchase and manage wetland sites, which are then valued in 'credits'. Banks may purchase and manage wetland sites with existing high wildlife value, restore degraded wetlands or (less frequently) create wetlands '*de novo*'. 'Credits' can be based on a range of wetland 'values' (protocols for site evaluation vary) and can be considered as 'deposits' in a conventional bank account. For developments affecting 'jurisdictional' wetlands, compensation is required for loss of wetland 'credits'. The banks have the option of selling the 'credits' earned by their mitigation projects or represented by the sites they own, to third parties (including developers), who can use them to offset any environmental impacts that their own projects may incur. Guidelines for the establishment of wetland mitigation banks were issued in 1993 by 'the Corp' and the US Environmental Protection Agency. Wetland mitigation banks were identified as the desired 'vehicle' for compensation with respect to water-dependent projects, projects involving small but unavoidable impacts on wetlands, linear projects such as pipelines and highways that involve numerous small impacts on wetlands, and routine public projects such as drainage ditches which can affect wetland hydrology.

Most commonly, a bank is created after a memorandum of agreement is signed between the bank sponsor and various federal and state agencies. Some mitigation banks are 'dedicated', being designed specifically to compensate for wetland damage or losses associated with particular types of construction activity. For example, a number have been sponsored by State transportation departments. Others have been sponsored by private developers to provide opportunities for compensation for wetlands potentially damaged by their own projects or have been set up privately to sell credits to the general public. One of the largest mitigation banks in the US, 'Ecobank', purchases and restores damaged wetlands in Florida. The bank receives mitigation credits from the State which are sold on to developers. Its projects have included a 9547-acre (3818.8-ha) parcel in Seminole County, that is pending approval, and a 5000-acre (2000-ha) parcel bordering the St Johns River, which may be worth as much as 300 mitigation credits upon completion.

One of the main advantages of mitigation banks is that they facilitate the employment of specialists with expertise in wetlands management, which greatly improves chances of restoration success. Individual developers, on the other hand, do not always have access to suitable expertise. The mitigation banks can take on the long-term responsibility of managing and maintaining wetland sites and absorb some of the costs of restoration failure. Mitigation banks also facilitate the restoration and protection of large parcels of land, as opposed to the fragmented tracts that often result when developers are required to mitigate their own development areas. Large numbers of small, disjointed wetland parcels are harder to manage than larger contiguous areas and often have less ecological value. One of the main difficulties remaining with wetland mitigation banking is the difficulty of valuing wetland resources consistently in both ecological and financial terms.

The theory and practice of mitigation banking is explored in depth in Marsh *et al.* (1996).

7.8 Implementing mitigation proposals

Mitigation plans should include the following components (after Salvesen 1990):
• a clear statement of the objectives of mitigation (i.e. is it intended to ensure no net loss of a specific habitat?);
• an assessment of the natural resource values that will be lost in comparison with those that will be replaced;
• a description of what action will be taken and when (e.g. planting proposals);
• a monitoring and maintenance plan;
• a contingency plan for problem issues/in the event of mitigation failure, etc.

In terms of compensatory mitigation through wildlife habitat creation, projects demonstrating the greatest success in practice tend to be those with the following characteristics (Parker 1995):
• a narrow range of objectives;
• correct substrate conditions;
• long-term, but flexible finance;
• management plans and management commitment;
• luck! (e.g. favourable weather).

7.9 Evaluating the effectiveness of mitigation proposals

Evaluating the effectiveness of ecological mitigation is essential to the pursuance of sustainability objectives through the EIA process. The magnitude and severity of any residual adverse impacts depends not only on the nature of a development proposal, but also on the effectiveness of mitigation. We must know the extent to which mitigation can be relied upon to offset damage or losses. This means knowing, for example, whether an equivalent area of land will be restored to compensate for habitat loss, and whether restored areas will have equivalent wildlife value. It is vital that decision-makers should be able to judge which impacts will be mitigated, how much mitigation will cost, whether it will work and how long it will take. Clarity about mitigation proposals is therefore very important.

An important first step is to determine the extent to which proposed mitigation measures match the impacts that have been identified. Without a clear indication of which impacts are to be mitigated, it is impossible to estimate the magnitude of any residual effects. Also, although mitigation measures are planned to reduce specific undesirable effects, they may have undesirable side-effects of their own, which should also be taken into account when estimating residual effects.

In the UK, there is often a gross mismatch between the impacts identified in ESs and the mitigation measures proposed. Failure to recommend mitigation measures for significant adverse ecological impacts is common. The review by Treweek and Thompson (1997) of UK ESs also highlighted many examples where mitigation proposals had little relevance to the impacts identified. In many examples, there was more emphasis on landscaping and tree planting as a 'mitigation measure' than on either the avoidance of impact or the restoration of particular ecosystems. This suggests an inability to tackle the mitigation of specific ecological impacts or an unwillingness to implement more expensive and complex mitigation proposals. The use of essentially cosmetic measures such as tree planting to substitute for loss of specific wildlife habitats is clearly unsatisfactory and does little to safeguard biological diversity. One reason for this is the lack of objective criteria that can be used to decide which impact types merit mitigation or the extent of mitigation required to offset loss or damage. In the absence of guidelines or legislative requirements for appropriate mitigation, the principles of mitigation can be loosely applied and inappropriate habitat substitutions are common.

Decision-makers therefore need to ask the following.
• Have recommendations been made for the mitigation of all significant adverse ecological impacts?
• Will important adverse impacts remain unmitigated?
• Will the mitigation measures themselves have any adverse side-effects?
• What will the overall residual effects be (e.g. will there be in-kind compensation or will habitat substitution occur)?
• Are residual losses or damages acceptable?
The next phase of evaluation relates to the likely effectiveness of mitigation proposals.

No mitigation measure should be proposed without some indication of its likely effectiveness. It is unreasonable to expect decision-makers to take recommendations for mitigation on trust, particularly with respect to impacts in specialist subject areas, such as ecology, where they may lack specific expertise. In addition, decision-makers need to understand the likely consequences of failure to implement appropriate mitigation measures.

> **Box 7.3** A checklist of questions for decision-makers to ask about the effectiveness of proposed mitigation measures.
>
> - Are mitigation proposals realistic?
> - Are mitigation proposals technically feasible?
> - Have they been tried and tested elsewhere?
> - How much will the proposals cost to implement?
> - How much follow-up management will be required to ensure effectiveness of mitigation proposals?
> - What are the chances of failure (has a risk analysis been done?)
> - What are the consequences of failure?
> - How long will mitigation take?

Key issues are the value of the resources that are to be damaged or lost, the extent to which they can actually be restored or replaced using available technology, the time this will take and the cost of achieving an acceptable degree of mitigation. Decision-makers therefore need to ask themselves the questions listed in Box 7.3. In addition, the effectiveness of mitigation through restoration or creation will depend on the spatial ecology or context of habitats that are lost or restored. Small, isolated patches of new habitat may deteriorate due to edge effects and external influences. Restored sites that are isolated from sources of colonizing species may fail to reach a satisfactory species complement without their deliberate introduction or re-introduction. Provision for the management of restored or newly created habitat is also very important.

7.9.1 The value of natural resources

Particularly rigorous measures might be expected to offset adverse impacts on designated areas or protected species. In the UK, however, the most important international and national designations for nature protection and conservation fail to act either as a deterrent to development or as a trigger for comprehensive mitigation proposals. Of the 112 SSSIs directly affected in the review by Treweek and Thompson (1997), specific mitigation measures were only recommended for nine (approximately 8%). There were other cases of potential impacts on internationally designated areas (including one UNESCO World Heritage Site, nine RAMSAR sites and four Special Protection Areas) where no specific mitigation measures were listed at all. The situation appeared to be little better with respect to protected species. Mitigation measures were proposed in 25% of cases where there were potential adverse impacts on birds protected under Schedule 1 of the Wildlife and Countryside Act, but in cases where there were potential adverse

Table 7.1 Mitigation of impacts on designated sites and protected species in the UK. (After Treweek & Thompson 1997.)

Impacts on	Number potentially affected	Specific mitigation measures
UNESCO World Heritage	1	0
RAMSAR	9	0
SPAs	4	0
National Park	4	0
SSSIs	Directly affected: 112	9 (avoidance: 8, reinstatement of habitat: 1)
NNRs	8	0
TPOs	8	0
TPOs	8	2
Protected species		
Amphibians	Great crested newts: 7	1
Mammals	Badgers: 17	Badgers: 3 (avoidance of setts: 1, tunnels: 2)
	Bats: 6	Bats: 1
Birds (Schedule 1, Wildlife and Countryside Act (WCA))	20	5 (avoidance of breeding site: 1, avoidance of breeding season: 4)
Nationally rare species		
Invertebrates	12	0
Plants	15	1 (translocation of bee orchid)

See Table 4.4, p. 113, for abbreviation definitions.

impacts on other protected species, including badgers, bats and great crested newts, mitigation was only proposed in 18%, 17% and 14% of ESs, respectively (Table 7.1).

It can be argued that environmental impacts should be measured in monetary terms, an equivalent value being invested in shadow projects to offset the environmental degradation caused by development (Munro & Hanley 1991). There have been many attempts to value natural systems in monetary terms in order to provide a common unit of measurement, but this has been fraught with difficulties. The concept of a uniform currency that can be used to 'trade' one habitat type for another founders due to the non-substitutability of most natural resources. For example, it is difficult to determine the extent to which chalk downland might substitute for lowland heathland. In a recent report by the statutory consultee for nature conservation in England, English Nature (Gillespie & Shepherd 1995), it was argued that irreplaceable resources constituting 'critical natural capital' should be regarded as inviolable. It is not possible to replace 'critical natural capital', so restoration could never compensate for its lost value, however this is measured. Clearly, it is crucial that we should understand which natural resources can be replaced and which cannot.

7.9.2 Replaceability

The issue of replaceability was introduced in section 6.3.15. Research suggests that the rehabilitation of damaged ecosystems or the restoration of habitats and species to sites they are known to have occupied in the past is far more likely to be successful than the creation of new or substitute habitats. In nearly all circumstances, newly created habitats lack certain structural and/or functional attributes and are incomplete copies of what has been lost. It is important to remember that all ecosystems are unique and it is never possible to recreate, down to the finest detail, the circumstances under which they evolved. This is particularly true for really old, mature habitats that have developed over centuries; the best that can be hoped for is to achieve something 'similar'. Table 7.2 summarizes some of the differences between ancient woodland and younger, replacement woodland.

However, habitat restoration or creation can play a very important part in compensating for past losses of natural and seminatural habitat or in circumstances where economic, social or political imperatives override ecological ones.

Table 7.2 Some differences between ancient and younger woodland. (After English Nature 1993.)

Ancient woodland	Younger 'replacement' woodland
Continuity of habitat over time	By definition impossible to recreate until woodland has aged to an equivalent degree
Soils undisturbed for long periods	Impossible to recreate. Effects of disturbance often irreversible
Long-standing mycorrhizal associations	May not be present. Poorly understood and difficult to reinstate
Possibility of locally evolved genetic variation	Can only be recreated if planting material (seeds, cuttings) is derived from same source
Specific associated herbaceous species ('ancient woodland indicators')	Tend not to colonize younger woodlands
Rich snail fauna, particularly in wet flushes and springs, but also in drier areas. Some very rare species which tend not to occur in other habitats	Poor colonizers. New habitat provides microclimates which are unsuitable
Rich woodland bird fauna	May take some time for birds to colonize. Hole-nesting species may not colonize for decades, when trees have matured sufficiently
A wide variety of habitats provided, including dead wood (important for many invertebrates)	Dead wood scarce in younger woodlands

Some ecosystems and wildlife habitats are more replaceable than others. Replaceability is sometimes simply a function of practical experience, but ecosystem complexity is also an important factor. This is an area where practice is often based largely on conjecture due to a lack of hard scientific evidence. There are a number of habitats for which restoration techniques are largely untested and for which there are no records of success in the literature. A review of habitat creation attempts in the UK (Parker 1995) showed some success in the creation of heathlands, salt marshes and some grasslands, but there was little long-term data for comparison with reference examples. Quantifiable definitions of 'habitat' are required to determine whether an acceptable degree of 'replacement' has occurred. In this context, it may be possible to evaluate success in terms of species composition. Fully effective replacement demands the reinstatement of viable populations of characteristic species in assemblages approaching their composition before damage occurred, i.e. in their 'pre-impacted' state. Although a characteristic species complement can be identified for most habitats, it is not always clear how many of these species need to be present and in what proportions, for a fully functioning system to be established through restoration or creation. More research is needed in this area, so that reference standards can be established.

'Reference' systems or habitats can be used to provide benchmarks for evaluating the effectiveness of any compensatory mitigation. Both structural and functional attributes have been used to characterize reference states.

Reference systems:
• provide evaluation standards with respect to key structures and/or functions;
• make the goals of compensatory mitigation explicit;
• provide templates from which restored/created habitats can be designed;
• establish a framework whereby declines in function resulting from adverse impacts (or recovery of function following restoration) can be estimated.

Explicit use of reference systems has been most common with respect to wetland ecosystems in the US. As the practice of compensating wetland losses through construction/restoration has become commonplace, the need for methods to assess the replacement of wetland function has grown. The 'evaluation for planned wetlands' (EPW) procedure is one example of a formalized approach to the evaluation of replacement wetlands (Bartoldus 1994). This rapid assessment procedure can be used to document and highlight differences between 'planned' and reference wetlands based on their capacity to perform certain functions, including control of shoreline bank erosion, stabilization of sediments, conservation of wildlife and fish stocks, and the maintenance of water quality. Richardson (1994) also refers to an evaluation method based on definition and scaling of wetland functions in reference systems for comparison with those in new or altered systems.

Using species/habitat affinities, for wetland habitats, Adamus (1995) developed an 'avian richness evaluation method' (AREM) for assessing lowland wetland and riparian habitats of the Colorado Plateau. This method scores habitats for avian richness based on simple observations of habitat characteristics. It was tested against data from 76 sites during the breeding season and gave consistent predictions of species composition in the wetland areas studied. The method could be used both for detecting losses of habitat quality (e.g. resulting in loss of species) and for evaluating the effectiveness of mitigation (is a full complement of species present in new/restored systems?).

Clearly, these approaches depend on the selection of reference examples using some sort of consistent classification. Brinson and Rheinhardt (1996), for example, selected reference wetland habitats from different classes of wetlands sharing similar hydrogeomorphic settings.

The use of reference standards is based on the premise that mitigation should achieve a certain degree of equivalence with what has been damaged or lost. It is necessary to quantify ecosystem attributes simply to be able to work out whether like has been substituted for like. Equivalence needs to be estimated in terms of both quantity and quality if wildlife habitats and distributions are to be safeguarded. Decision-makers should therefore ask the following.

• Will the quantity and quality of natural resources be maintained?
• Has all critical natural capital been safeguarded?
• Will key structures and functions be maintained, created or restored?
• Will the implementation and effectiveness of mitigation proposals be monitored?

By quantifying gains and losses in function it is possible to set minimum replacement ratios for restored impacted areas, the implementation of which can be monitored.

7.9.3 Issues of redistribution

It is impossible to evaluate the effectiveness of mitigation proposals without a clear idea of the intended outcome. In other words, is complete replacement required in all circumstances and what level of residual impact will be tolerated? In circumstances where 'no net loss' of natural resources will be accepted, damage must be avoided completely, or mitigation measures must guarantee the replacement of natural resources in their 'pre-impacted' condition. This might be demanded in the case of internationally protected wildlife habitat, for example. More commonly, a certain level of damage or loss is accepted, but the terms of acceptance are not always explicit. Mitigation to defined ecological standards is the exception rather than the rule. In the absence of reference standards for mitigation, it is difficult either to evaluate the success of mitigation or to regulate the extent of damage or loss accepted as a consequence of development. Under such circumstances issues of redistribution are more likely to come into play. This is particularly true for ecological impacts, many of which are cumulative and trans-boundary. Most mitigation proposals are evaluated individually, rather than on an integrated regional basis.

Where there is a strong dependence on compensatory mitigation, the distribution of ecosystems in the landscape may alter considerably. Habitats may be created or restored in one area to compensate for losses in another, or one habitat type may be substituted for another. Ecologists are beginning to document considerable differences (both quantitative and qualitative) between what is lost and gained (Zedler 1996). In evaluating the effectiveness of mitigation, the extent to which whole landscapes or regions are re-configured should therefore be considered (Bedford 1996). It is only by taking a broader perspective that it becomes possible to evaluate the potential cumulative effects of individual mitigation decisions on landscape-scale patterns of biodiversity.

Allen and Feddema (1996) made such a regional evaluation of the effectiveness of the '404 permit programme' in achieving 'no net loss' of wetland resources in southern California. They evaluated 75 Section 404 projects permitted between 1987 and 1989. They found that 80.47 ha of wetlands had been affected by Section 404 permits and 111.62 ha of wetland mitigation had been required to compensate for losses. Visits to the project sites revealed that only 77.33 ha of the mitigation area had actually been provided or completed, resulting in a net loss of 3.14 ha of wetlands. It also became clear that freshwater wetlands were experiencing a disproportionate loss, riparian woodland wetlands more often being used in mitigation efforts. The combined effect of development and mitigation was therefore to result in both a loss of wetland habitats and an overall substitution of types throughout the region. Allen and Feddema (1996) concluded that planning of mitigation on a regional or a watershed scale would be required to improve the effectiveness of the Section 404 permitting programme in ensuring 'no net loss' of wetlands and ensuring that replacement was appropriate in terms of quantity, quality and regional distribution.

There may also be redistributional consequences of relocating with respect to 'use values' placed on natural resources by people. For example, the notional present-day value of a habitat may derive, in part, from the importance placed on it by a local community for recreation or amenity: a 'use' value that would be lost by replacing the habitat elsewhere. There are similar problems for wildlife species. The use of habitat by associated species is influenced by its spatial ecology, or landscape context. Less mobile species may be unable to utilize replacement habitat if it is isolated in an inhospitable landscape 'matrix'.

Ecological mitigation for EIA frequently entails the loss of mature habitat and its replacement with a younger version, or even an earlier successional stage. For many of the seminatural wildlife habitats found in the UK, there appears to be a correlation between age, or maturity, and the presence of certain discriminating species that are unlikely to be found in younger examples of the habitat. Many of these species are rare and threatened, like many of the UK's 'ancient woodland indicator species', for example. For many species, 'young' and 'old' examples of the same habitat type are not equivalent. It is therefore necessary to take account of the time that may be needed to reach an equivalent stage in the succession of the habitat from a new or newly restored state and to attain a value equivalent to that lost. There are many examples where this might take decades. With respect to sustainability and the maintenance of natural capital for inheritance by future generations, the restoration of a complex habitat over decades may safeguard intergenerational, but not intragenerational, equity.

7.9.4 Technical feasibility

An indication of the likely success of proposed measures should be given, based either on experience elsewhere or on evidence drawn from the literature. In the absence of this requirement, proponents can make all sorts of elaborate undertakings to restore habitats without giving any indication of how likely they are to work in reality.

7.9.5 Prescriptions for mitigation

In many parts of the world, appropriate and consistent management is essential for the maintenance of important seminatural wildlife habitats. For habitats that have evolved under human intervention, the restoration and creation measures are prone to failure without follow-up management. Management plans for habitat restoration or creation should therefore always be attached to mitigation proposals. Some provision for monitoring is also important and this should also be prescribed. Without monitoring it may not be possible to tell whether or not mitigation has been successful.

Because monitoring of the implementation and effectiveness of mitigation measures has been rare in practice, it has been difficult to build up the knowledge and experience that is needed to improve the effectiveness of mitigation in the future.

7.9.6 Costs of mitigation

Cost–benefit analysis is commonly used to evaluate the social implications of development proposals. The consideration of adverse ecological effects can be included in such an analysis only where they imply some loss of welfare to society. However, it can be difficult to value the social costs of adverse effects on natural resources, particularly where there is no obvious or direct social use. In such circumstances, the 'replacement cost' technique is sometimes used. This is based on the monetary cost of restoring an asset to its original state and therefore has obvious applications in the evaluation of mitigation proposals. The costs of mitigation are interpreted as a proxy for the value or cost of environmental degradation.

The more costly replacement is, the higher the probability that development proposals will fail on efficiency grounds, as costs will exceed benefits. The cost of replacement tends to be higher for more complex habitats. Increased costs of mitigation for these habitats increases the probability that the benefit/cost ratio of developments affecting them would be less than unity. In such a situation, conservation of the original resource would be preferable on efficiency grounds. In fact, the real costs of replacing most habitats, or of restoring species to a location are unknown. In most restoration attempts, the focus has been on macro-organisms and techniques for the restoration of characteristic microfloras and faunas are hardly known. Nevertheless, the concept of 'replacement cost' offers considerable potential for evaluating the suitability of proposed mitigation measures.

The costs of research should not be ignored. There are many cases where research would be required simply to understand ecosystems sufficiently for economic values to be placed on lost functions. It is not inconceivable for the cost of carrying out such research to exceed the expected value of any predicted damage. There may also be insufficient time to undertake research (Unsworth & Bishop 1994). For many ecosystems, the existing valuation research base is inadequate to support defensible benefits transfer. As an alternative to more traditional valuation techniques, Unsworth and Bishop (1994) therefore proposed a simplified approach based on environmental annuities. The principal assumption behind the approach is that the public can be compensated for past losses in environmental services through the provision of additional services of the same type in the future.

7.10 The role of mitigation in sustainable development

Mitigation measures are generally implemented to reduce negative environmental impacts within reasonable environmental and economic constraints (Canter 1996). They therefore have a clear part to play in the achievement of 'sustainability'. Mitigation of adverse ecological impacts as part of the EIA process has an important part to play in maintaining a constant stock of 'natural capital' (Forbes & Heath 1990), which can be passed on to future generations (Barbier *et al.* 1990). However, there is little evidence to suggest that current policy recognizes its importance. Under the UK legislation, proponents of development are required only to recommend suitable mitigation measures, not to demonstrate that they can and will be undertaken. The actual implementation of proposed measures is not subject to any formal regulation, giving decision-makers (e.g. planning authorities) an important role in ensuring that ecological impacts are mitigated in accordance with the undertakings made in ESs. In the absence of any legislative requirement for evaluating the appropriateness and effectiveness of proposed mitigation measures, there is a considerable risk that significant adverse ecological impacts will remain unmitigated and that natural capital will be lost irrevocably.

If mitigation in EIA is to be consistent with the concept of sustainability and the maintenance of 'critical natural capital', the following conditions must be met:
• it should always be clear exactly which impacts are to be mitigated so that any residual impacts can be assessed;
• the need for mitigation should be determined in relation to the value of the resource affected and the severity of the impacts identified;
• mitigation should be more rigorous and comprehensive where potential impacts on designated sites and protected species have been predicted;
• mitigation proposals should be sufficiently detailed for their effectiveness to be evaluated;
• some indication should be given of the effectiveness of the proposed measures, based on similar experience elsewhere;
• where untested techniques are proposed, this should be made clear;
• the extent of residual impact with and without mitigation should be estimated;
• contingency measures should be included with respect to possible mitigation failures.

7.11 Recommended reading

Guidelines

The best guidance on ecological mitigation measures has been produced for road developments. Guidelines produced by the US Army Corps of Engineers (1983) include mitigation for different types of terrestrial and aquatic habitats, wildlife reserves and threatened and endangered species in the US. The Dutch Ministry of Transport, Public Works and Water has also produced excellent guidance on the design of wildlife crossings for roads and waterways Management (DWW 1995) and in the UK, useful information is available in the booklet produced by English Nature (1993).

Mitigation banking

Marsh, L.L., Porter, D.R. & Salvesen, D.A. (eds) (1996) *Mitigation Banking: Theory and Practice*. Island Press, Washington, DC.
Articles in Ecological Applications **6** (1) (1996) [A special issue on wetland mitigation.]

Role of ecological mitigation in biodiversity conservation

Treweek, J. & Thompson, S. (1997) A review of ecological mitigation measures in UK environmental statements with respect to sustainable development. *International Journal of Sustainable Development and World Ecology* **4**, 40–50.

8 Monitoring

8.1 Introduction

Fundamental to any testing of hypotheses or quantification of impacts is the need for testing and monitoring. There is a strong conviction amongst ecologists that individual proposals should be considered to provide an experimental context in which post-project monitoring is required to test the hypothesis: in this case, the impact predictions which have been made. According to Beanlands and Duinker (1983), this is the 'only concept of impact assessment in which the interdependencies of the various activities — baseline studies, predictions and monitoring — become coherent in a scientific sense'.

In the context of ecological impact assessment (EcIA), monitoring has two main roles. Firstly, monitoring has a vital part to play in developing the scientific basis for EcIA by:
- establishing the status of individuals, populations, communities or ecosystems to establish a baseline;
- enhancing understanding of 'natural variation';
- detecting trends in the status and distribution of ecosystems or their components;
- measuring departures from baseline status;
- providing a framework for evaluation at a variety of scales;
- enhancing future predictive ability through follow-up of EcIA predictions.

In other words, monitoring provides essential information. Without it, the ecological basis for impact prediction will remain severely limited due to high levels of uncertainty.

Secondly, monitoring is important for maintaining or enhancing the effectiveness of the EcIA process by providing a basis for:
- evaluating the validity of impact predictions;
- evaluating the success of mitigation measures;
- detecting departures from predicted states;
- generating consistent benchmarks for implementing remediation measures;
- deciding when alternative contingency measures should be implemented in the light of mitigation failure;
- evaluating overall methods, procedures and standards.

In this context, 'monitoring' is often referred to as 'follow-up' or 'audit'. It may be implemented on a case-by-case basis, or to evaluate national standards. Without it, the effectiveness and appropriateness of the EcIA process cannot be judged. Its value as part of an acknowledged commitment to the conservation of biological diversity, for example, will remain largely unknown. It will also remain difficult to determine the real benefit of effective EcIA in improving the quality of environmental decision-making or in contributing to better environmental outcomes. From the perspective of environmental regulators, post-project follow-up is important to check that recommended mitigation measures, for example, have been implemented and are effective. To date, environmental impact assessment (EIA) and EcIA have been used predominantly as part of the process of acquiring development consent and have not been extended into the actual lifetime of approved projects. There has therefore been relatively little opportunity to check that actions are actually undertaken in the manner specified. Furthermore, the motivation to use rigorous, quantitative approaches in EcIA has been lacking because there has been no formal requirement to measure the impacts that actually do take place.

8.2 Ecological monitoring

Ecological monitoring entails the systematic observation and measurement of ecosystems (or their components) to establish their characteristics and changes over a period of time. Monitoring is essential to check the validity of predictions and provide a rational basis for precautionary action 'in the face of uncertainty and ignorance' (Barlow *et al.* 1992). No system can be managed effectively without an understanding of causal relationships. As noted earlier, it is impossible to attribute observed changes to any particular cause without some knowledge of inherent or 'natural' variation. In theory, 'the more uncertain the expected outcome of a proposed management design and the more that is at stake, the greater the investment in monitoring should be' (Salwasser 1988). There are many examples to illustrate the inadequacy of short-term datasets to explain ecosystem 'behaviour' in the longer term. Long-term studies of alligators in Lake Apopka, Florida (Semenza *et al.* 1997), for example, showed that pollution by a spill of persistent organochlorine pesticides in 1980 resulted in a population crash several years after the spill, followed by a slow recovery, but then further declines in the mid-1990s. Variation in the population was considerable, and some of the effects of the pollution incident were considerably delayed. Short-term studies would have been likely to result in erroneous conclusions and would not have been adequate as a basis for predicting the impacts of the pollution incident on the longer-term status of the alligator in the lake.

8.2.1 Requirements for monitoring standards

If a primary goal of ecological monitoring is to provide comparative data for analyses, for example biodiversity, examination of population trends, local extinctions or the impact of human activities on populations, then studies must employ standard techniques (Coddington *et al.* 1991; Heyer *et al.* 1994; Stork 1994). The importance of standardizing techniques cannot be over-emphasized (Stork & Davies 1996). If different methods are used in two separate studies then their results will not be comparable and the value of one or both of the studies will be greatly diminished. 'It is therefore very important to ensure that the methods used are the 'industry standards' (Stork & Davies 1996). There are a number of texts that are valuable in providing standards for biodiversity assessment in particular. The Smithsonian Institution, for example, has started to publish a series of texts outlining standard methods for different groups of animals and plants, the first of which (Heyer *et al.* 1994) is a manual for measuring and monitoring amphibians. Others on mammals and birds are due to be published shortly and one on insects is being prepared. Some standard methods for biodiversity assessment are also outlined in HMSO's (1996) publication *Biodiversity Assessment: a Guide to Good Practice*, and a guide to the standard floras of the world has been produced by Frodin (1990).

In some countries, standard protocols have been established for monitoring certain groups. In the UK, for example, there is the well-established Butterfly Monitoring Scheme and also the national Breeding Bird Survey (BBS). Such surveys, carried out using consistent methods and repeated over relatively long periods, can provide valuable information about trends in bird distribution and populations (Tucker *et al.* 1997) and constitute a necessary baseline for evaluating local impacts in EcIA. For example, adverse impacts on populations of species that are in continuous, prolonged decline in status from numerous to rare, might be regarded as unacceptable.

8.2.2 Focusing ecological monitoring effort

Collection of data 'for the sake of it' or 'just in case' it proves useful in future is unlikely to be either helpful or cost-effective. 'Useful ongoing review is best ensured by designing the monitoring to challenge hypotheses' (Barlow *et al.* 1992). In other words, monitoring should be structured to address clearly defined assessment questions, fundamental questions about the behaviour and status of ecosystems or to redress specific imbalances or deficiencies in the level of our knowledge. A good monitoring programme must provide for both repeatability and control. It is therefore necessary to consider the required frequency of monitoring and its timing with respect to both the nature of affected ecosystems and intended (or already implemented) actions. Monitoring should be targeted to where the most useful environmental information can be gained and should be performed with sufficient frequency that trends can be identified accurately.

In some circumstances continuous monitoring may be required (e.g. to check for unplanned discharges or to check whether pollutants are persisting in the environment (Department of the Environment 1995). In general, however, continuous monitoring is likely to be prohibitively expensive. Because of the high costs of maintaining monitoring pro-grammes, the costs of monitoring should always be taken into account as one of the inherent costs of implementing an intended action.

It is important to structure ecological monitoring so that potential delayed or chronic effects can be recognized. Sublethal or chronic effects are generally more likely to be detected in older, well-established organ-isms such as perennial plants or longer-living fish. There may often be a considerable delay before the effects of an activity are manifested in the actual absence or death of organisms, so some way of detecting chronic damage is crucial, for example it might be possible to recognize stunted growth, reduced reproductive potential, morphological abnormalities (e.g. those observed in molluscs following contamination by tributyltin), increased susceptibility to disease or reduced longevity in selected indicators.

The use of indicators, assessment and measurement endpoints and ecological evaluation criteria applies to the design of ecological monitoring programmes just as it applies to the EcIA process. Ideally, EcIA should be considered as just one stage in a structured, longer-term monitoring process that permits comparison of impacts against clearly established ecological baselines.

One of the main barriers to effective EcIA is the shortage of reliable and up-to-date biodiversity information. Our knowledge of different groups of organisms varies considerably. For some groups, such as birds, almost all species have been inventoried and we know a fair amount about their biology, distribution and status (Bibby *et al.* 1992b), but for some other groups we know hardly anything. An all-biota taxonomic inventory (ABTI) has now been proposed (see section 8.2.5). This would make some of the less well studied groups like termites, butterflies and fishes the focus of concerted collecting and study over the next few years so that our level of understanding of their biology and taxonomy will match that of birds and some other vertebrate groups. The objective of the ABTI is to focus on specific taxa and to inventory all species in those groups on a global basis. As part of this initiative, the world's major taxonomic institutes have therefore had to set priorities about which groups of organisms to target for special attention. The directors of some of the major entomological systematic collections, for example, have reached an agreement about which insect taxa should be the focus of increased inventorying effort (Stork & Davies 1996).

Focusing effort is particularly important for large-scale monitoring programmes due to their potentially enormous cost and the organizational challenges they entail. Monitoring programmes may be implemented on a site-specific, regional, national or international basis. National monitoring programmes in particular should be planned so that the information they generate is likely to remain germane to management issues in the longer term. At the very least, national (or large-scale) monitoring programmes should provide information that is gathered using consistent, replicatable techniques and which can be applied effectively in a variety of potential management scenarios, some of which may not yet have been envisaged.

While the need for ecological monitoring and EcIA follow-up is widely acknowledged, there has been considerable reluctance in many countries to make monitoring a formal or legal requirement of the EIA process. At the same time, however, some countries have invested in large-scale national ecological monitoring programmes intended to strengthen the knowledge base and improve predictive ability by providing up-to-date information on the status of ecosystems. Implementing such large-scale monitoring programmes is no mean task.

The following section introduces the US Environmental Protection Agency's Environmental Monitoring and Assessment Program (EMAP). This is probably the most ambitious national ecological monitoring programme to be attempted and serves to illustrate many of the benefits and challenges associated with large-scale ecological monitoring.

8.2.3 The US Environmental Monitoring and Assessment Program

The US Environmental Protection Agency initiated the EMAP in 1988 to provide information on the status, extent and condition of ecological resources in the US. When fully implemented, EMAP was intended to be an integrated, multi-resource programme that could be used to determine the status of ecological resources at various geographical scales over long periods of time and to detect changes in the status of these resources on a regional and national basis (Messer et al. 1991).

One of the most important features of EMAP is its use of an 'indicator framework' to monitor ecological attributes. One of the main motivations for developing this approach was a desire for consistency in terminology and compatibility of data with the Environmental Protection Agency's Ecological Risk Assessment Program. The following definitions were therefore agreed.

- *Environmental value*: a characteristic that contributes to the quality of life provided to an area's inhabitants, i.e. the ability of an area to provide desired functions or services, for example production of food, provision of clean water and air, aesthetic experience, recreation and the support of desired plant and animal species.
- *Assessment endpoint*: an explicit expression of the environmental value that is to be protected.
- *Measurement endpoint*: a measurable ecological characteristic that is related to the valued characteristic chosen as the assessment endpoint. Measurement endpoints are often expressed as the statistical or arithmetic summaries of the observations that comprise the measurement.
- *Indicator*: an environmental characteristic that can be measured to assess the status and trends of environmental quality, i.e. ability to support a desired human or ecological condition.

EMAP covers seven broad resource categories: near-coastal waters, the Great Lakes, surface waters, wetlands, forests, arid lands and agro-ecosystems, a coordinated monitoring network and series of indicator measurements being developed independently for each category (Novitzki 1995). The strategy used to select indicators for EMAP is illustrated in Fig. 8.1.

EMAP also addresses the condition of landscape ecological conditions under EMAP-L. This particular program is developing indicators that can be measured from remotely sensed images (Cain *et al.* 1997), primarily land-cover maps derived from satellite images (US Environmental Protection Agency 1994b). Monitoring can only be effective if it is based on meaningful classification of the whole environment into relevant subcategories and if suitable measurements can be made that reflect environmental or ecosystem state. For EMAP, it was considered important to retain scope for combining assessments of individual resource categories to carry out landscape-level assessments of ecological resources overall. However, this does not appear to be facilitated by setting up resource categories that seem to overlap in some cases. For example, the Great Lakes could fall within the category of 'surface waters'.

In terms of implementing monitoring, the different resource categories have been handled separately, although the approach taken is similar for all categories. For the EMAP-Wetlands programme, responsible for the 'wetlands' category, quantitative assessment is based on:

- definition of the current wetland resource;
- selection of indicators of wetland resource condition;
- establishment of a monitoring framework.

A wetlands classification is used as a basis for inventorying and

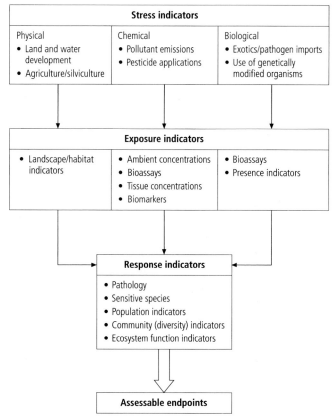

Fig. 8.1 Conceptual diagram of the Environmental Monitoring and Assessment Program (EMAP) indicator development strategy. (After Messer *et al.* 1991.)

mapping the current wetland resource. The US Fish and Wildlife Service runs a National Wetlands Inventory (NWI), which reports on the status of wetlands and trends in wetland area and this uses a wetland classification system described by Cowardin *et al.* (1979). The focus of EMAP-Wetlands is primarily on monitoring trends in condition, but wetland area and condition are not independent and NWI data on wetland area changes was concluded to be important for EMAP-Wetlands assessments of wetland condition. To facilitate the integrated use of available data in this way, it was concluded that the EMAP-Wetlands monitoring programme should use the same classification system as the NWI (Leibowitz *et al.* 1993). The NWI classification is based on three levels: wetland systems that are based primarily on hydrogeomorphic characteristics (e.g. 'marine', 'estuarine'), subsystems that are more based on hydrological characteristics (e.g. 'subtidal', 'intertidal') and classes based primarily on substrate and vegetative life-form (e.g. 'rock bottom' or 'emergent'). The classification generates 56 classes, which can be further subdivided if necessary.

EMAP-Wetlands uses a sample-based approach to inventory wetlands. To survey the whole of the land area of the US would clearly be an overwhelming task. As it is, the annual field survey required to maintain the programme is considerable. EMAP-Wetlands set out to assess the condition of 'estuarine emergents' (e.g. salt marshes), 'palustrine emergents' (e.g. prairie potholes) and 'palustrine forested wetlands' (e.g. bottomland hardwoods) for the whole country by 2004. It was estimated that about 200 samples would be required for each class for each assessment region, resulting in a possible overall total of 2400 sample sites. It was intended to sample one-quarter of these sample sites (600) each year: a considerable field programme.

EMAP as a whole has a uniform, systematic sampling grid that covers the whole of the US. The point frame is hierarchical, consisting of a nested series of grids at increasing densities. The reference-level density (the basis for regional sampling) consists of 12 500 sampling grid points located approximately 27.1 km apart. The nested nature of the sampling frame allows different indicators, or indicators in different ecosystem categories to be measured at different levels of resolution. Ecosystem distributions can be measured from satellites or aircraft at relatively low cost and can therefore be measured at a higher grid density, whereas field surveys are usually based on measurements taken at lower grid densities. Sampling density can also be adjusted according to the scale of variation of ecosystems, for example geographically restricted types need more intensive sampling than widespread ecosystem categories (Messer *et al.* 1991).

As a general rule, EMAP describes resources in an area centred around grid points in two 'tiers'. Sampling within a hexagonal area of 39.7 km^2 around each grid point results in 1/16th of the land area of the US being sampled overall. For 'Tier 1' sampling, existing information is used, or information is collected using remote sensing (aerial photography, satellite imagery). This generates a Tier 1 probability sample of the resources of the US from which regional estimates of areal extent of all landscape entities can be generated as well as regional estimates of more discrete features like lakes or wetlands. Tier 1 wetlands can then be sampled at random to provide a 'Tier 2' sample, so the network of sample sites visited or surveyed using remote sensing to obtain measurements of wetland condition is a subset of the Tier 1 EMAP sample (Novitzki 1995). There are cases where specific resource types are distributed too sparsely for the standard approach to generate enough survey sites, in which case Tier 1 sampling can be intensified.

EMAP-Wetlands demonstrates a fundamental problem with sample-based monitoring programmes: the fact that evaluation of 'status' and monitoring of 'trends' can demand different approaches to the allocation of limited survey resources. To evaluate status it is preferable to visit as many sites as possible overall, but to monitor trends it is better to re-visit a limited number of sites. EMAP-Wetlands reached a compromise by revisiting sites every 4 years instead of annually (Novitzki 1995).

Measurements or observations (indicators) of wetland characteristics can be used individually or in combination (e.g. through indices) to compare any wetland to reference wetlands in their region. To be effective, indicators must discriminate successfully between 'good' and 'degraded' wetlands. They must also be linked to the types of assessment questions that will be tackled using monitoring data. By basing monitoring on measurements of wetland resource condition relative to reference wetlands, it is possible to assess the success of regional 'no net loss' policies for wetlands and to make objective decisions on wetland mitigation, the establishment of mitigation banks or other activities that might affect the survival of wetlands and their condition. Some examples of EMAP-Wetlands assessment questions and associated indicators are given in Table 8.1. for estuarine emergent wetlands. In effect the indicators bridge the gap between the processes of wetland assessment (for regulation or decision-making purposes) and wetland evaluations to compare the relative values or qualities of wetland ecosystems.

Indicators are chosen to reflect certain wetland 'values' and are selected by panels of local experts. Potential indicators are evaluated in pilot studies carried out to test whether indicators are successful in discriminating between what are regarded as 'good' or 'degraded' examples of local wetland types. Table 8.2 shows the relationship between

Table 8.1 Examples of EMAP-Wetlands assessment questions and associated indicators. (After Novitzki 1995.)

Assessment question	Relevant indicators
What portion of salt marshes exhibit plant and animal communities similar to reference wetlands?	Total number of species Relative abundance of species Number/proportion of native, threatened, endangered and 'nuisance' species
What portion of salt marshes provide a degree of shoreline protection similar to reference wetlands?	Plant species presence Plant density Wetland area Tidal range
What portion of salt marshes accumulate sediments in a manner similar to reference wetlands?	Accretion rates Percentage of organic matter Bulk density
What portion of salt marshes provide habitat for fish and wildlife similar to that of reference wetlands?	Shellfish production Finfish production Invertebrate species composition and abundance

Table 8.2 EMAP-Wetlands proposed values, subvalues and indicators for estuarine emergent wetlands in the Gulf of Mexico. (After Novitzki 1995.)

Wetland value	Subvalues	Indicators
Biological integrity	Plant diversity	Total number of species. Number of native, rare, threatened, endangered and nuisance species
	Animal diversity	Total number of species. Number of native, rare, threatened, endangered and nuisance species
	Plant abundance	Percentage cover/stem density, stem height and width index, spectral reflectance
	Animal abundance	Biomass or number of each species
Harvestable productivity	Plant abundance	Percentage cover/stem density, stem height and width index, spectral reflectance
	Animal abundance	Biomass or number of each species, wetland extent
Hydrology	Shoreline protection	Hydraulic conductivity, percentage cover or stem density
	Water regime	Wetland area, tidal amplitude, tidal range, water depth, sulphides, soil salinity
	Salinity	Soil salinity
Water-quality improvement	Sediment retention	Accretion rate, bulk density, percentage organic matter, tissue and soil analysis for contaminants
	Nutrient processing	Percentage cover/stem density, dead vegetation, sulphides, plant tissue nutrient analysis, C/N ratio

indicators and wetland values for estuarine emergent wetlands as used in EMAP-Wetlands pilot studies in the Gulf of Mexico (Novitzki 1995).

There are certain characteristics that are shared by 'good' indicators of ecological status or condition. The criteria used by the US Environmental Protection Agency (1994a) to select indicator criteria for EMAP are summarized in Table 8.3. A vital aspect in the success of any large monitoring programme is the management of data, its interpretation and the reporting of results. All too often, ecological monitoring generates

Table 8.3 Indicator selection criteria for EMAP. (After US Environmental Protection Agency 1994a.)

Critical criteria	
Regionally responsive	Must reflect changes in ecological condition, pollutant exposure or habitat condition and respond to stressors across most pertinent habitats within a regional resource class
Unambiguous	Must be related unambiguously to an assessment endpoint or relevant exposure or habitat variable that forms part of the resource group's overall conceptual model of ecological structure and function
Simple	Quantifiable by synoptic monitoring or by cost-effective automated monitoring
Stable	Must have low management error and be stable (show low temporal variation) during an assessment period
High signal/noise ratio	Sufficiently high signal strength when compared with natural variation to allow detection of ecologically significant changes within a reasonable timeframe
'Low impact'	Sampling must have minimal environmental impact
Desirable criteria	
Available method	Should have a generally accepted, standardized measurement method
Historical record	Has an existing historical database or one can be generated from accessible data sources
Retrospective	Can be related to past conditions through retrospective analyses
Anticipatory	Provides an early warning of changes in ecological condition or processes
Cost-effective	Has low incremental measurement cost relative to information provided

information that is inaccessible and therefore never used. As emphasized by Messer *et al.* (1991), 'monitoring programs often live or die on the basis of their data management structure'. For a large-scale, integrated monitoring programme such as EMAP, for example, it may be necessary to ensure that data are accessible from research centres throughout the country. Constant quality checks are necessary to ensure consistency of sampling and interpretation. At the same time, the end-users of the information generated by the programme will have highly variable interests and technical abilities, making it necessary to develop innovative approaches to the analysis, interpretation and display of results.

8.2.4 International guidelines for biodiversity monitoring

The Convention on Biological Diversity (UNEP 1992) established certain priorities for nations in the task of inventorying and monitoring biodiversity. The main intention (as specified in Annex 1 of Article 7) is to identify and monitor ecosystems, habitats, species, communities, genomes and genes that satisfy certain criteria.

Priority is to be placed on ecosystems and habitats that satisfy the criteria summarized in Box 8.1.

The Convention on Biological Diversity recognizes the potential role of legislative processes like EIA in delivering sustainable development and the conservation of biological resources. National programmes for inventory and monitoring can only improve the chances of EIA and EcIA to realize this potential. However, it is important that national biodiversity monitoring programmes should be structured to provide relevant information. UNEP (1993) has drawn up a list of general priorities for the types of data that should be collected (Table 8.4). When these key parameters have been addressed, it is then possible to develop the information base and provide more detailed information. Clearly, if all countries were able to collect up-to-date information about all the parameters listed in Table 8.4, the task of performing effective EcIA would be greatly facilitated.

A need for up-to-date information on various components of biological diversity is generated by a number of pieces of international legislation.

Box 8.1 Criteria for prioritizing ecosystems, habitats, species and communities for monitoring.

Monitoring should prioritize ecosystems and habitats which:
• contain high diversity, large numbers of endemic or threatened species or wilderness
• are required by migratory species
• are of social, economic, cultural or scientific importance
• are representative, unique or associated with key evolutionary or other biological processes

Monitoring should prioritize species (or communities) which are:
• threatened, wild relatives of domesticated or cultivated species
• of medicinal, agricultural or other economic value
• of social, scientific or cultural importance
• of importance for research into the conservation and sustainable use of biological diversity, such as indicator species

Emphasis is also to be placed on 'described genomes and genes of social, scientific or economic importance' (Stork & Davies 1996)

Table 8.4 Key parameters for monitoring biodiversity at the country level. (From *Guidelines for Country Studies on Biological Diversity* (1993), United Nations Environment Programme, Nairobi.)

Monitoring genetic diversity

In situ plant and animal genetic resources, including land races and non-improved crop varieties, medicinal plants and wild ancestors of domestic breeds and cultivars

Numbers of varieties of crop (or livestock) grown *in situ* in sample localities

Coefficient of genetic diversity or kinship of crop (or livestock) grown *in situ* in the same sample localities

Number of accessions of crops and livestock held in *ex situ* storage within the country

Percentage of these accessions regenerated over a specific time period

Numbers of microorganisms, microbial strains and other cultures held in *ex situ* storage in the country

Full data on the propagation and release of any genetically modified organisms into the wild

Release, introduction and re-introduction into the wild of any captive-bred and cultivated plant or animal from *ex situ* collections

Species monitoring

Species threatened at the national level

Trends in numbers of nationally threatened species measured as changes in the proportion of species listed in national threat categories

Country-specific endemic species

Species of commercial value for sustainable use at both the national and community levels

Flagship species that serve as a focus for the conservation of a diversity of other species and habitats

Indicator species that can serve to monitor ecosystem disturbance, particularly predators and invasive colonizing species

Alien or introduced species that threaten indigenous biological diversity

Percentage of species with declining, stable or increasing populations, particularly amongst the categories of species listed above

Time-interval data on changes in population of a few key species of particular ecological or economic significance

Percentage of threatened and country-endemic species held in national and international *ex situ* collections

Changes in species composition and yields of important food resources for human consumption such as fisheries and game meat production for local use and export

Agents identified as direct threats to species diversity

Implementation of priority actions for species conservation and use identified in the latest strategic plan, assessed against specified success indicators

Habitat monitoring

Mapped distributions of natural habitats of conservation concern, using natural classification system

Rates of change of habitat areas

State or condition of habitats of conservation concern using some simple measure of biodiversity quality

Rates of change of habitat condition

Percentage of remaining natural habitats occurring in patches greater than $10\,km^2$, $100\,km^2$ and $1000\,km^2$ as a measure of habitat fragmentation

Rates of change of ecologically sensitive areas

Percentage of remaining habitats within protected areas

Percentage of identified centres of species diversity and endemism within protected areas

Changes in the area and identified benefits of the mapped distributions of categories of biodiversity function

Protected areas monitoring

Agents identified as direct threats to habitat diversity

Areas of habitat restored or reinstated to a more natural condition

Implementation of priority actions for habitat conservation and use identified in the latest strategic plan, assessed against specified success indicators

Number, area and location of protected areas

Percentage of terrestrial and marine ecosystems under conservation management

Protected areas that have management plans

Continued on p. 254

Table 8.4 (*Continued*).

Protected areas monitoring (*continued*)
Protected areas inventories of species and habitats
Effectiveness of management of protected areas to achieve specified objectives
Agents identified as threats to protected areas
Protected areas infrastructure such as staff numbers and budgets
Sustainable use benefits to local communities generated by protected areas
Economic values of protected areas including visitor numbers and income generation

Table 8.5 Requirements to collect data under various EC Directives and international conventions. (From HMSO 1996. Crown copyright is reproduced with the permission of the Controller of Her Majesty's Stationery Office.)

Legislation requiring biodiversity information	EC Birds Directive	EC Habitats Directive	Bonn Convention	Bern Convention	Biodiversity Convention
Information required on	Wild birds	Habitats and species	Migratory species	Threatened habitats	All components of biodiversity
Requirement to monitor	Bird population levels	Habitats and species (extent and status)	Migratory species	Threatened habitats	Key components of biodiversity
Information required on designated sites for biodiversity conservation	Special Protection Areas	Special Areas for Conservation			
Data required on sustainable use of biological diversity	Yes	Yes	Yes	Yes	Yes
Data required on threats to biological diversity	Yes	Yes	Yes	Yes	Yes

Table 8.5 summarizes the main legislative requirements for monitoring of biological diversity in the UK. This monitoring is required to:
• establish baselines;
• provide regular and systematically obtained data that enable detection of change in status;
• establish the reasons for change, both desirable and undesirable.
Clearly, it is advisable to tackle these requirements in an integrated manner, as many countries, including the UK, are attempting to do.

Table 8.6 Current status of monitoring of species on the UK's biodiversity 'long list'. (From HMSO 1996. Crown copyright is reproduced with the permission of the Controller of Her Majesty's Stationery Office.)

Group	Species on list	Species currently with 'biodiversity action plans'	Species with 'species action plans' now or soon to begin	Species lacking adequate status assessment	Species with some form of assessment programme in place
Algae	18	0	0	13	5
Fungi	21	4	6	Unknown	Unknown
Lichens	81	7	6	26	44
Liverworts	32	5	0	1	30
Mosses	79	6	6	20	52
Stoneworts	13	1	0	4	9
Vascular plants	230	28	35	9	185
Ants	5	3	0	1	4
Bees	19	1	1	0	18
Beetles	72	12	3	19	50
Butterflies	25	6	25	0	0
Caddis flies	2	0	0	2	0
Crickets/grasshoppers	7	1	3	2	2
Dragonflies	7	1	0	2	5
Two-winged flies	54	3	0	32	22
Mayfly	1	0	0	1	0
Millipedes	7	0	0	2	5
Molluscs	45	11	2	28	15
Moths	122	3	8	11	103
Spiders	44	0	2	0	42
Stonefly	1	0	0	1	0
True bugs	5	0	1	4	0
Wasps	7	0	1	1	5
Other invertebrates	40	4	4	23	13
Amphibians	7	2	2	0	5
Birds	200	9	43	15	140
Fish	25	4	6	1	16
Mammals	66	9	7	17	24
Reptiles	10	1	5	0	5
Total	1245	116	166	235	799

However, the collection, organization and dissemination of the information needed to protect key habitats and species is a formidable task. Data published by the UK's Biodiversity Steering Group indicates that current monitoring coverage of species on the 'biodiversity long list' is markedly uneven. Table 8.6 summarizes the number of species in different groups for which adequate information is available to assess their status. For some groups, like the fungi, our knowledge is so limited that we cannot reach any conclusions about species status or even the extent of current monitoring. Overall, of the 1245 species currently included on the UK's biodiversity 'long list', there are 235 that are poorly known and for which there are no current, systematic assessment schemes.

8.2.5 Biodiversity inventories

Our knowledge about biological diversity varies considerably between groups of organisms. It also varies between sites, regions and countries, some of which have been studied in much more depth than others. Two types of biodiversity inventory have emerged to tackle these two aspects of biodiversity knowledge provision. They are the ABTI (see section 8.2.2) and the 'all-taxa biodiversity inventory' (ATBI).

The ABTI is intended to focus on specific taxa and to inventory all species in those groups on a global basis. Priority is to be given to certain taxa in the first instance, to ensure that more recording effort is invested in currently under-recorded groups. ATBI, on the other hand is intended to carry out systematic inventory of all taxa within a certain area. As Stork (1994) pointed out, no single site has ever been fully inventoried for all taxa. This has made it even more difficult to tease out the roles of different organisms in ecosystem processes. Janzen and Hallwachs (1994), cited in Stork and Davies (1996) recommended that an ATBI should take the form of a 5-year, species-level, total inventory of a site. Sites chosen for ATBI should be sufficiently large and biologically complex to incorporate a large proportion of a nation's or region's species, for example sites of 50 000–100 000 ha containing hundreds of thousands of species. Sites for ATBI should also include a variety of habitats, preferably with altitudinal gradient. Another important requirement is for formal protection status, accessibility and for an organizational structure that will permit the straightforward implementation of a major biological recording scheme. Clearly ATBIs are expensive and time-consuming. This, together with the fact that there is a world-wide shortage in expertise for many groups of organisms, means that full ATBIs are only likely to be possible at a limited number of sites (Stork & Davies 1996).

Full biodiversity inventories take a long time to generate results that can be used to support decision-making. 'Rapid biodiversity assessments' (RBAs) have emerged to provide a more responsive and easily implemented approach to biodiversity inventory. They generally focus on certain taxa and provide an estimate (rather than a total inventory) of biological richness or the uniqueness of a particular area. They can never substitute entirely for full, systematic and scientific inventory backed up by the collection of voucher specimens, but they are considerably cheaper and generate results much more quickly. The groups most commonly surveyed are birds, large mammals, reptiles, amphibians and trees and, less commonly, invertebrates such as butterflies and dragonflies. Often, 'visual encounter survey' (VES) is used to record presence/absence, followed by more detailed sampling as appropriate, using transects, quadrats or random walks.

From a purely pragmatic point of view, RBAs are clearly beneficial. However, their reliability depends crucially on the availability of skilled field recorders. RBAs also suffer from some fundamental drawbacks, including:

- recorder bias;
- failure to survey all microhabitats and biotopes equally;
- inability to estimate animal densities (unless carried out in conjunction with mark–recapture);
- focus on better known and more 'visible' or 'apparent' groups;
- failure to provide voucher specimens for longer-term validation.

For EcIA, all these types of inventory have a potential role and should be regarded as complementary: ABTI and ATBI in providing longer-term, baseline information on the status of species, communities, habitats and ecosystems, RBAs in providing up-to-date, locally relevant information for use in specific situations. Whatever the method used for biodiversity inventory, its value will depend on how reliable and how exhaustive it is. In other words, on whether the elements of biodiversity (usually species) have been correctly identified, named and distinguished from one another and the extent to which the inventory represents a complete census of all species present. As with any ecological survey, the methods used for biological inventory should be able to account for temporal changes in species presence (Stork & Davies 1996).

8.2.6 The UK's Countryside Survey 1990

The emergence of united, global concern about environmental change and biodiversity loss has been relatively recent, Likewise, in many individual countries, the negative impacts of land use on biodiversity have often gone unnoticed until they have been made obvious through dramatic species decline. Often, the extent of species decline is impossible to document, except from snippets of historical and anecdotal information. Even in countries such as the UK, which have a well-developed tradition of record-keeping, it can be difficult to find relevant information about wildlife status or decline. More recently, however, a number of large-scale survey schemes have been implemented to provide necessary information about the status of wildlife and wildlife habitat in the British countryside. One of the most comprehensive is the 'Countryside Survey' coordinated by the UK Natural Environment Research Council's Institute of Terrestrial Ecology (ITE). ITE first developed a 'land classification system'. This was designed to provide a framework for ecological sampling at the national level (Bunce et al. 1996). Its constituent 'land classes' describe the character of 32 'land types' in the UK. These could be used to provide a sampling frame to carry out more detailed field surveys, the results of which may be amalgamated to generate national estimates of environmental 'stock and change' (Potter *et al.* 1996). In 1990 an extensive 'Countryside Survey' (CS 1990) was carried out. This rested on field survey of a stratified random sample of 508 × 1 km squares selected to represent each of the 32 land classes in the 'land classification system'. For each square, data were collected on land cover, landscape features, wildlife habitats and vegetation. Estimates of change were calculated by comparing results from CS 1990 with results from an earlier survey in 1984 when 381 of the same squares had also been visited and a pilot survey in 1978 when 256 × 1 km squares had been visited. Analysis of changes between these dates produced the most comprehensive assessment of recent land cover and ecological change in the British countryside to date (Barr *et al.* 1993). For example, CS 1990 estimated a cumulative loss of some 28 000 km of hedgerows (valuable wildlife habitat in farmed landscapes) in England and Wales (Potter *et al.* 1996), with a slight acceleration in the rate of loss from 1984 to 1990 (Barr *et al.* 1994). CS 1990 quantified a general impression held by nature conservationists that biodiversity had been declining on agricultural land in the UK over the period of the survey. CS 1990 also demonstrated the importance of linear features such as hedges, streams and roadside verges in preserving botanical diversity in the lowland landscapes of the UK (Potter *et al.* 1996) and provided estimates of rates of loss for the UK's main seminatural habitats. Such information has obvious benefits in EcIA, where it is often necessary to make decisions about the relative value of different landscape features in supporting wildlife, or to evaluate the regional or national implications of local habitat losses. A repeat survey is planned for the year 2000. It is to be hoped that it will be possible to continue the survey well into the future, providing much-needed baseline information about the causes of species decline in the countryside and the relative values of different countryside 'features' in preserving biological diversity.

8.3 EcIA audit and review

The need for expert review is indicated by a number of studies that have highlighted shortcomings in the EcIA process and serious deficiencies in the quality of ecological reports submitted in support of development applications (Spellerberg & Minshull 1992; Treweek 1996; Thompson *et al.* 1997). In some countries, such as the Netherlands, ecological studies carried out under EIA legislation are formally audited or reviewed by officially appointed ecological experts. In countries where there is no formal review process, it can be difficult to judge the effectiveness of EcIA. Review of environmental statements (ESs) submitted with applications for development consent may be the only practical way. If nothing else, the opportunity to comment on ESs is valuable in ensuring that ecological issues are given full consideration in the decision-making process. However, it is always important to distinguish between review of ESs as legal documents and review of proposals to determine their acceptability. It is difficult to review ESs effectively without a consistent approach. For this reason, 'review packages' have been developed to assist decision-makers and official consultees in the planning process.

One of the best known is that developed by Lee and Colley (1990) and adopted in abridged form by the UK's Institute of Environmental Assessment. ESs are graded according to the completeness with which particular issues are addressed and an overall grade is then given for the ES as follows:

A excellent, no tasks left incomplete;

B good, only minor omissions and inadequacies;

C satisfactory, despite omissions and inadequacies;

D parts well attempted, but must as a whole be considered unsatisfactory because of omissions and/or inadequacies;

E poor, significant omissions or inadequacies;

F very poor, most tasks left incomplete.

This package provides a consistent framework for review, but reaching a final grade requires subjective assessment and the package does not

---

Box 8.2 Checklist of factors to consider in reviewing ESs with respect to identification of ecological impacts and the survey methods used.

Impacts
- What are the main ecological impacts?
- Are there any obvious omissions?

Have impacts been quantified?

Would it be possible to quantify them using the data available or provided?

Have probabilities (risks) and timescales been specified?

Can suitable and effective mitigation measures be identified for these impacts?

Survey methods

Have the results of previous surveys been referenced?

Have suitable literary sources been used?

For any new surveys carried out:
- Were surveys carried out at an appropriate time of year?
- Were surveys of sufficient duration and/or intensity to produce meaningful results?
- Were suitable techniques used?
- Was the scale of the study appropriate to the range or distribution of impacts?
- Are the results quantified?
- Are results presented in full and have appropriate analytical techniques been used to interpret them?

address the quality of ESs with respect to coverage of ecological issues *per se*. It may be sufficient simply to produce checklists of issues that should be considered in assessing the adequacy of ESs with respect to ecological concerns. These issues might relate to:
- compliance of the ES with the relevant legislative requirements with respect to nature conservation; and
- the appropriateness and adequacy of coverage of ecological issues given existing knowledge and techniques.

Box 8.2 includes some of the factors that might be included in such a review checklist with respect to identification of impacts and the suitability of ecological survey methods. More site-specific criteria might be required to reflect current knowledge and the particular conditions of any particular case.

Clearly, the criteria used to review ESs will vary considerably between countries, according to legislative requirements and the special characteristics of ecosystems. The important thing is to ensure that some sort of check is maintained on the quality of EcIA and on the adequacy of consideration of ecological issues in the decision-making process. Without any formal or consistent form of review, shortcomings in EcIA are likely to be perpetuated and important 'natural resources' will continue to be destroyed at an unacceptable rate.

8.4 Recommended reading

Principles of ecological monitoring
Spellerberg, I.F. (1991) *Monitoring Ecological Change*. Cambridge University Press, Cambridge.

Biodiversity inventories
Stork, N. & Davies, J. (1996) Biodiversity inventories. In: *Biodiversity Assessment: a Guide to Good Practice*. Field Manual 1. *Data and Specimen Collection of Plants, Fungi and Microorganisms*. HMSO, London.

Survey and monitoring techniques
Bibby, C.J., Burgess, N.D. & Hill, D.A. (1992) *Bird Census Techniques*. BTO and RSPB, Academic Press, London.

9 Geographical information systems for ecological impact assessment

9.1 Introduction

Geographical information systems (GIS) are 'integrated systems of computer hardware and software for entering, storing, retrieving, transforming, measuring, combining, subsetting and displaying spatial data that have been digitised and registered to a common coordinate system' (Johnston 1998). More simply, a GIS is a tool for managing information of any kind according to where it is located. GIS can be hosted on a range of computing systems, from desktop pcs to workstations and mainframe computers. As GIS have become more accessible and adaptable (and to some extent, cheaper), so their potential applications to ecological impact assessment (EcIA) have mushroomed.

Three of the fundamental questions that lie behind effective EcIA lend themselves perfectly to GIS-based analysis.

1 What is where?
2 Why is it there?
3 What happens if …?

As discussed earlier in the book, it is not possible to describe, explain or predict ecosystem 'behaviour' without knowing how ecosystem components are distributed in time and space or with respect to each other ('what is where?') and understanding the relationships and processes that explain their distribution and behaviour ('why is it there?'). As well as requiring knowledge of spatial distributions and relationships, the ability to make reliable predictions ('what happens if?') often demands knowledge about temporal trends ('what has happened in the past?').

Tackling these questions demands the very same spatial, analytical and predictive procedures that GIS are designed to facilitate, so GIS have a particular relevance to EcIA.

9.1.1 The GIS database

In simple terms, GIS store data according to two main attributes.

1 *Where?* (The 'locational attribute'.)
2 *What?* (The 'thematic attribute'.)

Locational attributes are defined using coordinates, which are stored
digitally. GIS are therefore structured around a straightforward *X*, *Y* coordinate system to which any number of thematic 'layers' of information
can be referenced.

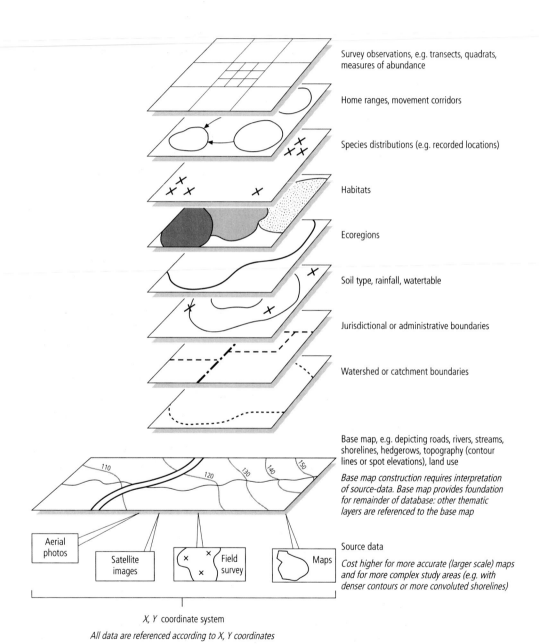

Survey observations, e.g. transects, quadrats, measures of abundance

Home ranges, movement corridors

Species distributions (e.g. recorded locations)

Habitats

Ecoregions

Soil type, rainfall, watertable

Jurisdictional or administrative boundaries

Watershed or catchment boundaries

Base map, e.g. depicting roads, rivers, streams, shorelines, hedgerows, topography (contour lines or spot elevations), land use

Base map construction requires interpretation of source-data. Base map provides foundation for remainder of database: other thematic layers are referenced to the base map

Aerial photos

Satellite images

Field survey

Maps

Source data

Cost higher for more accurate (larger scale) maps and for more complex study areas (e.g. with denser contours or more convoluted shorelines)

X, Y coordinate system

All data are referenced according to X, Y coordinates

Fig. 9.1 A hypothetical geographical information systems (GIS) database structure.
(After Korte 1994.)

Figure 9.1 shows how different layers of thematic information relevant to EcIA might be built up, all referenced to a common X, Y coordinate system.

Map features are defined using 'topologic data elements': points, lines and areas or polygons. Points are located by pairs of X, Y coordinate values. Lines are described by series of coodinate pairs (two for straight lines and more for curves). Areas or polygons are bounded by lines, but also have a uniquely numbered centroid (situated anywhere within the area) defined by an X, Y coordinate pair. Attribute data are associated with topologic data elements and provide descriptive information about them. They are stored separately from the graphics data that are used to structure the system. GIS can therefore be used to store, manage and retrieve data describing what ecological entities or components are present in a study area and where exactly they are located in space and in relation to each other. This databasing function: the ability to organize information according to a common reference system, is of great value in EcIA, which often demands the integration of many different types of data.

Creating the GIS database is likely to account for more than half of the total cost of a GIS project, and the base map is usually the most expensive of all map 'topics' included in the database. Data for other thematic layers are positioned by reference to the base map. In other words, the scale and accuracy of the base map are fundamental to the positional accuracy of the remainder of the system (Korte 1994). One of the most important developments to have accompanied the development of GIS is the greatly increased accessibility of remotely sensed data. Satellite imagery, radar and even conventional aerial photographs are now readily accessible as sources of data for EcIA and are routinely used by some practitioners to generate base maps of study areas. There are standards for horizontal and vertical accuracy that specify certain limits for deviation of GIS map points from true positions 'on the ground', such as the National Map Accuracy Standards (NMAS) used in the US.

9.1.2 Sources of data

Sources of data for GIS are incredibly varied, including existing paper or digital maps, results of field survey, aerial photographs, radar and satellite images. In fact, one of the main advantages of using GIS is the ability to tap into a variety of data sources and interpret them on a 'common platform'.

Aerial photography has been in use for a long time, but the first environmental remote sensing satellites only came into use in the 1970s. Technological developments in this area are incredibly rapid and the spatial, spectral and temporal resolutions of satellite imagery are increasing all the time. Environmental mapping and monitoring from space is now a practical reality and in many ways it has changed the face of EcIA. Interpreted satellite imagery has proved a valuable source of information on topography, land use, vegetation cover and habitat distribution. It has also enabled us to quantify global rates of habitat destruction: otherwise an incredibly difficult task (Veitch *et al.* 1995). Relatively recent provision of remote sensing data from space is outlined by Plummer *et al.* in Danson and Plummer (1995) and CEOS (1992) provides a comprehensive survey of the remote sensing systems planned between 1992 and 2002. Key landmarks in the provision of remotely sensed data are summarized in Fig. 9.2. Well-known sources of solar reflective data used by ecologists include the American Landsat series of satellites (the Multispectral Scanner (MS) and Thematic Mapper (TM)) and the French SPOT (Satellite Pour l'Observation de la Terre) High Resolution Visible (HRV) sensor. The most advanced operational satellite system from a spectral point of view at present is Landsat-TM, which has seven bands spread between the visible and thermal infrared. However, the launch of Landsat-6 failed and Landsat-5 is already 7 years beyond its design lifetime, so it cannot be relied upon as an ongoing

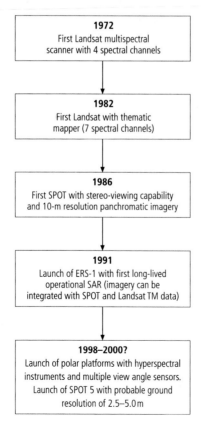

Fig. 9.2 Landmarks in remote sensing. ERS-1, European Remote Sensing satellite; SAR, Synthetic Aperture Radar. (From Wilkinson 1996.)

source of ecological information (Wilkinson 1996). The SPOT-5 satellite, to be launched towards the end of this decade, is planned to carry a sensor with a ground resolution in the order of 2.5–5 m (Novajosky 1993). If successful, this will greatly enhance the resolution of data available to ecologists.

Decisions about what sources of data to use are fundamental to the development of GIS for EcIA. In particular, it is often necessary to decide what relative emphasis to place on field survey or remotely sensed data. The main advantage of using remotely sensed data is the ability to acquire spatially referenced information, which can be updated quickly, straight-forwardly and often relatively cheaply. There are particular advantages for studies at the regional level, in areas where access for field survey is limited or difficult (remote 'wilderness' areas, for example) or where future monitoring is intended. Ready access to ecological information without the need to implement site or field surveys has removed one of the main barriers preventing early consideration of ecological constraints in the EIA process (Treweek 1995). Access to remotely sensed data has also made it considerably easier to develop 'regional ecosystem approaches' to natural resource management, which take account of wider geographical and temporal trends in distribution and abundance. Continuing problems in interpretation and classification of satellite imagery are reviewed by Wilkinson (1996). For ecologists in particular, there may be problems because the spatial and temporal scales of available remotely sensed data differ from those of interest in any particular case (Mack *et al.* 1997), whereas field surveys can be tailored to provide the information required (except where there is a historical dimension of course). Aerial photography remains an important source of information for ecological studies, particularly where historical information about land cover or land use is necessary. Some specialized data suppliers can provide interpreted aerial photography for incorporation into GIS that will not necessarily be more expensive than carrying out field surveys from scratch.

Remotely sensed data will only be appropriate if they do actually permit detection of important habitat characteristics, which depends on the spatial and spectral resolution of sensors and on the spatial scale of 'behaviour' of the system or species under study. One problem is that ecological studies often require high-resolution data, which are invariably more expensive to obtain than low-resolution data.

Few cost–benefit analyses have been carried out to establish ground rules for selecting data for particular ecological applications. However, a paper by Mack *et al.* (1997) describes an approach that could be used to compare datasets with different resolutions according to their suitability for a particular ecological purpose. Their study compared results of higher-resolution field survey with land-cover data taken from the

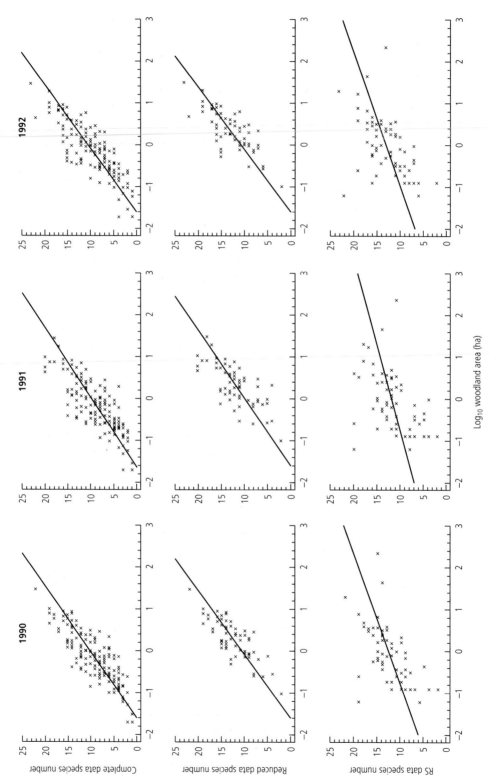

Fig. 9.3 Species–area relationships for birds in UK woodlands in 3 years. RS, remotely sensed. (From Mack *et al.* 1997.)

Institute of Terrestrial Ecology (ITE) Land Cover Map of Great Britain (Fuller *et al.* 1994), which had a resolution of 25×25 m. The study was based on 151 woodlands ranging in size from 0.02 to 30 ha. Use of these woodlands by birds was studied over a 3-year period (Hinsley *et al.* 1995) and used to derive species–area relationships. Figure 9.3 shows regression curves for species number against $\log_{10}$ woodland area (in hectares) where woodland area estimates were derived from field survey, remotely sensed data, or a combination of the two (a 'reduced' dataset based on ground-based area estimates but only for those woodlands detectable using remote sensing). The goodness-of-fit of the species–area models derived from the remotely sensed data was less than that obtained using ground-based data (not surprising as in this case the analysis was close to the limits of the remotely sensed data). Even so, the remotely sensed data permitted coarse estimates of species–area relationships.

This study demonstrated that, where there are existing ecological models available with an appropriate scale, the suitability of two datasets with different resolutions can be assessed by fitting the model to both datasets, along with a control in which the higher-resolution dataset is applied only to those landscape units which are detectable using both datasets. In this way it is feasible to evaluate the benefits of increased resolution both in terms of improved detection of habitat units and the goodness-of-fit of models that relate species distribution to habitat units. It is possible to envisage the use of such an approach in EcIA in cases where species–habitat relationships are already relatively well understood and post-implementation monitoring is envisaged. The approach also has advantages when one area has been studied quite intensively and there is a need to extrapolate to areas that are less well known. The benefits of intensive field survey in one area can then be realized in others through the use of remotely sensed data, which is relatively cheaper per unit area. What this example emphasizes most of all, however, is that the benefits of using remotely sensed data can usually only be realized following a considerable investment in field survey and detailed ecological investigation, even if this does not always appear as a direct cost in any particular EcIA.

Most importantly, using GIS, there is no reason why field data and remotely sensed data cannot be integrated, thereby combining the benefits of each.

9.1.3 Exploring relationships

Once data are entered, GIS often enhance data-handling capability and speed up access to information for analytical purposes. For this reason, they can play a useful part in conceptualization and in EcIA scoping. Visual presentation of data 'layers' can facilitate recognition and understanding of potential relationships between different ecological attributes thereby ensuring that all relevant avenues of enquiry are explored and that suitable data are collected in field study programmes. Using a GIS platform for EcIA can also help to ensure that potential impacts are studied at an appropriate scale. In EcIA, GIS have proved particularly effective for studying the relationships between proposed road or pipeline routes, for example, and areas of high nature conservation importance.

As well as performing straightforward databasing functions, GIS can also be constructed to explore functional relationships by querying data in different ways. For example, in addition to storing data describing plant species distributions in a watershed (or catchment), a GIS might also be used to store information on the depth to watertable recorded at the same (and different) locations within the watershed. Watertable depth may be a key factor explaining the observed distribution of plant species. GIS can be used to combine relevant thematic data layers and explore the possible relationships between them, often using overlaying functions ('how do these layers of information relate to each other?'). Statistical testing of observed relationships or correlations between thematic layers is a function that may be performed internally or externally and depends crucially on the quality of data (often field data) collected. Once ecological relationship have been confirmed and responses can be quantified (e.g. the responses of plants to watertable change), GIS can be used to tackle the 'what if?'-type questions, which are the crux of the EcIA process. For example: 'what happens to currently observed plant species distributions if watertables are raised or lowered?' Theoretically, GIS can be used to generate any number of potential 'impact scenarios', the relative significance of which can then be tested using relevant data to tackle the 'so what' stage of the EcIA process. In practice, the use of GIS for EcIA prediction has been hampered by the lack of undisputed rules of ecosystem behaviour. GIS are only as good as the data and conceptual models on which they are based.

9.2 GIS applications in EcIA

GIS can be used as a tool in scoping or conceptualizing the EcIA, deriving suitable study limits and generating appropriate impact scenarios and mitigation strategies as well as simply handling the relevant data and making them accessible. GIS can therefore play an important part in managing the EcIA process. If used as a platform for EcIA from the outset, GIS can help to ensure that:
- study areas capture all ecological parameters falling within the potential 'impact zone';
- study areas incorporate the full range of ecological parameters affected;

- sources of digital data can be readily accessed (e.g. remotely sensed data);
- data collection programmes are structured to 'capture' variation at a suitable scale (in other words intensity and range of sampling reflects apparent scale of variation);
- data from different sources can be used on a common interpretive 'platform';
- alternative impact scenarios can be generated and addressed;
- alternative mitigation strategies can be tested;
- surveys are repeatable (follow-up monitoring is a realistic option).

However, GIS can also play a part in tackling the more detailed ecological questions or problems that must be addressed before ecologically sound decisions can be made. Examples of the kind of ecological question that might be tackled using GIS are listed in Box 9.1.

Despite its obvious potential, the use of GIS to tackle ecological problems has a relatively recent history. The International Association for Landscape Ecology (IALE), for example, first formed a GIS working group in 1988 (Johnston 1998). Field ecologists in general have been relatively slow to adopt GIS, but this has been partly due to the difficulties of collecting spatially explicit field data. Availability of cheap and effective global positioning systems (GPS) has helped to remove this barrier and there are now some very good examples of GIS that have combined spatially referenced field data with data derived from other sources, including satellite imagery. A review of GIS use for EIA by environmental consultants (Joao & Fonseca 1996) confirmed that GIS

Box 9.1 Ecological questions amenable to GIS analysis.

- Where is plant species or community A?
- Is the distribution of plant species A linked or correlated with any particular abiotic variable (e.g. altitude or rainfall)?
- Where is plant community A in relation to plant community B?
- How has the distribution of plant community A changed over time?
- How is the distribution of plant community A likely to change in future if environmental conditions remain as they are at present (what is its baseline status)?
- How is the distribution of plant community A likely to alter if environmental factor X is altered (what will be the impact of altering environmental factor X)?
- What vegetation types are included in the 'home range' of animal J?
- Are there any other factors that appear to influence the shape or extent of J's home range (e.g. presence of roads, levels of human disturbance)?
- Has the amount of habitat available to animal J changed over time (baseline habitat availability) and will a predicted reduction cause carrying capacity to be exceeded?

are used more for presentation of data and results than they are for data analysis. Even the more simple analytical tools provided by GIS have tended to be neglected by practitioners, for example map overlaying to derive simple measurements of area of habitat, to measure habitat loss or to carry out constraint mapping. However, for large-scale impact assessments that involve massive data-collection programmes and have sufficiently long 'lead times', GIS are becoming too useful to ignore, even if they are only used to generate base maps and structure field surveys. In some countries, notably Canada and the US, ecological impact assessments have been managed around GIS databases for years.

The following sections refer to areas of EcIA in which use of GIS has had demonstrable benefits. These include assessment of cumulative ecological effects, regional ecological studies, conflict resolution in land-use planning, strategic ecological assessment, ecological mitigation planning and assessment of landscape-scale effects.

9.2.1 GIS for cumulative effects assessment

Cumulative habitat loss is a problem for many species in many countries. It is well known that the geographical location and spatial organization of habitat has a bearing on its quality and accessibility to associated species. GIS technology has made it considerably easier to quantify those spatial attributes of habitat distribution and organization which affect the value of an area to associated species and therefore to recognize when unacceptable thresholds of habitat loss and fragmentation may have been reached (Wadsworth & Treweek 1999). These might include:

- habitat dimensions;
- habitat isolation;
- edge/interior ratio;
- size of 'core' areas;
- habitat fragmentation;
- average patch size in fragmented habitats;
- position of habitat relative to same or other habitat;
- distance between habitat patches;
- overall amount of one habitat in a study area relative to other types.

One of the reasons why GIS have a particular role in assessment of cumulative ecological effects is because they facilitate mapping and modelling of ecological impacts expressed over large geographical scales using remotely sensed data. For example, one recent study used satellite imagery to quantify rates and patterns of change in land cover between two dates in a study area covering 967 847 ha in two different countries, China and North Korea (Zheng *et al.* 1997). This study also demonstrated the usefulness of GIS for modelling temporal changes or trends and therefore their role in establishing more reliable baselines for tracking cumulative change.

Studies by Sebastini *et al.* (1989) and Johnson *et al.* (1988) also demonstrate how GIS can facilitate studies of cumulative ecological changes expressed over large areas or long time-periods. Sebastini *et al.* (1989) used a GIS-based approach to trace the cumulative loss of coastal wetlands as a result of development in a large study region. Johnson *et al.* (1988) used GIS-based analysis to assess the effects of wetland loss on water quality and flood attenuation using land-use maps spanning many years to quantify the magnitude, location and rate of wetland loss as a result of development and to project possible future rates of loss. The relationship between water quality and the extent and position of wetlands was then modelled using empirical data from 37 watersheds. Studies of this kind benefit from the ability to address spatial and temporal trends in habitat distribution and also the ability to study correlations between habitat presence and other features of interest, such as water quality. Without GIS, such studies would be incredibly laborious, time-consuming and costly (Wadsworth & Treweek 1999).

The use of GIS for assessment of cumulative ecological effects demands:
- selection of a mapping scale that provides sufficiently detailed information about sources and receptors of impacts;
- availability of data on potential sources of cumulative impact — their spatial and temporal range and magnitude;
- local, ecological landscape mapping undertaken to fit within broader, regional mapping frameworks (e.g. to facilitate understanding of a project area's contribution to regional habitat supply and allow accurate assessment of regional project impacts);
- use of a landscape classification and mapping system that captures landscape features and ecosystem processes at both local and regional levels.

9.2.2 GIS for mapping ecosystems

GIS have played a key role in ecosystem classification and mapping, particularly for regional and national applications. Ecosystem classifications are used to derive homogeneous map units with predictable characteristics. Ecosystem mapping stratifies the 'land base' into these map units based on information about environmental factors such as climate, geology, soils and vegetation. Map units are displayed as polygons, and data (attributes) associated with each polygon can be stored in the map database. Interpretive products such as habitat suitability maps use polygon attributes for assigning landscape values. Digital maps generated from such GIS databases provide a spatial expression of ecosystem classifications that can be used to depict habitat, wildlife and other ecological resources in a standardized and directly comparable fashion. For example, a fundamental part of habitat evaluation procedure (HEP) (US Fish and Wildlife Service 1980) is the accurate delineation of vegetation cover types. Ideally these are generated on an accurate base map, as measurements of vegetation cover type and structure form the basis for quantification of habitat availability. Satellite imagery, aerial photography or radar images may be suitable to derive distribution maps, together with colour infra-red photography where separation of vegetation structure is required, for example to distinguish species' feeding and reproductive guilds and select suitable evaluation species.

Accurate vegetation mapping is useful for:
- selecting evaluation species for further study;
- extrapolating from sampled to unsampled areas;
- comparing baseline habitat provision with future habitat availability.

Mapping scale is very important, as it influences the types of decisions that can be made. For example, Table 9.1 summarizes the hierarchy of ecological units used in British Columbia's system for ecosystem mapping and indicates the main planning or decision-making applications of units mapped at different scales. For local decision-making, more detailed information is generally required than for broad, regional planning. The US Fish and Wildlife Service (1980) suggests that 'maps generated from remotely sensed data at a scale of 1:20000–1:60000 will usually permit acceptable resolution for terrestrial habitat evaluations'.

Ecosystem mapping can be greatly enhanced at the regional level by drawing on a number of data sources to stratify the landscape. Stohlgren et al. (1997) used remotely sensed data, aerial photography and field data to differentiate ecological units in the Rocky Mountain National Park in Colorado. They demonstrated that the identification of gaps in the regional protection of species of conservation concern ('gap analysis') could be improved by reducing the size of 'minimum mapping units' (MMUs). 'Methods to identify gaps in the protection of biodiversity are typically applied to large areas with coarse MMUs' in the region of 100 ha, or sometimes 2 ha. This is clearly ineffective for species dependent on habitats with much smaller scales of variation, and there is a risk that areas of high or unique diversity might be missed due to the poor resolution of mapping data. These might include, for example, narrow riparian zones that are often concealed by over-storey trees and cover relatively small areas. Stohlgren et al. (1997) therefore used MMUs of 0.02–0.09 ha for more detailed ecosystem mapping. They were able to do this by using Landsat Thematic Mapper data (30-m pixels based primarily on bands 3, 4 and 7) and colour aerial photography to classify land cover and land-use types.

Table 9.1 Ecological units used in ecosystem mapping in British Columbia. (After Ecosystem Working Group 1998. Copyright of the Province of British Columbia.)

Ecological unit	Common level of mapping	Common scale of presentation	Purpose, objectives and general use
Ecoregion units: Stratification based on broad differences in climate			
Ecodomain	1 : 7 000 000	1 : 30 000 000	Area of broad climatic uniformity, relevant for international and national planning
Ecodivision	1 : 2 000 000	1 : 7 000 000	Area of broad climatic and physiographic uniformity, relevant for international and national planning
Ecoprovince	1 : 2 000 000	1 : 7 000 000	Area with consistent climate, used in provincial and regional planning
Ecoregion	1 : 100 000	1 : 2 000 000	Areas with major physiographic and minor climatic variation
Ecosection	1 : 100 000	1 : 2 000 000	Areas with minor physiographic and macroclimatic variation
Biogeoclimatic units: Represent classes of ecosystems under the same regional climate: describe variation in vegetation and site conditions occurring within ecosections. The basic unit is the subzone, which may be grouped into zones or divided into variants or phases based on differences in regional climate. Biogeoclimatic subzones and variants are the units mostly used in ecosystem mapping			
Zone	1 : 250 000	1 : 2 000 000	Group of subzones: large geographical area with broadly homogeneous macroclimate, used in provincial and regional planning
Subzone	1 : 100 000	1 : 250 000	Climax or near-climax plant associations on zonal or typical sites, related to regional climate. Used in regional, operational and local planning
Variant	1 : 100 000	1 : 250 000	Reflects differences in average regional climate which result in climax plant subassociations on zonal sites
Phase	1 : 50 000	1 : 250 000	Reflects regional climatic variation resulting from local relief, e.g. extensive grasslands on steep south-facing slopes in an otherwise forested subzone
Ecosystem units			
Site series: site modifiers structural stage seral association	1 : 5000	1 : 20 000–1 : 50 000	Regional, operational and local planning

By developing consistent approaches to ecosystem mapping and incorporating ecosystem maps into GIS, integration with other resource inventories is facilitated and regional baselines of ecological information can be established which greatly improve ecological analysis and decision-making.

9.2.3 GIS for mapping habitat potential

Wildlife species that have typically low population density and 'secretive lifestyles' can be very difficult to survey in the field. Direct counts or population censuses are therefore more time-consuming and expensive than for less elusive species. For these species in particular (but also for others), it can be more straightforward to infer possible distributions by combining knowledge of relationships between species and habitat (derived from the literature, for example) with remotely sensed data on the distribution of landscape or ecological units. In other words, to derive 'potential distribution maps' from established knowledge of relationships between species, their life requisites and their associations with defined units of ecosystem classification. For example, to map potential distributions of grizzly bear habitat in a remote, forested region it might be possible to:

- derive forest cover maps from satellite images or aerial photographs (e.g. 1 : 15 000 scale);
- digitize existing surficial geology maps and soils maps;
- obtain digital elevation data for slope/aspect mapping (e.g. 1 : 20 000 scale);
- use GIS overlays to produce map (landscape) units with unique combinations of vegetation cover, slope, aspect, surficial geology and soils;
- use existing ecosystem information to predict plant communities associated with defined landscape units;
- refine GIS landscape units using field data;
- interpret suitability of ecological land units for provision of grizzly bear food and cover (e.g. from the literature or field studies);
- produce a final base map as a basis for wildlife habitat evaluations.

This kind of approach lends itself well to the design of stratified random field sampling, to ensure that limited resources for field survey are invested to best effect. For example, a minimum of sampling plots can be collected per homogeneous landscape unit. In remote areas with few landmarks, accurate geo-referencing of sampling locations can be a problem, but sampling sites can be located by compassing from known map features (e.g. intersection of two seismic lines) or through use of 'global positioning systems' (GPS) (e.g. Usher & Kansas 1994).

Standards for wildlife habitat capability/suitability ratings in British

Columbia (Wildlife Interpretations Subcommittee 1998) summarize critical steps in the development of wildlife habitat capability and suitability maps using GIS (see Fig. 9.4).

The usefulness of ecosystem maps for assessing habitat suitability depends on how well the attribute information stored in the GIS map database relates to key habitat requirements. Consider the amount of detail that will be required for impact assessment or evaluation and ensure that the mapping scale permits capture of relevant habitat features or variables. If only limited information is available about habitat requirements, mapping may be restricted to small scales (e.g. $1:250\,000$–$1:500\,000$), which cannot be used to assess suitability. Whereas grizzly bear denning habitat can be readily mapped at scales greater than $1:20\,000$, for example, it is not distinguishable at a scale of $1:250\,000$. It is therefore necessary to consider the minimum size of map polygons needed to define meaningful habitat units (Wadsworth & Treweek 1999). Table 9.2 compares the usefulness of different mapping scapes for studying habitat use. Smaller scales reveal less detail of mapped ecosystem units than larger-scale maps, but may not be so useful for gaining overviews of habitat suitability, availability or quality for larger, more mobile species.

The suitability of a map polygon for a species can be assessed in

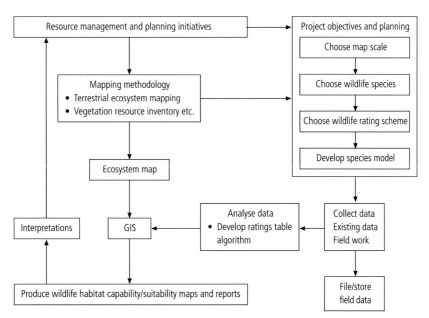

Fig. 9.4 Critical steps in the development of wildlife habitat capability and suitability maps using GIS. (From Wildlife Interpretations Subcommittee 1998.)

Table 9.2 Ecosystem mapping scales appropriate to different levels of information about habitat use. (After Wildlife Interpretations Subcommittee 1998. Copyright of the Province of British Columbia.)

Scale of map	Information on habitat use
1 : 5000 ('very large')	Permits wildlife suitability to be delineated at the micro-habitat level
	Appropriate for most species at risk because they often have limited distributions and very specific habitat requirements (e.g. suitable breeding ponds for amphibians)
	Also useful for mapping habitats important to a large diversity of species, using habitat units interspersed in close proximity
	Mapping at this scale targets small areas with high values
1 : 10 000–1 : 20 000 ('large')	Enables differentiation of ecosystem units and classification of level of use by study species
	Generally used for land capability planning, e.g. for forestry
1 : 50 000–1 : 100 000 ('medium')	Maps at this scale show general habitat distribution and quality. Most useful for large vertebrate species like ungulates or bears, which use several ecosystem units, or have seasonal habitat use
	Medium-scale ecosystem mapping is useful for relatively local or regional planning, where there is sufficient habitat knowledge to warrant detailed analysis at the landscape level
1 : 250 000–1 : 500 000 ('small')	Small map scales depict general ecological boundaries, e.g. belts of seasonal habitat use by migratory species
	Useful at regional or provincial planning levels to highlight areas that require more attention
	Approximate abundance of important habitat types can be assessed and used to generate relative provincial values and distributions for assessment species, especially larger and more mobile species such as the grizzly bear or caribou

various ways, for example a rating can be assigned based on professional experience or knowledge. If more detailed information is available, algorithms may be developed that rate habitats based on explicitly stated relationships between a species' life requisites and ecosystem units. Habitat suitability can also be estimated by linking GIS data (e.g. on soils, vegetation, topography and land use) with species population or ecosystem process models.

9.2.3.1 *Expert knowledge*

Brown *et al.* (1994), for example, combined geophysical mapping with expert knowledge to resolve resource conflicts between commercial forestry and wildlife interests (availability of suitable habitat) in the trans-boundary areas of two National Parks (Mount Revelstoke and Glacier in British Columbia). One of the species of concern was the caribou *Rangifer tarandus caribou*, which has traditionally migrated out of the park in winter to seek better habitats in lower-elevation forests that are managed commercially. In addition to the limits on winter habitat imposed by climate, heli-skiing, snowmobiling and other recreational activities further restrict the overall area of habitat available to caribou.

A geophysical database was constructed, consisting of topography, hydrology, forest inventory and road network information. A 1 : 20 000-scale digital topographic map was used to provide a common reference base and an individual watershed was used as a case study. It was divided into blocks ('polygons') based on topography, forest cover and management. A dominant forest type was assigned to each forest polygon, and growth and yield projections were calculated. Harvesting costs were estimated as well as habitat suitability for caribou. High, medium, low and 'nil'-use suitability categories were identified. The habitat suitability model was developed for woodland caribou based on species–habitat relationships observed by local wildlife experts over the past 10 years. Habitat requirements for caribou include food, cover, reproduction, space and security. The spatial arrangements and seasonal distributions of these components also play a role in determining the overall area and quality of habitat needed to maintain a viable herd of caribou. This model incorporated environmental conditions and food resources that are critical for over-wintering of caribou. Because these animals move according to environmental conditions and food availability, their habitat varies seasonally with forage availability, the degree of snow accumulation, etc. Critical seasonal habitats were identified (early winter, late winter and spring) and key features selected for each. By constructing habitat suitability maps it was then possible to arrive at a combined suitability classification where a high-suitability rating in any one or all critical seasons gave a 'highly suitable' combined rating for areas of caribou habitat (Table 9.3).

Using this habitat suitability model, the amount of habitat in the high, medium and low classes could be estimated. In this particular watershed,

Table 9.3 Criteria used to differentiate caribou habitat in early winter. (After Brown *et al.* 1994.)

Habitat variables	High suitability	Medium suitability	Low suitability	Not suitable
Elevation	0–1199 m	1200–1499 m	1500–1799 m	>1800 m
Slope	<29%	30–39%	40–46%	>70%
Slope position	Lower	Mid	Upper	Crest
Dominant tree species	Spruce, true fir, cedar, hemlock	Spruce, true fir, cedar, hemlock	Douglas fir, lodgepole and whitebark pine	Others
Height class	>3	>2	2	1
Crown closure class	6–10	4–10	4–5	0–3

spring habitat was the most limiting, only 14% of the watershed being classified as moderately to highly suitable in spring. At the end of the day, it was possible to identify those areas of low-moderate value as caribou habitat that would yield positive economic benefits from harvesting and also those areas where harvesting would be uneconomic. In this way, the use of GIS provided an objective summary of differing resource needs and the ability to identify solutions that would optimize overall benefits for wildlife and commercial forestry.

9.2.3.2   *A Bayesian rule-based approach with GIS*

Rather than using expert knowledge to assign habitat ratings, or values, Tucker *et al.* (1997) used a combined GIS and Bayesian rule-based approach to model bird distributions in north-east England. They were particularly interested in predictive distribution models that could be used to anticipate the likely effects of changes in land use on selected bird species: the coal tit *Parus ater*, the golden plover *Pluvialis apricaria* and the snipe *Gallinago gallinago*. Bayes Theorem for conditional probability was used to modify initial estimates of the probability of encountering the species in a landscape, using their known preferences for individual habitat characteristics and information on the distribution of those characteristics in the landscape. Habitat suitability models linked detailed information on habitat preference for nesting with spatially referenced environmental information stored in a GIS database. Figure 9.5 is a flowchart of the Bayesian-GIS modelling system used. Model complexity varied between species. For the coal tit, satellite land cover alone could be used as a habitat variable. For the snipe, satellite cover and altitude were necessary, while the golden plover model required satellite cover, altitude and gradient. The golden plover only breeds at gradients less than 10 degrees, so this could be used as a filter and the three-tiered model could be reduced to two tiers. This approach made it possible to predict how changes in land use might affect future distributions of the study species and also to quantify those distributions. (One cautionary note: when using such approaches for EcIA, specialists should always be consulted to avoid the use of sophisticated and expensive approaches to draw conclusions that would be obvious to an experienced field ecologist.)

9.2.3.3   *Gap analysis*

Gap analysis has emerged as a method for identifying gaps in the protection of species richness. The Gap Analysis Project of the US National Biological Service (Scott *et al.* 1993) seeks to map the spatial distribution of vertebrate species richness in order to identify potential gaps in protection of vertebrate species. Key elements are vegetation maps characterizing the pattern and distribution of vegetation across

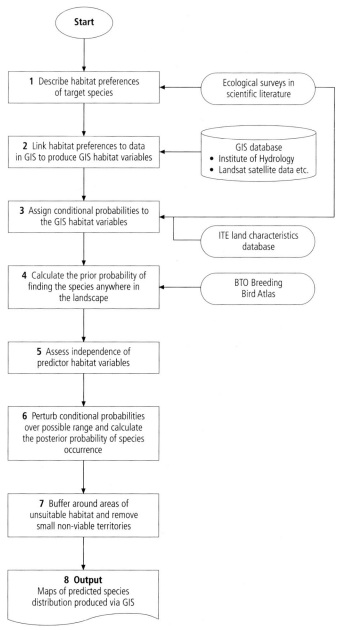

Fig. 9.5 Flowchart of the Bayesian-GIS modelling system used by Tucker *et al.*
(1997) to model bird distributions in north-east England. (Reproduced with kind
permission from Kluwer Academic Publishers.)

the US, and maps of predicted vertebrate species richness. Vertebrate
species richness is predicted from the aggregation of individual species
distributions, which in turn are based on the vegetation maps and
on models relating vegetation to species occurrence (Williams 1996).

Outside the US, Savitsky *et al.* (1995) used GIS for gap analysis in Costa Rica.

9.2.4   GIS for strategic ecological assessment

Strategic ecological assessment has been promoted as a means of resolving conflicts in natural resource use at an early enough stage in land-use planning for adverse ecological impacts to be avoided or reduced 'at source'. This often requires consideration of large amounts of information and a variety of complex and interconnected issues. However, if all data are geo-referenced, they can be integrated on a 'unifying' GIS platform and alternative land-use scenarios can be developed and displayed. This makes it possible to evaluate the likely consequences of different land-use strategies before they are implemented and also to test the effectiveness of alternative approaches to ecological mitigation ('ecological mitigation planning'). Impacts operating at different geographical and temporal scales can also be captured, quantified and compared relatively easily.

Overall, ecological mitigation is unlikely to work unless it is planned on a regional ecosystem basis. Work on wetland mitigation in the US has demonstrated that, if wetland losses are compensated on a case-by-case basis, an imbalance in the regional distribution of wetland types can result. Associated imbalances in the distributions of wetland species are likely, with some of the more 'discerning' species becoming increasingly rare. Ecosystems are not 'closed' and do not operate in isolation. It is therefore inappropriate to manage them as if they are isolated. Wetlands, for example, are greatly influenced by external land-uses that affect hydrology and water quality. Cumulative ecological effects are particularly difficult to compensate for. In the Netherlands, the government has decided to attempt to halt the decline in native plants and animals by restoring their habitat on farmland: in other words to implement a strategic approach to compensation for the adverse effects of past, present and future development on wildlife habitats. Starting in 1994, staff from the Ministry of Agriculture, Nature Conservation and Fisheries drew up a typology of 150 possible 'nature types' to describe all the ecosystems that could possibly occur in the Netherlands.

They categorized these according to levels of management and selected target ecosystems, which included:
• wooded, undulating-to-hilly landscapes on rich, loamy soils;
• wooded, undulating-to-hilly landscapes on poor, sandy soils;
• two types of drift-sand inland dunes;
• managed alluvial wooded landscapes;
• non-managed alluvial wooded landscapes;
• fen marsh landscape;

- reed marsh landscape;
- two types of semidynamic coastal dunes.

The next step was to identify those abiotic conditions which would suit each ecosystem. Information on soils, geomorphology, hydrology and biogeography was imported from an existing database of national ecological data into an Arc/Info GIS. For each target ecosystem it was then possible to derive values indicating how suited each ecosystem type was to the abiotic conditions found in different parts of the country. In other words, the GIS could be used to define the abiotic potential for restoration of target ecosystems and therefore to select suitable areas. The Dutch government particularly wanted to locate areas of 500–1000 ha where 'nearly natural' ecosystem could be developed, as land reclamation and intensive agricultural use since the 1860s has reduced most natural sites in the Netherlands to small islands within large stretches of farmland. By using GIS to plot 'high potential' areas for each of the target ecosystems, it became clear that some of the ecosystems could only be restored in very limited areas, whereas others could be restored much more widely (Fry 1996).

The need to ensure that development is compatible with national and international obligations to conserve biological diversity demands methods for strategic ecological assessment to assess the overall effects of development on national and international wildlife stocks. Relatively few studies have been carried out to address this problem, due to the difficulties of obtaining reliable data at a suitable resolution. Treweek *et al.* (1998) studied the coincidence of new road proposals in England with areas of high-value lowland heathland habitat using remotely sensed land-cover data and species-distribution maps to derive habitat distributions. However, they were unable obtain reliable data at a fine enough resolution to study the local spatial relationships between habitats and road proposals. Bina *et al.* (1997) measured the proximity of routes in the trans-European transport network to sites designated for nature conservation. Again, it was not possible to predict the actual impacts of the proposed routes, due to uncertainties in the locations of routes and the inability to differentiate between sites in terms of habitats and species. Nevertheless, in both cases, the use of a GIS platform made it possible to estimate the overall potential extent of ecological impacts due to national and international proposals: something that would otherwise have proved very difficult. Figure 9.6 shows the results of the GIS analysis in terms of the percentage of important bird areas (IBAs) located within 10 km of proposed trans-European networks (TENs) in 12 European Union countries. Plate 9.1 (facing p. 82) is the same output produced as a map. This could later be combined with data describing the range and magnitude of stressors (such as barrier effects, noise disturbance or pollutant deposition) to estimate the overall levels of

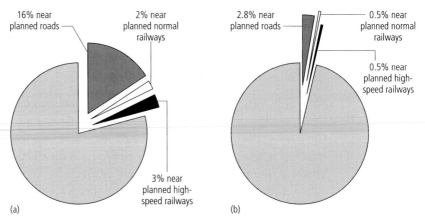

16% near planned roads

2% near planned normal railways

2.8% near planned roads

0.5% near planned normal railways

0.5% near planned high-speed railways

3% near planned high-speed railways

(a)   (b)

Fig. 9.6 Percentage of important bird areas (IBAs) falling within (a) 10 km and (b) 2 km of proposed trans-European networks (TENs) in 12 European Union countries. (From Bina *et al.* 1997.)

impact on specific habitats and species. At the very least, such an approach makes it possible to redesign new routes for greater compatibility with internationally agreed conservation objectives.

9.3 Weighing up the benefits of GIS for EcIA

For EcIAs addressing wide-ranging effects in relatively large or inaccessible areas, the cost of laborious field survey can be prohibitive. In such circumstances, a GIS-based approach can be cheaper and more efficient than a traditional approach based on field inventory and hand-drafting of maps to present results. It can also open up opportunities for using alternative source data such as aerial photography or satellite imagery. Most importantly, automated ecological study using GIS comes into its own as a basis for monitoring programmes. Satellites can repeat land-cover surveys much more frequently, efficiently and accurately than people on foot. When considering the use of GIS for EcIA it is therefore important to ask whether or not GIS will make the process more efficient, cost-effective and productive than more traditional methods. It is worth noting that there are some ecological problems that were far too intractable to tackle before the advent of GIS. It is undoubtedly a technology that has opened up new approaches to EcIA as well as new sources of data (Wadsworth & Treweek 1999).

At present, the lack of readily available and affordable digital, ecological data means that a certain amount of hand digitizing is very likely to be necessary: a laborious and often soul-destroying exercise. Johnston (1998) cites one example where the average time taken to interpret aerial photographs, carry out field checks and draft inventory maps for wetlands was 58.3 hours per 36 square miles. Manual digitizing can be considered as a project 'overhead', which may or may not be worthwhile, depending on the use that is made of the data. If complex relationships need to be considered and, particularly if iterative analyses are likely to be performed, the initial costs of data entry may well be justified as each layer of data is likely to be used more than once (McAuley 1991). However, for more straightforward analyses, which are unlikely to be repeated, the costs of developing a GIS may be prohibitive. Existing digital datasets are often very expensive and for commercial use, new licences may be required for each new application. This is a significant barrier to the use of GIS for commercial EcIA.

In terms of general provision for wider access to digital data for environmental impact assessment (EIA), it is worth mentioning the National Spatial Data Infrastructure (NSDI) in the US (US Executive Office of the President 1994), the National System for Geographical Information (SNIG) in Portugal (Henriques 1996) and the National Geospatial Database (NGD) in the UK (Nanson *et al.* 1995). The European Environment Agency has also devoted considerable resources to the development of digital datasets, which can be used to address environmental problems on a GIS-platform. Of particular relevance to EcIA are Europe-wide data on land cover, biotopes and natural areas important for nature conservation. In the UK, it is possible to purchase digital versions of Ordnance Survey maps as well as digital land cover, soils and geological data. However, worldwide, access to reliable, up-to-date, affordable digital data for EcIA remains relatively limited.

In summary, consider using GIS for EcIA if:
• you find you are spending a lot of time and money hand-drafting maps for reports;
• your project has a long lead time;
• you have access to necessary software and hardware;
• you have, or have access to, relevant GIS expertise and experience;
• your study area is large, remote or inaccessible;
• there are no reliable existing maps to assist in survey design or EcIA scoping;
• you need to tackle complex issues requiring access to multiple datasets;
• you need to model a variety of impact scenarios;
• likely impacts require spatial modelling;
• you need to address cumulative impacts;
• target species are difficult to census directly;
• habitat or species distributions can be predicted from other data;
• relevant digital data are available at an appropriate scale and affordable price.

On the other hand, common problems with the use of GIS for EcIA include:

- lack of time for data collection and entry;
- lack of existing digital data;
- lack of experience and familiarity with software;
- false precision (obscuring sources of error);
- a technology-led approach;
- 'using a hammer to crack a nut';
- over-investment in data irrelevant to decision-making.

To conclude, Johnston (1998) gives the following useful advice to anyone considering a GIS-based approach:

- keep it simple;
- ask whether a GIS is really necessary to tackle key questions;
- use existing data where possible rather than developing new databases;
- plan ahead—conceptualize, for example using data management systems or flowcharts to guide GIS development;
- keep good records, including a description of source data and analysis performed for each step in the GIS process;
- check results—is GIS output logical?
- consult with experienced individuals for advice on database management, data needs and GIS procedures.

9.4 Recommended reading

Introductory GIS for ecologists

Johnston, C.A. (1998) *Geographic Information Systems in Ecology*. Blackwell Science, Oxford.

Wadsworth, R. & Treweek, J. (1999) *Geographical Information Systems for Ecology: an introduction*. Addison Wesley Longman, Harlow.

Remotely sensed data for ecological study

Danson, F.M. & Plummer, S.E. (eds) (1995) *Advances in Environmental Remote Sensing*. John Wiley and Sons, Chichester.

10 Ecological impact assessment design and analysis

10.1 Introduction

Ecological impact assessment (EcIA) relies on a number of tools and techniques, many of which are common to all kinds of ecological study. This book is not intended to provide an exhaustive account of practical ecological survey techniques, as these are described elsewhere. Accepted techniques and survey standards can vary between countries and are often highly species or habitat specific. However, it is possible to identify some guiding principles for the design of EcIA surveys and analyses. EcIA is all about making scientifically defensible decisions. It therefore raises issues of predictive reliability and questions of how to handle uncertainty. As discussed throughout this book, it is a process often constrained in ways that make it difficult to adhere to fundamental scientific principles.

This chapter explores some of the particular scientific, statistical and practical considerations that should be borne in mind when designing and undertaking EcIAs. It summarizes some of the shortcomings commonly identified and considers their practical implications.

10.2 Common shortcomings

A number of shortcomings have been identified in EcIA since the US National Environmental Policy Act (NEPA) first created a requirement for impact assessment of proposed development actions in 1969. Reviews of the quality of ecological input to EIA can be found in Beanlands and Duinker (1984), Spellerberg and Minshull (1992), Treweek *et al.* (1993), Thompson (1995), Treweek (1996), Thompson *et al.* (1997) and Warnken and Buckley (1998). A summary of the shortcomings most commonly identified is given in Box 10.1. Some can be reduced through good practice (see section 10.9), but others are a result of legislative loopholes (see section 10.3) or lack of investment in suitable institutional structures. Some of the difficulties of adopting a scientifically defensible approach to EcIA are explored in more detail in sections 10.5–10.8.

Box 10.1 Common criticisms of EcIA.

Problem	Possible solution
Legislative	
Indicative thresholds take no account of ecological impacts, are set too low, or allow potential cumulative effects to slip through the net	Indicative thresholds should be reviewed regularly in the light of ongoing monitoring of biodiversity loss
Minimalist interpretation of legislation resulting in neglect of key issues	Tighter wording of legislation to remove potential loopholes. Independent review by ecological experts
Failure to audit effectiveness of predictions or mitigation measures	Introduce requirements for follow-up and monitoring
Failure to take regional ecosystem approaches or to consider potential cumulative ecological effects	Introduce SEA with explicit requirements to consider status of ecological resources at regional, national and international levels
Lack of formal requirements for independent review means that many poor EcIAs are allowed to stand. Ecologists are often under pressure to omit or downplay potentially adverse impacts	Introduce requirements for independent review
Institutional	
Poor standards and inconsistent methodology	Publish guidance on suitable methods for different categories of development and different classes of ecological receptor
EcIA undertaken by unqualified personnel	Introduce professional standards and accreditation schemes
EcIAs are reviewed by unqualified personnel (decision-making authority has no ecological expertise). Decision-makers are unaware of potentially serious omissions	Publish guidance on review of ecological content. Ensure EcIAs are reviewed by personnel with ecological expertise. Introduce requirements for independent review
Practitioners are under pressure to carry out EcIA on shoestring budgets (developer pays)	Decision-makers and regulators should reject poor examples of EcIA. Introduce requirements for independent review

Continued

Box 10.1 (*Continued*).

Problem	Possible solution
Failure to evaluate the significance of local impacts in a wider context	Increase investment in national monitoring programmes to provide up-to-date data on ecosystem/habitat/species distributions and status

Practical

Problem	Possible solution
Inappropriate survey intensity, timing or duration. (Can be due to lack of awareness of behaviour of receptors or due to inadequate provision of time/resources for survey)	Undertake pilot surveys to guide more intensive study. Ensure lead times are adequate. Ensure duration of EcIA tallies with possible seasonal fluctuations. Early consideration of potential ecological effects and early inclusion of ecological expertise in EIA
Over-reliance on vague verbal forecasts, usually due to lack of knowledge or understanding about ecological receptors	Increase investment of resources in EcIA. Introduce formal requirements for ecological monitoring to develop knowledge base and enhance predictive ability
Failure to indicate levels of uncertainty, often due to pressure from clients. May also be due to inability to quantify levels of uncertainty due to inadequate study	Introduce formal requirements to specify levels of confidence in predictions or uncertainty about outcomes. If nothing else, back predictions and recommendations up with anecdotal examples
Failure to provide data needed to predict ecological impacts, usually due to under-resourcing of EcIA	Enforce legislative requirements to provide necessary data. Increase investment in EcIA
Failure to quantify impacts (sometimes due to lack of data, but complexity of ecosystem is also a factor)	Enforce legislative requirements for testable impact prediction and introduce requirements for follow-up. Increase investment in EcIA
Bias towards easily surveyed and more charismatic taxonomic groups	Peer review or quality audit of EcIA. Development of professional standards and accreditation of EcIA practitioners
Inadequate replication (access to 'control' areas is often limited)	Peer review or quality audit of EcIA. Introduce requirements for ecological monitoring

Continued on p. 288

Box 10.1 (*Continued*).	
Problem	**Possible solution**
Scientific Weak predictive ability due to complexity of ecosystems and lack of clear ecological theories and laws	Invest in relevant research. Introduce requirements for post-development monitoring of ecosystem responses
Avoidance of statistical method in EcIA design and analysis, often due to lack of opportunity to test outcomes through monitoring	Build EcIA case histories: accumulated experience can substitute for formal replication in some cases. Invest in research on techniques for measuring cause–effect relationships between actions and receptors. Introduce requirements for monitoring and testing of impact predictions

10.3 Legislative barriers

Individual practitioners cannot be expected to solve problems that derive primarily from limitations in existing legislation. However, legislative loopholes are not an adequate excuse for departures from good practice. For example, scientifically defensible approaches to EcIA are hampered by the widespread lack of legislative requirements for follow-up or monitoring. Lack of follow-up means it is possible to implement ecological mitigation measures, the effectiveness of which has never been demonstrated. This is clearly not something a scrupulous ecologist would recommend, but review of environmental statements suggests it is not uncommon.

The most common legislative barriers to good practice in EcIA are:
• lack of requirements for strategic environmental assessment;
• lack of requirements for follow-up or monitoring;
• lack of requirements for independent review;
• failure to include ecological considerations in indicative thresholds for environmental impact assessment (EIA).

10.4 Institutional barriers

In the majority of jurisdictions the developer pays for EcIA, resulting in frequent under-resourcing of EcIA and pressure on practitioners to omit or downplay potentially significant adverse effects. To some extent, this tendency can be counter-balanced if decision-makers or regulators ensure that poor EcIAs are rejected or referred back for improvement. There are many examples where failure to address ecological issues adequately from the outset has resulted in costly delays in obtaining development consent: early dialogue between proponents, practitioners and decision-makers is therefore to be encouraged. Many decision-making authorities do not have adequate time or resources to play as active a role as they might like in the scoping or review of EcIAs. Nevertheless it is increasingly common for decision-making authorities to demand follow-up monitoring as a condition on development consent despite the lack of clear legislative requirements. Also, international pressure to consider impacts on biodiversity has resulted in many new national and regional initiatives that should make it easier to evaluate the wider significance of local impacts in future.

10.5 Scientific barriers

Many of the scientific barriers to good practice in EcIA have been considered in depth elsewhere in the book. They derive from the inherent complexity of most ecosystems, the immaturity of ecology as a science and a lack of investment in relevant research. Subsequent sections focus on the inherent difficulties of making reliable ecological predictions and problems with the development of suitable statistical techniques for EcIA design and analysis.

10.6 Accounting for natural variation

It is necessary to account for natural variation before action-induced impacts can be identified, but often the necessary data are lacking and there is limited scope for studying affected ecosystems in sufficient detail.

Variability represents heterogeneity across some dimension (e.g. time or space) that can generally be represented by a frequency distribution. Natural variation cannot be reduced, but it may be possible to stratify potentially affected populations into more homogeneous subpopulations to rationalize subsequent analyses. However, some variables may be on non-horizontal trajectories in the long term, in other words they may be on 'moving baselines' (Christensen *et al.* 1976). Also, natural fluctuations may be so great that it becomes difficult to detect small changes against the background 'noise' of natural variation. Cowell (1978) suggested that shifts in the abundance of populations of many rocky shore species, for example, would need to exceed 25% to be detected with confidence. Pielou (1981) used a hypothetical example to emphasize the tremendous difficulties of sampling in the field, for example to estimate populations. Figure 10.1 shows a typical ecological dot pattern. It consists of 478

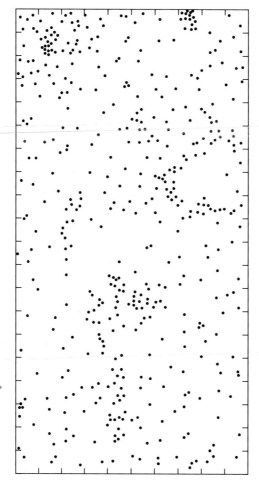

Fig. 10.1 Hypothetical sampling 'population'. (From Pielou, E.C. (1981) The usefulness of ecological models: a stock-taking. *The Quarterly Review of Biology* **56**, 17–31; published by the University of Chicago.)

dots that could represent 'shrubs in a desert, earthworms in a pasture, deermice in a forest, seals on an ice floe, or anything else you care to envisage'. Divide this into 200 square sampling units by connecting the check marks on the boundaries to those opposite. In Pielou's example, dots were counted in 60 of these sampling units and the results were used to estimate the size of the total population. The expected 95% confidence limits were found to be 421 and 535, the width of the interval being 24% of the true population size. This example reinforces the fact that, even with a relatively straightforward sampling task and a sampling fraction as high as 0.3 (considerably higher than would generally be used in EcIA surveys), the precision of the resulting estimate may be very low.

In a practical sense, there has been relatively little research on the intensity of sampling effort required to reduce uncertainty to acceptable levels in different situations. However, a recent paper by Reading (1997) experimented with different approaches to surveying reptiles on lowland heathland in the UK. Different arrays of refuges were established and compared with the mean number of captures of individuals per year. For two species, the smooth snake *Coronella austriaca* and the sand lizard *Lacerta agilis*, increased refuge density resulted in an initial increase in the number of captures, but this levelled off when a refuge density of $80–120\,ha^{-1}$ was reached. For other species, including the slow-worm *Anguis fragilis* and the grass snake *Natrix natrix*, the number of captures continued to rise beyond this density. This difference may be attributable to the fact that grass snakes are able to use a variety of habitats and may simply have been 'passing through' the study area. Slow-worms, on the other hand, are a relatively sedentary species with small home ranges. It is therefore possible that even the higher densities of refuges used in this study failed to encompass all the territories occupied in the study area. This study not only provided useful information about suitable densities of artificial refuges for estimating total populations of certain species in their preferred habitat, it also illustrates how important it is to take account of species behaviour and habitat fidelities when designing survey strategies. It is only by matching sampling approaches and intensities to species characteristics that it is possible to account for natural variation in population numbers or distribution.

Where relatively detailed information is required about species abundance or the dynamics of populations, a pilot study of some kind is advisable to ensure that field data gathering will be sufficiently intensive to permit meaningful impact prediction.

10.7 Dealing with uncertainty

Connected with the problems of characterizing natural variation are those connected more generally with levels of uncertainty. Levels of uncertainty are rarely analysed explicitly in EcIA, making it difficult to evaluate the significance of results and generating potential errors of interpretation.

Uncertainty in ecological analysis generally has three main sources. These are discussed in considerable detail in the book on ecological risk analysis by Suter (1993a). They include:
- inherent randomness or variability (stochasticity);
- incomplete knowledge (ignorance);
- mistakes in execution of assessment activities (error).

Whatever its source, uncertainty can be attributed to a lack of information, which may or may not be rectifiable (Rowe 1994). Stochasticity, for example, results in uncertainty that is difficult to reduce because it is an inherent characteristic or property of the system being assessed. Incomplete knowledge, on the other hand can, theoretically, be rectified through more detailed study. Knowing whether or not knowledge gaps can be filled is very important in EcIA. There are many examples where detailed studies have failed to produce useful information because the systems under study are so complex. In such cases it can be better to take a more pragmatic approach and to spend limited funds on practical mitigation measures.

There are three main categories of uncertainty (after Rowe 1994):

1 uncertainty in future or past states (temporal uncertainty);
2 uncertainty due to complexity;
3 uncertainty in measurement.

10.7.1 Temporal uncertainty

Uncertainty in future states is the most familiar category of uncertainty. The future is always uncertain because (Rowe 1994):

• behaviour of non-linear dynamic (chaotic) systems makes it very difficult to predict future behaviour;
• sparse, rare events with low probability but high or serious consequences are difficult to predict;
• timeframes for observation are often inappropriate (e.g. long-term trends may be missed or short-term excursions filtered out if timeframes for observation are too short or long, respectively);
• internal system parameters change with time (e.g. genetic drift).

The traditional approach to decision-making under uncertainty deals with possible future states without measuring their likelihood of occurrence, but simply attempting to optimize the consequences should any of them occur. In EcIA it is more helpful to estimate the likelihood of possible future states, so that decisions can be made about appropriate avoiding or remedial action. However, attempts to model the probability of future events or states are usually based on knowledge about the past. While historical data can be used to characterize past patterns of behaviour, these data may be incomplete or 'incorrect', in that they were not necessarily recorded for the purpose now sought.

Sources of uncertainty about past states are likely to include (Rowe 1994):

• incomplete data;
• limited validation (lack of external data references);
• changing system parameters, which prevent identical conditions from being re-visited;
• inappropriateness of measured parameters for current purpose;
• lack of knowledge about how data were collected.

When predicting the probability of future events it is worth bearing in mind that many ecological 'events' are both extreme and rare (in other words they have significant consequences but occur with low probability). In such cases it may not be enough to know the mean of a long-term frequency distribution. Instead, we might want to characterize a distribution for the 95th percentile (as in the case of designs for flood protection, for example) and more samples or replicates are likely to be needed to characterize upper percentiles with an acceptable confidence level than to characterize means.

The intensity of data gathering required for EcIA will always depend on the degree of uncertainty about past and future states of ecosystems. Field survey programmes should therefore be designed following an assessment of the ability to predict future states of an ecosystem, given existing data and understanding of ecosystem behaviour.

10.7.2 Uncertainty due to complexity

Uncertainty also derives from complexity, which influences the number of degrees of freedom in a system and how the parameters expressing those degrees of freedom interact. When systems become too complex to deal with all parameters directly, simplification of one or more parameters becomes necessary. In other words, a model, or abstraction of the system, is required. If the system being modelled is a real system, it is possible to measure the validity of the model empirically. Confidence in the model depends on the degree to which it is considered to represent reality. Models (conceptual or otherwise) that are over-simplified are an obvious source of uncertainty.

Sources of uncertainty due to complexity (Rowe 1994) include:
- inherent random processes;
- degree of complexity;
- interdependence of parameters;
- initial conditions in chaotic systems;
- incomplete information about all parameters.

Most ecosystems are inherently complex and we have only limited information about them. Ecosystem models are therefore often over-simplified and predictions are inherently uncertain.

10.7.3 Uncertainty due to measurement

The techniques used to measure parameters or effects do not always give perfect information capture. Information may be missed because measurement techniques have inappropriate detection limits or cannot discriminate sufficiently between measurement units. Measurements are made to gain information using a variety of scales (e.g. nominal, ordinal, cardinal or ratio scales). There is always a minimum unit of measurement for which it is possible to discern one unit from another, using whatever measurement tools are available. This determines the precision of measurement scales. Imprecise measurement is one obvious source of uncertainty. 'Accuracy' refers to the correctness with which measurements of scale values have been made. Inaccuracy in measurement is another possible source of uncertainty, but one that should theoretically be rectifiable. It is critically important to understand the measurement error and its magnitude relative to the variability in attributes being measured. Ideally, measurements should be carried out using some sort of quantitative basis for characterizing the random and systematic error of any measurement technique (e.g. through the use of replicate samples or blanks) before using it to measure variables of interest. For multiple observations of scale values, statistical models may be used to describe results, in which case increased sample sizes help to reduce uncertainty and increase confidence in our understanding of underlying processes.

Rowe (1994) lists the following sources of uncertainty in measurement.

1 *Empirical observations*:
  • precision of measurements (resolution of measurement tools);
  • accuracy of measurements (quality of measurement);
  • measurement interaction;
  • systemic measurement errors (measurement bias).

2 *Interpretation of observations*:
  • judgements about the inclusion or exclusion of data;
  • sample size adequacy;
  • objective vs. subjective methods for sampling;
  • objective vs. subjective methods for analysis and reduction.

3 *Interpretation of measurements*:
  • differences in judgement or interpretation by experts (is the glass half empty or half full?);
  • biases, beliefs and dogma of experts (e.g. bias error in making observations, or bias in interpreting results).

In ecological assessments, there is often a lack of empirical information on which to base analyses. Sample sizes tend to be too small and there is an over-dependence on subjective methods. Interpretations are therefore prone to bias. While subjective judgements have their own part to play in EcIA, lack of transparency about the level of subjectivity applied can result in decisions based on misleading evidence. Uncertainty is something we have to live with, but it helps to know how uncertain we are in any particular situation. As standard practice, all models or analyses should therefore be accompanied by a measure of variability.

10.7.4 Errors under uncertainty

One way to study ecosystem behaviour is to test null hypotheses. An example of a null hypothesis might be that removal of habitat due to road construction will not result in local extinction of a protected species. It is not always practical to test null hypotheses for EcIA because of time constraints, but if more attempts were made even just to formulate appropriate and uncontroversial null hypotheses, less time and effort might be wasted on studies that provide irrelevant information.

When a decision is made about a null hypothesis under conditions of uncertainty, at least two different kinds of error can occur (Shrader-Frechette & McCoy 1992). Type 1 errors occur if one rejects a null hypothesis that is true, resulting in a 'false positive' (e.g. in assertions of effects where none exists). Type 2 errors on the other hand occur if one fails to reject a false null hypothesis (actual effects are not recognized).

- Type 1 error: false positive — predicted effects do not exist.
- Type 2 error: ignorance of possible effects.

Under uncertainty, it is necessary to decide how to deal with these types of error and to use statistical tests that are based on tolerance limits for them. Shrader-Frechette and McCoy (1992) indicate how the concept of significance in statistics is often defined in terms of a 'type-1 risk of error of either 0.01 or 0.05, where there is not more than a 1 in 100 or a 5 in 100 chance of committing the error of rejecting a true hypothesis'. Thresholds for type 2 errors on the other hand are set less often in ecology. In EcIA it is particularly important to consider the nature of the hypothesis being tested and the possible consequences of tolerating different types of error. While there is a justifiable reluctance to make type 1 scientific errors, the consequences of type 2 errors can in fact be just as serious. Hypotheses might be constructed positively (hypothesizing no harm will occur) or negatively (hypothesizing that there will be no benefit). Shrader-Frechette and McCoy (1992) summarize the possible consequences of type 1 and type 2 errors for these two types of hypothesis:

Positive null hypothesis (no harm):
- Type 2 error: ignorance (possible harm).
- Type 1 error: false positive.

Negative null hypothesis (no benefit):
- Type 2 error: ignorance (possible loss of benefit).

They use the example of drug testing to illustrate the fact that a type 2 error for a positive null hypothesis (no harm) can result in cases where adverse effects are missed and ignored, whereas the consequences of type 1 error would be much less serious (although perhaps more expensive). In EcIA also, the consequences of a type 2 error for a positive null hypothesis can result in failure to identify potentially serious ecological impacts. In cases where there are potential risks to important ecological receptors (e.g. protected species) it is entirely appropriate to expect a development proponent to demonstrate that there will be no harm. Regulators or decision-makers would then benefit from setting thresholds for type 2 errors.

10.7.5 Sensitivity analysis

Whatever the source of uncertainty, it helps considerably if we can quantify it in some way. Sensitivity analyses can be used to evaluate the range and impact of variability in an assessment, by presuming that outcome values are known (no uncertainty) but have different ranges. By holding variability fixed, using only one range to represent it, it is possible to assess the impact of uncertainty in the same way (Haimes *et al.* 1994).

After an outcome redistribution has been determined, sensitivity analysis can help in judging the relative importance of an input distribution based on sparse data. If it transpires that the choice of input distribution does not account for much variation in the outcome, then the lack of data can be assumed not to have that much impact. On the other hand, if input distribution is a major contributor to variation in the outcome, lack of data generates more uncertainty in the outcome, which must be taken into account when reaching conclusions. Otherwise, it may be decided that more data are required to 'fill the gaps'. In this case, a cost–benefit type of analysis might be justified to decide whether the expected improvements in the overall assessment justify the additional efforts.

10.8 Statistical solutions

One common criticism of the EIA process in general and of EcIA in particular, is the widespread failure to set or test hypotheses (Beanlands & Duinker 1983). In fact EcIAs are generally more likely to generate hypotheses than to test them. The limited time and resources generally available for EcIA mean that 'it may not be possible to establish true experimental controls under field conditions, nor to undertake the sampling programmes required to meet normally accepted confidence limits in statistical analyses' (Beanlands & Duinker 1983). Access to suitable control sites is often limited and lead times are often too short to set up adequate sampling programmes. Not only may it be inappropriate to apply classical inferential methods to 'pre- and post-operational data on one impacted site' (Eberhardt 1976), but many ecologists never even have access to post-operational data due to the lack of monitoring.

Eberhardt (1976) does identify some situations where it may be possible and appropriate to draw on traditional methods, formulating hypotheses that can be subjected to straightforward statistical analyses. For example, where cooling water is released into a river it might be possible to compare densities of freshwater organisms in two different locations because 'many samples can be taken over time and/or space to compare two reasonably well defined populations'. In circumstances where a 'readily measured gradient of 'insult' exists and persists in time', it may be possible to rely on regression analyses in which 'various measurements on sessile (or sedentary) elements of the biota are regressed on measurements on the stressing element'. This is an example where clearly defined relationships between a stressor and a population-level response can be established by drawing on experience of other, similar situations and extrapolating to new ones. More often, our understanding of the ecological processes likely to 'drive' responses and our experience of other similar situations is too limited for such an approach to be taken. In situations where the behaviour of potential ecological receptors is poorly understood, however, it is perfectly reasonable to demand resources for studying similar situations elsewhere.

Beanlands and Duinker (1983) include a schematic diagram modelled on work by Christensen *et al.* (1976) which illustrates gradients of increasing difficulty in either measuring or predicting ecological effects (Fig. 10.2).

Although it is not always possible to adopt classic experimental designs for impact assessment studies, much greater use 'could, and should, be made of hypotheses and statistically based designs' (Beanlands & Duinker 1983). Another barrier to this approach is that ecological

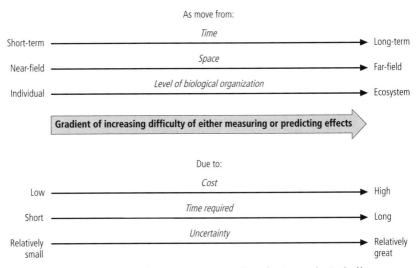

Fig. 10.2 Gradients of difficulty in measuring and predicting ecological effects. (After Christensen *et al.* 1976.)

concepts such as species composition, trophic structure and stability
have rarely been operationally defined to allow comparisons to be made
between impacted and un-impacted systems (La Point 1995). This is
something that could be undertaken on a project-by-project basis, but
not within the timeframes normally set for EcIAs.

There is still a tendency to omit from EcIA evidence that is hard
to quantify. It can also be very difficult to develop reliable statistical
techniques that permit 'attribution of cause', i.e. that quantify the
evidence that changes have actually been caused by the human activity
under study (Stewart-Oaten 1993). To overcome such problems, it is
advisable to ensure that:

• conclusions are not based solely on hypothesis tests;
• results for individual species are not considered in isolation.

In the absence of a clear causal chain, a convincing case requires that:

• results for several species follow a consistent pattern;
• plausible mechanisms for an ecological impact can be identified;
• reasonable alternative mechanisms have been explored and ruled out
(Schroeter *et al.* 1993).

One of the main barriers to the development of statistical techniques
for use in EcIA is the scarcity of post-development monitoring and
therefore of opportunities for testing of either impacts or predictions.

Because of uncertainty, we need to design ecological analyses with great
care. Although statistical techniques are greatly under-used in EcIA, a
certain amount of research has been devoted to techniques of particular
relevance to EcIA. Some of these are outlined in following sections.
Firstly, however, it is useful to consider certain key considerations in
statistical design: notably the need for effective controls and the need for
adequate replication.

10.8.1   The need for effective controls

'Controls' are required in order to be able to compare 'treated' or
'affected' ecosystems with systems that have not been treated or affected.
Statistical designs should permit estimates of the variation associated
with either 'treatment' or 'control' so that 'treatment effects' or 'impacts'
can be distinguished from variation that derives from other sources.
Otherwise it is impossible to predict ecological impacts with any degree
of confidence.

'True' controls (those that differ only in one characteristic, namely the
presence or absence of a specific treatment or impact) are rarely, if ever,
achieved in ecology (Peters 1991) and are certainly unusual in EcIA.
Particularly at the ecosystem level, it may be more appropriate to refer
to 'reference system' instead of 'control' (Likens 1985). This is certainly
the case for most examples of EcIA, where replication of conditions in

both time and space is rarely possible. In particular, EcIA is generally carried out to tight deadlines. It may be difficult to identify suitable 'control' sites in cases where potentially affected ecosystems are limited in extent or rare. Even when suitable sites can be identified, access to them for experimental purposes may be restricted. As Peters (1991) points out, the incorporation of a 'proper control' is often precluded simply for financial reasons.

10.8.2 The need for replication

Effects or impacts can only be distinguished if they are large relative to the inherent variability of replicates and the number of samples taken. Ideally, variability is estimated in pilot studies and then used to estimate the number of samples required to test for effects of a given magnitude (Elliott 1977). The results of such exercises can be startling. Post-hoc study of the statistical representation achieved in a number of vegetation surveys in the UK showed that the sample sizes that would have been required to estimate the total number of plant species represented were usually considerably greater than the sample sizes actually used. In one example, sampling effort would have to have been increased about 10-fold to detect 90% of the species present (Sparks *et al.* 1997).

If the expected magnitude of impacts is so small that it is unlikely to be detected with realistic sampling effort, there is no point proceeding with sampling. Insufficient sampling can result in type 2 errors as described earlier in section 10.7.4 (in which no significant difference is found, although it may nevertheless exist). Likewise, if sampling effort is seriously constrained by limited time and/or funding, it is unlikely that it will be possible either to characterize underlying variation, to detect change or to make the connection between changes and their potential causes. Limited survey can be just as misleading as carrying out no survey at all.

10.8.3 Pseudoreplication

Pseudoreplicates (*sensu* Hurlbert 1984) are repeated samples that misrepresent the range of possible values that treatments or controls might take because the replicates are not statistically independent and the number of independent measurements are over-estimated. In short, interdependence of samples makes it difficult to extrapolate from their 'behaviour' to that of the larger population from which they are selected (Peters 1991). Pseudoreplication can be extremely difficult to avoid in EcIA. In streams, for example, it is very difficult to find independent control sites, and discrete locations upstream of a disturbance really represent no more than 'pseudocontrols'. Furthermore, undisturbed or 'control' sites are always likely to be physically separated from impacted sites, for example they will tend to be upstream of an impact, whereas impacted areas will be downstream (Humphrey *et al.* 1995). Most importantly, most ecological processes within streams are linked. Similar problems are encountered in many other ecosystems, which make it impossible to assign treatment effects randomly and independently to replicates.

10.8.4   BACIP design

The 'Before, After, Control, Impact, Paired differences' (BACIP) design was proposed by Stewart-Oaten *et al.* (1992). In this design, samples are collected at the same time from single 'impact' and 'control' sites before and after an impact has occurred. The difference between sampled abundances at impact and control areas at any one time is regarded as a replicate observation. The means of sets of differences between the two areas before and after impact are compared by a *t*-test or the equivalent. In effect, this method uses a form of temporal replication as a solution to the problem of pseudoreplication (Humphrey *et al.* 1995).

The BACIP design requires that data satisfy certain assumptions, including additivity of time and location effects and independence of temporal replicates over time. Tests for checking the necessary assumptions are provided by Stewart-Oaten *et al.* (1986, 1992). The independence assumption is likely to be violated whenever 'large, local and long-lasting' influences affect only one site and are therefore dependent on the site location and the type of organism and response under investigation (Stewart-Oaten *et al.* 1986). This is a situation commonly encountered in EcIA. As Stewart-Oaten *et al.* (1992), pointed out, the assumption of independence is more likely to be valid for short-lived and relatively mobile species than it is for sedentary, long-lived ones. There are a number of other situations in which a design such as this may prove inadequate to demonstrate causal relationships. For example, Underwood (1993) has argued that multiple control sites are essential to infer causal relationships between an impact and observations of change in 'natural systems'. However, in EcIA, access to multiple control sites is rarely possible.

The method does suffer from the possibility of confounding effects due to a large unpredictable event or perturbation just happening to be coincident with the onset of the potential impact. For this reason, it is necessary to consider the plausibility of observed effects and their possible causes. There are a number of inferential techniques that could and should be used to support the design, for example incorporating an element of spatial modelling using 'near and far' impact sites to observe whether effects wane or change with increasing distance from a source.

Again, it has more relevance to the design of EcIAs for which subsequent monitoring will take place. In other situations, opportunities to assess post-development impacts will be limited.

10.8.5 State space analysis

As Suter (1993a) points out, in cases where one wishes to use heterogeneous measures to characterize an ecosystem relative to others of the same type, state space analysis may be a useful tool. This describes the location of a system of interest as 'a point in a Cartesian space with n dimensions corresponding to the n properties measured. If multiple measures are taken over time or space, then the system is defined as a vector or a cloud of points in the n space'. It is possible to estimate both the likelihood (risk) that a disturbed or altered system belongs to a different state space from a set of nominally undisturbed reference systems (Johnson 1988) or to calculate the deviation of a disturbed system from reference systems in the chosen state space (Kersting 1988). There are therefore obvious applications for these techniques in EcIA.

Other ordination techniques that may be useful include principal components analysis (PCA) and similarity indices (Suter 1993a). These have the advantage of:

• retaining the identity of the individual, measured properties;
• showing which measured properties account for most differences among systems;
• having non-arbitrary statistical and mathematical properties;
• not purporting to represent a property of the system, but simply showing how they differ.

An annotated bibliography of canonical correspondence analysis and related constrained ordination methods referred to in the ecological literature between 1986 and 1993 has been produced by Birks *et al.* (1993). Common applications are to analyse vegetation–environment relationships, to study the effects of disturbances (fire or trampling, for example) or to study the effects of pollution on aquatic communities (see papers by Snoeijs 1991 and Snoeijs & Prentice 1989). The analysis of vegetation–environment relationships by canonical correspondence analysis is described by ter Braak (1987a,b).

10.9 Closing checklists

There is no, one correct way to go about undertaking EcIAs. In fact, one of the main challenges to EcIA practitioners is knowing when to expect the unexpected. No amount of scientific understanding can substitute for practical knowledge and experience. However, there are some points it may be helpful to consider when planning a development with potential ecological effects, when planning to undertake EcIAs or when reviewing them. These are summarized in the following checklists.

1 Proponents:
- consult with a qualified ecologist as early as possible: failure to take account of ecological effects can result in serious and expensive delays;
- provide detailed information about the nature, magnitude, spatial extent and duration of activities associated with your proposal;
- allow long lead times for ecological studies;
- provide scope for pilot studies to enhance the focus and effectiveness of EcIA;
- expect EcIAs to span seasons;
- invest in arrangements for mitigation as soon as possible — considerable forward planning may be necessary to put effective measures in place;
- adopt a flexible approach, incorporate ecological criteria into the planning and design of proposals;
- remember that good EcIAs are generally cheap at the price!

2 Practitioners:
- start with what is known already, research existing literature and data, locate any existing information about the distributions and status of any ecological receptors with protected status;
- think spatially from the outset, collect or draft suitable maps of the receiving area(s), including habitat distributions, range maps, etc.;
- consider using geographical information systems (GIS) in cases where the study area is large, where numerous overlaying or interpolation procedures appear necessary to work out spatial relationships between stressors and receptors, there is a long lead time and relevant expertise, hardware and software are available;
- contact statutory consultees as early as possible and initiate constructive dialogue;
- carry out pilot studies if possible to ensure that the scope of EcIA is appropriate (e.g. that all key receptors have been identified) and to establish an appropriate intensity of survey or data gathering (e.g. the UK's Phase I Habitat Survey). Set thresholds for detailed survey and analysis;
- identify valued ecosystem components (VECs) (to include protected habitats or species, suitable indicators of change, etc.), only carry out detailed surveys for those parameters that can be used to support decision-making;
- establish assessment and measurement endpoints (e.g. limits of acceptable change);

- check that proposed field surveys will coincide with key periods of receptor presence, activity or visibility;
- identify suitable taxonomic experts where necessary;
- survey on more than one occasion if possible;
- provide evidence to support conclusions about potential impacts — estimate their likely magnitude, severity, probability of occurrence, duration and permanence, reversibility;
- provide evidence to demonstrate the effectiveness of proposed mitigation measures — in cases where effectiveness is unknown, recommend trials to provide suitable evidence;
- recommend monitoring/follow-up where possible to evaluate the accuracy of predictions and the effectiveness of mitigation.

3 Decision-makers:
- consult widely and enable/engage public consultation where possible;
- establish limits of acceptable change;
- establish criteria for reviewing the acceptability of both environmental impact statements (EISs) and development effects;
- ensure availability of relevant ecological expertise for the review stage;
- consider cumulative effects (how will single proposals affect the overall status of ecological resources?);
- reject inadequate EISs;
- consider cross-sectoral issues.

All the stakeholders in the EcIA process have an important part to play in delivering information that really is valuable for decision-making.

10.10 EcIA and decision-making

As pointed out in Chapter 1, EcIA should be an iterative process: a cycle rather than a stepwise procedure. Its purpose is to ensure the full and appropriate consideration of ecological issues in environmental assessment, environmental management, environmental planning and decision-making. Methods by which consideration of different sectoral interests can be integrated in decision-making are the subject of ongoing debate. This book has outlined the factors that should be taken into account when attempting to predict ecological outcomes of proposals and evaluate their ecological significance *before* carrying these conclusions forward into a decision-making process where other factors (social, economic and political) come into play. However, considerably more thought needs to be given to the ways in which decisions are reached concerning ecological and environmental integrity. As Mapstone (1995) pointed out, we tend not to consider what levels of ecological impact we wish to detect (the 'critical effect size') or the types of uncertainty or error we are prepared to tolerate when designing EcIAs.

Stipulation of a critical effect is not simply a statistical or procedural decision: it involves a multitude of judgements about the ecological consequences of an effect of a nominated magnitude or greater, many of which will be social, aesthetic or economic. It becomes essential in cases where impact monitoring is proposed: the essential ingredient in the EcIA of the future! Effective monitoring depends on finding answers to the following questions for each variable being monitored.

• How much anthropogenic disturbance is acceptable (used to define 'limits of acceptable change')?

• What amount of development-induced change should precipitate management action?

Mapstone (1995) emphasizes that the point at which action to reverse adverse impacts is considered desirable may well vary according to how 'importance' is assessed. Aesthetic or economic considerations might demand more stringent critical effect sizes than those required to safeguard ecological function. However, we still need to know what the ecological consequences of proposed actions will be, and we need to determine these consequences using objective, transparent and defensible methods. Most importantly, EcIA should be carried out intelligently and carefully to ensure that decisions are based on reliable information.

10.11 Recommended reading

Ecological risk assessment (including assessment under uncertainty)
Rowe, W.D. (1994) Understanding uncertainty. *Risk Analysis* **14** (5), 743–750.
Suter, G.W. II (1993) *Ecological Risk Assessment*. Lewis Publishers, Boca Raton, FL.

Statistical techniques for impact assessment
Eberhardt, L.L. (1976) Quantitative ecology and impact assessment. *Journal of Environmental Management* **4**, 27–70.
Shrader-Frechette, K.S. & McCoy, E.D. (1994) What ecology can do for environmental management. *Journal of Environmental Management* **41**, 293–307.
Stewart-Oaten, A., Murdoch, W. & Parker, K. (1986) Environmental impact assessment: 'pseudoreplication' in time? *Ecology* **67**, 929–940.
Stewart-Oaten, A., Bence, J.R. & Osenberg, C.W. (1992) Assessing the effects of unreplicated perturbations: no simple solutions. *Ecology* **73**, 1396–1404.

Introductory statistics for biologists and ecologists
Watt, T.A. (1997) *Introductory Statistics for Biology Students*, 2nd edn. Chapman and Hall, London.

Glossary

Abundance: The number of organisms in a population.

Anthropogenic: Caused by humans.

Assessment endpoint: A criterion or benchmark against which to evaluate an outcome. (*See also* Measurement endpoint.) Assessment endpoints reflect management goals and link decision-making criteria with the specific ecological measures made in an ecological assessment.

BACIP design: The Before, After, Control, Impact, Paired differences design requires samples to be collected at the same time from single impact and control sites before and after an impact has occurred. The difference between sampled abundances at impact and control sites at any one time is regarded as a replicated observation. The means of sets of differences between control and impact sites before and after impact are compared by a *t*-test or the equivalent.

Baseline conditions: The conditions that would pertain in the absence of a proposed action. To characterize baseline conditions it is necessary to quantify natural variation.

Biodiversity: Biological diversity. The Convention on Biological Diversity defines biodiversity as 'The variability among living organisms from all sources including, *inter alia*, terrestrial, marine and other aquatic ecosystems and the ecological complexes of which they are part, this includes diversity within species, between species and of ecosystems'. Biodiversity is a measure of genetic variation between and within species.

Biogeoclimatic units: Classes of ecosystem under the influence of the same regional climate.

Biomass: The weight of living material per unit area, usually expressed as weight of dry matter per unit area (e.g. $kg\,ha^{-1}$).

Biota: A general term for all living organisms.

Biotope: The smallest unit of a landscape that can be classified according to the biota present (usually defined using floristic characteristics).

Capability (habitat capability): The ability of a habitat, under optimal natural (seral) conditions to provide life requisites of a species, irrespective of its current habitat conditions.

Carrying capacity: The maximum number of organisms or amount of biomass that can be supported in a given area.

Checklist: Lists of factors or components to be considered in an impact assessment.

Community: Populations of different species inhabiting the same area or habitat at the same time.

Compensation: A form of mitigation in which loss or degradation of a natural resource is compensated for by its creation or recreation at an alternative site.

Compensatory restoration: Restoration projects designed to compensate the public for the interim loss of an injured natural resource before it is returned to its pre-impact condition.

Component interaction technique: A technique in which the environment is modelled as a list of components ranked in order of their ability to initiate secondary impacts, based on known chains of dependence between them.

Constant natural assets: Those elements of biodiversity for which maintenance of the overall stock is desirable, but which can be created on alternative sites (*see also* Critical natural capital).

Contingent valuation: A technique that can be used to estimate economic values for commodities that cannot be traded conventionally in markets. It works by soliciting willingness to pay for or accept a change in the levels of environmental service flows in a carefully structured hypothetical market.

Critical natural capital: Those elements of native biodiversity that cannot be readily replaced (such as ancient woodlands) and will undergo potentially irreversible declines if any depletion occurs (*see also* Constant natural assets).

Cumulative effect: An umbrella term for effects that accumulate over space or time. *See also* Synergistic effect; Interactive effect; Time-crowding; Space-crowding.

Cumulative effects analysis (CEA): A formal process in which cumulative effects are assessed explicitly.

Density dependence: The tendency for the death rate in a population to increase or for the birth or growth rate to decrease, as the density of the population increases.

Density independence: The tendency for the death, birth or growth rate in a population neither to rise nor fall as the density increases.

Developer: The initiator or proponent of a project.

Development: A specific action undertaken to satisfy human needs.

Direct impact: An outcome that is directly attributable to a defined action.

Disturbance: Disruption of normal process or behaviour. In community ecology, disturbance is defined as an event that displaces organisms, opening up space, which can be colonized by individuals of the same, or different species.

Diversity: A measure of variety and abundance.

Ecological impact assessment: The process of defining, quantifying and evaluating the potential impacts of defined actions on ecosystems or their components.

Economic impact assessment: The process by which changes in employment, income and levels or business activity are assessed with respect to a proposed project or programme.

Ecosystem: Communities of organisms interacting with abiotic factors and with each other as a distinct unit (i.e. a bounded system).

Effective population size: The size of a genetically idealized population with which an actual population can be equated genetically.

Elasticity: A measure of the rapidity of restoration of a stable state following ecosystem disturbance.

Endangerment: The degree to which a species is threatened by, or at risk from, extinction.

Environment: A general term, variously defined, which describes all components of human surroundings. The term must be applied to a particular subject (environment of what or whom?) and its limits are determined by the limits of perception or experience of that subject.

Environmental impact statement: Report summarizing the findings of an environmental impact assessment. Used interchangeably with environmental statement.

Exclusive threshold: A limit applied to screen out projects or actions for formal environmental assessment automatically.

Exposure assessment: The process of locating and measuring the coincidence between key receptors and defined activities.

Extinction: The condition that arises when the last remaining individual of a species, group or gene, ceases to exist. Extinction may be defined locally or globally.

Focusing: The process by which ecological impact assessment is refined, by selecting suitable 'valued ecological components' (VECs) for in-depth study.

Fragility: The inverse of ecosystem stability.

Gene flow: The consequence of cross-fertilization between members of species across boundaries between populations, or within populations, resulting in the spread of genes across and between populations.

Geographical information systems (GIS): Integrated systems of computer hardware and software for entering, storing, retrieving, transforming, measuring, combining, subsetting and displaying spatial data that have been digitized and registered to a common coordinate system.

Guild: A group of species that exploit similar environmental resources in a similar way.

Guild indicator: A species selected to represent a guild, which can be relied on to reflect the likely behaviour and responses of other guild members.

Habitat: The place and resources occupied and used by a population of organisms.

Habitat equivalency analysis: A methodology used to determine compensation for natural resource injuries in natural resource damage assessments.

Habitat evaluation procedure (HEP): A formal procedure developed by the US Fish and Wildlife Service to assess the consequences of habitat loss for wildlife.

Habitat potential: A measure of the ability of a given habitat to support a certain species.

Habitat specificity: The degree to which a species is associated with one habitat, compared with its occurrence in all habitats.

Habitat suitability index (HSI): Used in habitat evaluation procedure. Derived by comparing habitat conditions in a study area with optimum habitat conditions for the same evaluation species.

Health impact assessment: The process of predicting and evaluating the potential effects of proposed actions on human health.

Hedonic pricing: A technique in economics used to impute a price for an environmental asset by examining the effect its presence has on a relevant market-priced asset.

Home range: The area habitually used by a species to fulfil its requirements for food, shelter and a place to breed. Excursions beyond this area are rare.

Hysteresis: The degree to which the path of restoration or ecosystem recovery is an exact reversal of the path of ecosystem degradation.

Impact range: The area likely to be affected by a proposed action or stressor.

Indicative threshold: Limits used to assess the eligibility of an action for formal environmental assessment.

Indicator: Any representative component, used to provide surrogate measurements reflecting the likely behaviour of other components.

Indirect impact: An impact that is attributable to a defined action or stressor, but that affects an environmental or ecological component via effects on other components. Indirect effects are often, but not necessarily, time-delayed or expressed at some distance from their source.

In-kind replacement: Replacing one kind of ecosystem or habitat with another of equivalent type and condition.

Insularization: The combined process of habitat fragmentation, isolation and loss.

Integrated pollution prevention and control (IPPC): A formal process of identifying, prescribing and regulating industrial processes that may pollute or harm the environment.

Intergenerational equity: A term used to describe the fairness of distribution of resources between generations.

Interim lost value: The diminution in value of natural resources following injury and pending their recovery to baseline but for the injury (US Oil Pollution Act 1990).

Inventory: An exhaustive survey of species.

Keystone (species): Originally used to denote a species that preferentially consumes and holds in check another species that would otherwise dominate the system. Also used to describe the effect of a change in one species on some characteristic of its community or ecosystem. Keystone species have an impact that is out of proportion to their proportional abundance.

Map units: Map units are established as a result of applying a classification to map polygons.

Matrix: Table of potential interactions between actions and receptors.

Measurement endpoint: An assessment endpoint for which a numerical value can be assigned.

Metapopulation: A population perceived to exist as a series of subpopulations, linked by migration between them. However, the rate of migration is limited, such that the dynamics of the metapopulation should be seen as the sum of the dynamics of the individual subpopulations (Begon *et al.* 1996).

Minimum dynamic area: A species might occupy habitat patches within a larger 'biotope' at different times. Within each patch, the species might go extinct, but re-establish later from other patches. The 'minimum dynamic area' is the area of biotope patch necessary to retain enough habitat patches to prevent overall extinction.

Minimum viable habitat: The minimum area and quality of habitat required to support a given population.

Minimum viable population: The smallest isolated population having a 99% chance of remaining in existence for 100 years despite the foreseeable effects of demographic, environmental and genetic stochasticity, and natural catastrophes.

Mitigation: The process of preventing, avoiding or minimizing adverse impacts by: (i) refraining from a particular action; (ii) limiting the degree of an action; (iii) repairing, rehabilitating or restoring the affected environment; (iv) providing substitute resources.

Mitigation bank: The creation, restoration or enhancement of habitats (usually wetlands) to be sold or exchanged to compensate for future losses. Values of wetlands created, restored or enhanced are determined and the habitats are assigned credits, which can be sold or 'withdrawn' to compensate for the loss of habitat elsewhere.

Natural variation: Variation attributable to non-anthropogenic causes.

No net loss: The point at which habitat or biodiversity losses equal their gains, both quantitatively and qualitatively.

Policy assessment: A process concerned with assessing all the intended and unintended outcomes of policies being planned, proposed, implemented or reviewed.

Polygons: Delineation that represent discrete areas on a map, bounded by a line.

Population density: The numbers in a population per unit area.

Population dynamics: The variations in time and space in the size and densities of populations.

Population viability analysis: The structured, systematic and comprehensive examination of the interacting factors that place a population or species at risk.

Primary restoration: Restoration designed to return an injured natural resource to its baseline, pre-injury condition.

Programme: A set of projects such as a transport programme.

Project: An individual development scheme.

Pseudoreplicates: Repeated samples that misrepresent the range of possible values that treatments or controls might take because the replicates are not statistically independent and the number of independent measurements are over-estimated.

Rarity: A measure of relative abundance.

Receptor: Any ecological component affected by a particular action or stressor.

Reference system (reference habitat): An ecosystem or habitat selected to provide benchmarks against which to compare or measure the degradation of other, similar systems or habitats.

Replaceability: A measure of the extent to which a habitat or ecosystem can be restored or reconstructed.

Replacement cost: The financial cost of replacing an ecosystem or habitat or restoring it to pre-impact condition.

Resilience: The tendency of a system to return to its former state following a disturbance.

Resource: That which may be consumed by an organism thereby becoming unavailable to other individuals of the same or different species.

Restoration: The re-establishment of a damaged or degraded system or habitat to a close approximation of its pre-degraded condition.

Scoping: The process by which study limits are set. In environmental impact assessment, the process by which key issues are identified.

Site of Special Scientific Interest (SSSI): A UK nature conservation designation. SSSIs are privately owned but are regarded as essential to maintain the UK's full variety of native habitats and species. They are managed under agreements between the national nature conservation agencies and land owners.

Social impact assessment: The process of assessing the social consequences of specific policy actions or project developments.

Social impacts: Social and cultural consequences for human populations associated with actions that alter the ways in which people live, work, play, relate to one another, organize to meet their needs, and cope as members of society.

Space-crowding: Effects that are concentrated in a defined area.

Species-centred environmental analysis (SCEA): A procedure for diagnosing species-specific environmental factors that limit the size of a population.

Species composition: A qualitative measure of the range of species present.

Species diversity: A measure of species richness and the relative abundance of species.

Species richness: The number of species present in an area.

Stability: The ability of an ecosystem to maintain some sort of equilibrium in the presence of perturbations.

Stepped matrix: A matrix in which environmental factors are set against others in order to display the consequences of initial changes in some factors for other factors.

Stochastic processes: Random processes.

Strategic ecological assessment: Evaluating the ecological impacts of policies, plans and programmes.

Strategic environmental assessment: Evaluating the environmental impacts of policies, plans and programmes. 'The formalized, systematic and comprehensive process of evaluating the environmental impacts of a policy, plan or programme and its alternatives, the preparation of a written report on the findings and the use of the findings in publicly accountable decision-making' (Therivel *et al.* 1992).

Suitability (habitat suitability): The ability of a habitat in its current condition to provide the life requisites of a species.

Sustainable development: Development that meets the needs of the present without compromising the ability of future generations to meet their own needs.

Technology assessment: The systematic study of the effects on society that may occur when a technology is introduced, extended or modified.

Territory: A defended 'home range'.

Threat: A measure of the degree to which a species or habitat is exposed to potential damage (anthropogenic) within a defined geographical area.

Time-crowding: The temporal concentration of impacts.

Translocation: The process by which an individual, a population, a community or a habitat is moved from one location to another.

Travel cost method (TCM): A method in economics used to place a value on non-market environmental goods by using consumption behaviour in related markets. The costs of consuming the environmental asset are used as a proxy for price.

Type 1 error: A type 1 error is made if there is an incorrect conclusion that an impact will occur.

Type 2 error: A type 2 error is made if there is an incorrect conclusion of 'no impact'.

Umbrella species: Umbrella species are those for which targeted conservation management will also benefit other species using the same habitat.

Zone: A large geographical area with a broadly homogeneous climate.

References

Adamus, P.R. (1995) Validating a habitat evaluation method for predicting avian richness. *Wildlife Society Bulletin* 23 (4), 743–749.

Allaby, M. (1994) *The Concise Oxford Dictionary of Ecology.* Oxford University Press, Oxford.

Allen, A.O. & Feddema J.J. (1996) Wetland loss and substitution by the Section 404 Permit Program in southern California, USA. *Environmental Management* 20 (2), 263–274.

Allen, T.F.H. & Starr, T.B. (1982) *Hierarchy: Perspectives for Ecological Complexity.* The University of Chicago Press, Chicago, IL.

Anderson, I. (1997) Damage limitation. *New Scientist* 20/27 December 1997, 16–17.

Angold, P.G. (1997a) Edge effects in fragmented habitats: implications for nature conservation. In: *Proceedings of the International Conference on Habitat Fragmentation and Infrastructure* (ed. K. Canters). Dutch Ministry of Transport, Public Works and Water Management, Delft.

Angold, P.G. (1997b) The impact of a road upon adjacent heathland vegetation: effects on plant species composition. *Journal of Applied Ecology* 34, 409–417.

Anon. (1979) *Record Repositories in Great Britain.* HMSO for Royal Commission on Historical Manuscripts, London.

Anon. (1989a) Mammals. In: *Guidelines for Selection of Biological SSSIs*, pp. 232–241. Nature Conservancy Council, Peterborough.

Anon. (1989b) Reptiles and amphibians. In: *Guidelines for Selection of Biological SSSIs*, pp. 265–268. Nature Conservancy Council, Peterborough.

Anselin, A., Meire, P.M. & Anselin, L. (1989) Multicriteria techniques in ecological evaluation: an example using the analytical hierarchy process. *Biological Conservation* 49, 215–231.

Arrow, K., Bolin, B., Costanza, R., Dasgupta, P., Folke, C., Holling, C.S., Jansson, B., Levin, S., Maler, K., Perrings, C. & Pimentel, D. (1995) Economic growth, carrying capacity and the environment. *Science* 268, 520–521.

Arrow, K., Bolin, B., Costanza, R., Dasgupta, P., Folke, C., Holling, C.S., Jansson, B., Levin, S., Maler, K., Perrings, C. & Pimentel, D. (1996) Economic growth, carrying capacity, and the environment. *Ecological Applications* 6 (1), 13–15.

Ashley, M.V., Melnick, D.J. & Western, D. (1990) Conservation genetics of the black rhinoceros (*Diceros bicornis*), 1: evidence from the mitochondrial DNA of three populations. *Conservation Biology* 4 (1), 71–77.

Atkinson, S.F. (1985) Habitat-based methods for biological impact assessment. *The Environmental Professional* 7, 265–282.

Bailey, J.M. & Saunders, A.N. (1988) Ongoing environmental impact assessment as a force for change. *Project Appraisal* 3 (1), 37–42.

Bailey, R.G. (1980) *Description of the Ecoregions of the United States.* Miscellaneous Publication No. 1391. US Forest Service, Washington, DC.

Baker, A.J.M (1987) Metal tolerance. In: *Frontiers of Comparative Plant Ecology* (eds I.H. Rorison, J.P. Grime, R. Hunt, G.A.F. Hendry & D.H. Lewis). *New Phytologist* 106 (Suppl.) 93–111. Academic Press, London.

Barbier, E.B., Markandya, A. & Pearce, D.W. (1990) Environmental sustainability and cost–benefit analysis. *Environment and Planning A* 22, 1259–1266.

Barlow, S.M., Bridges, J.W., Calow, P., Conning, D.M., Curnow, R.N., Dayan, A.D. & Purchase, I.F.H. (1992) Toxicity, toxicology and nutrition. In: *Risk analysis, perception, management,* pp. 35–65. Report of a Royal Society Study Group. The Royal Society, London.

Barr, C.J., Bunce, R.G.H., Clarke, R.T., Fuller, R.M., Furse, M.T., Gillespie, M.K., Groom, G.B., Hallam, C.J., Hornung, M., Howard, D.C. & Ness, M.J. (1993) *Countryside Survey 1990: Main Report (Countryside 1990,* Vol. II). Department of the Environment, London.

Barr, C., Gillespie, M. & Howard, D. (1994) *Hedgerow Survey 1993 (stock and change estimates of hedgerow lengths in England and Wales, 1990–1993).* Contract Report to the Department of the Environment, London.

Bartoldus, C.C. (1994) A procedure for the functional assessment of planned wetlands. *Water, Air and Soil Pollution* 77, 533–541.

Baskerville, G. (1986) Some scientific issues in cumulative environmental impact assessments. In: *Cumulative Environmental Effects: a binational perspective* (eds G.E. Beanlands, W.J. Erckmann, G.H. Orians, J. O'Riordan, D. Policansky, M.H. Sadar & B. Sadler). The Canadian Environmental Assessment Research Council and the US National Research Council. Ministry of Supply and Services, Ottawa.

Bateman, I. (1993) Valuation of the environment, methods and techniques: revealed preference methods. In: *Sustainable Environmental Economics and Management. Principles and Practice* (ed. R.K. Turner), pp. 193–265. Belhaven Press, London.

Batten, L.A., Bibby, C.J., Clement, P., Elliott, G.D. & Porter, R.F. (1990) *Red Data Birds in Britain.* Poyser, London.

Beanlands, G. (1992) *Cumulative Effects and Sustainable Development.* United Nations University Conference on the Definition and Measurement of Sustainable Development: the Biophysical Foundation. June 22–25, Washington, DC.

Beanlands, G.E. & Duinker, P.N. (1983) *An Ecological Framework for Environmental Impact Assessment in Canada.* Institute for Resource and Environmental Studies, Dalhousie University, Halifax; in cooperation with the Federal Environmental Impact Assessment Review Office, Canada.

Beanlands, G.E. & Duinker, P.N. (1984) An ecological framework for environmental impact assessment. *Journal of Environmental Management* 18, 267–277.

Bedford, B.L. (1996) The need to define hydrologic equivalence at the landscape scale for freshwater wetland mitigation. *Ecological Applications* 6 (1), 57–68.

Bedford, B.L. & Preston, E.M. (1988) Developing the scientific basis for assessing cumulative effects of wetland loss and degradation on landscape functions: status, perspectives and prospects. *Environmental Management* 12 (5), 751–771.

Bedward, M. & Pressey, R.L. (1991) Scores and score classes for evaluation criteria: a comparison based on the cost of reserving all natural features. *Biological Conservation* 56, 281–294.

Bedward, M., Pressey, R.L. & Keith, D.A. (1992) A new approach for selecting fully representative reserve networks: addressing efficiency, reserve design and land suitability with an iterative analysis. *Biological Conservation* 62, 115–125.

Begon, M., Harper, J.L. & Townsend, C.R. (1996) *Ecology: Individuals, Populations and Communities*, 3rd edn. Blackwell Science, Oxford.

Berger, J. (1990) Persistence of different-sized populations: an empirical assessment of rapid extinctions in Bighorn sheep. *Conservation Biology* 4 (1), 91–99.

Bewley, D. (1997) Southern birds soak up poisons. *New Scientist*, 31 May 1997, 7.

Bibby, C.J., Burgess, N.D. & Hill, D.A. (1992a) *Bird Census Techniques*. BTO and RSPB, Academic Press, London.

Bibby, C.J., Collar, N.J., Croosby, M.J., Heath, M.F., Imboden, Ch., Johnson, T.H., Long, A.J., Stattersfield, A.J. & Thirgood, S.J. (1992b) *Putting Biodiversity on the Map: Priority Areas for Global Conservation*. International Council for Bird Preservation, Cambridge.

Billings, W.D. (1952) The environmental complex in relation to plant growth and distribution. *Quarterly Review of Biology* 27, 251–265.

Bina, O., Briggs, B. & Bunting, G. (1997) Towards an assessment of the trans-European transport networks' impact on nature conservation. In: *Proceedings of the International Conference on Habitat Fragmentation and Infrastructure* (ed. K. Canters). Dutch Ministry of Transport, Public Works and Water Management, Delft.

Birks, H.J.B., Peglar, S.M. & Austin, H.A. (1993) *An Annotated Bibliography of Canonical Correspondence Analysis and Related Constrained Ordination Methods 1986–1993*. Botanical Institute, University of Bergen, Bergen.

Block, W.M., Brennan, L.A. & Gutierrez, R.J. (1987) Evaluation of guild-indicator species for use in resource management. *Environmental Management* 11 (2), 265–269.

Bratton, J.H. (1991) *British Red Data Books, 3, Invertebrates Other Than Insects*. Joint Nature Conservation Committee, Peterborough.

Bridson, G.D.R., Phillips, V.C. & Harvey, A.P. (eds) (1980) *Natural History Manuscript Resources in the British Isles*. Mansell, London.

Brinson, M.M. & Rheinhardt, R. (1996) The role of reference wetlands in functional assessment and mitigation. *Ecological Applications* 6 (1), 69–76.

Brookshire, D., Thayer, M.A., Schulze, W.D. & d'Arge, R.C. (1982) Valuing public goods: a comparison of survey and hedonic approaches. *American Economic Review* 72, 165–178.

Brown, K. & Moran, D. (1993) *Valuing Biodiversity: The Scope and the Limitations of Economic Analysis*. Centre for Social and Economic Research on the Gobal Environment. Global Environmental Change Working Paper 93-09.

Brown, K.S. (1991) Conservation of neotropical environments: insects as indicators. In: *The Conservation of Insects and their Habitats* (eds N.M. Collins & J.A. Thomas), pp. 350–404. Academic Press, London.

Brown, S., Schreier, H., Thompson, W.A. & Vertinsky, I. (1994) Linking multiple accounts with GIS as decision support system to resolve forestry/wildlife conflicts. *Journal of Environmental Management* 42, 349–364.

Buckley, G.P. & Fraser, S. (1998) *Locating new lowland woods*. English Nature Research Reports No. 283. English Nature, Peterborough.

Buckley, R.C. (1991) How accurate are impact predictions? *Ambio* 20, 161–162.

Bullock, J.M., Hodder, K.H., Manchester, S.J. & Stephenson, M.J. (1995) *A review of information, policy and legislation on species translocations*. Report to the Joint Nature Conservation Committee, Institute of Terrestrial Ecology, Huntingdon.

Bunce, R.G.H., Barr, C.J., Gillespie, M.K. & Howard, D.C. (1996) The ITE Land Classification: providing an environmental stratification of Great Britain project. *Environmental Monitoring and Assessment* 39 (1–3), 39–46.

Burdge, R. & Vanclay, F. (1995) Social impact assessment. In: *Environmental and Social Impact Assessment* (eds F. Vanclay & D.A. Bronstein), pp. 31–67. John Wiley and Sons, Chichester.

Burhenne-Guilmin, F. & Glowka, L. (1994) An introduction to the Convention on Biological Diversity. In: *Widening Perspectives on Biodiversity* (eds A.F. Krattiger, J.A. McNeely, W.H. Lesser, K.R. Miller, Y. St Hill & R. Senanayake), Chapter 1.4, pp. 15–19. IUCN/IAE, Gland.

Cain, D.H., Riitters, K. & Orvis, K. (1997) A multi-scale analysis of landscape statistics. *Landscape Ecology* 12, 199–212.

Canter, L.W. (1996) *Environmental Impact Assessment*, 2nd edn. McGraw-Hill, New York.

Carpenter, R.A. (1976) The scientific basis of NEPA — is it adequate? *Environmental Law Reporter* 6, 50014–50019.

Carpenter, S.R., Chisholm, S.W., Krebs, C.J. & Schindler, D.W. (1995) Ecosystem experiments. *Science* 269, 324–327.

CEC (1993) Implementation of Directive 85/337/EEC, COM (93) 28 Final. Commission of the European Communities, Brussels.

CEOS (1992) *The relevance of satellite missions to the study of the global environment.* Committee on Earth Observation Satellites report to the UNCED Conference, Rio de Janeiro. British National Space Centre, London.

Chesser, R.K., Smith, M.H. & Brisbin, I.L. Jr (1980) Management and maintenance of genetic variability in endangered species. *International Zoological Yearbook* 20, 146–154.

Chesser, R.K., Rhodes, O.E. Jr, Sugg, D.W. & Schnabel, A. (in press) Effective sizes for subdivided populations. *Genetics.*

Chiras, D.D. (1988) *Environmental Science: A Framework for Decision Making,* 2nd edn. Benjamin Cummings, Menlo Park, CA.

Christensen, S.W., Van Winkle, W. & Mattice, J.S. (1976) Defining and determining the significance of impacts: concepts and methods. In: *Proceedings of a Workshop on the Biological Significance of Environmental Impacts* (eds R.K. Sharma, J.D. Buffington & J.T. McFadden), pp. 35–59. Publ. EE-5, Institute for Environmental Studies, University of Toronto, Toronto.

Clark, J.R., Lewis, M.A. & Pait, A.S. (1993) Pesticide inputs and risks in coastal wetlands. *Environmental Toxicology and Chemistry* 12, 2225–2233.

Clark, W.C. (1986) The cumulative impacts of human activities on the atmosphere. In: *Cumulative Environmental Effects: a binational perspective* (eds G.E. Beanlands, W.J. Erckmann, G.H. Orians, J. O'Riordan, D. Policansky, M.H. Sadar & B. Sadler). The Canadian Environmental Assessment Research Council and the US National Research Council. Ministry of Supply and Services, Ottawa.

Clements, F.E. (1920) *Plant Indicators.* Carnegie Institute of Washington Publication 290, Washington, DC.

Cocklin, C., Parker, S. & Hay, J. (1992) Notes on cumulative environmental change I: concepts and issues. *Journal of Environmental Management* 35, 31–49.

Coddington, J.A., Griswold, C.E., Silva, D., Peneranda, E. & Larcher, S.F. (1991) Designing and testing sampling protocols to estimate biodiversity in tropical ecosystems. In: *The Unity of Evolutionary Biology* (ed. E.C. Dudley), pp. 44–60. Proceedings of the Fourth International Congress of Systematic and Evolutionary Biology. Dioscorides Press, Portland, OR.

Collins, R.J. & Barrett, G.W. (1997) Effects of habitat fragmentation on meadow vole (*Microtus pennsylvanicus*) population dynamics in experiment landscape patches. *Landscape Ecology* 12 (2), 63–76.

Collins, S.L. & Benning, T.L. (1996) Spatial and temporal patterns in functional diversity. In: *Biodiversity: A Biology of Numbers and Difference* (ed. K.J. Gaston), pp. 253–280. Blackwell Science, Oxford.

Connell, J.H. & Sousa, W.P. (1983) On the evidence needed to judge ecological stability or persistance. *American Naturalist* **121** (6), 789–824.

Contant, C.K. & Wiggins, L.L. (1989) Toward defining and assessing cumulative impacts: practical and theoretical considerations. In: *Environmental Analysis: The NEPA Experience* (eds S.G. Hildebrand & J.B. Cannon), pp. 336–356. Lewis Publishers, Boca, FL.

Contant, C.K. & Wiggins, L.L. (1991) Defining and analysing cumulative environmental impacts. *Environmental Impact Assessment Review* **11**, 297–309.

Corbet, G.B. & Harris, S. (eds) (1996) *The Handbook of British Mammals*, 3rd edn. Blackwell Science, Oxford.

Cowardin, L.M., Carter, V., Golet, F.C. & LaRoe, E.T. (1979) *Classification of Wetlands and Deepwater Habitats of the United States*. US Fish and Wildlife Service, Washington, DC.

Cowart, R.H. (1986) Vermont's Act 250 after 15 years: can the permit system address cumulative impacts? *Environmental Impact Assessment Review* **6**, 135–144.

Cowell, E.B. (1978) Ecological monitoring as a management tool in industry. *Ocean Management* **4**, 273–285.

Dahuri, R. (1994) Incorporating biodiversity objectives and criteria into environmental impact assessment laws and mechanisms in Indonesia. In: *Widening Perspectives on Biodiversity* (eds A.F. Krattiger, J.A. McNeely, W.H. Lesser, K.R. Miller, Y. St Hill & R. Senanayake), pp. 319–325. IUCN/IAE, Gland.

Danson, F.M. & Plummer, S.E. (eds) (1995) *Advances in Environmental Remote Sensing*. John Wiley and Sons, Chichester.

Davis, R. (1963) Recreation planning as an economic problem. *Natural Resources Journal* **3** (2), 239–249.

Dee, N., Barker, J., Drobuy, N. & Duke, K. (1973) An environmental evaluation system for water resources planning. *Water Resources Research* **9**, 523.

Demarchi, D.A. (1993) *Ecoregions of British Columbia*, 3rd edn. 1 : 2 000 000 Map. BC Ministry of Environment, Lands and Parks, Victoria.

Demarchi, D.A. & Lea, E.C. (1989) Biophysical habitat classification in British Columbia: an interdisciplinary approach to ecosystem evaluation. In: *Proceedings — Land Classification Based on Vegetation: Applications for Resource Management, Moscow* (compiled by D.E. Ferguson, P. Morgan & F.D. Johnson), pp. 275–276. General Technical Report INT-257, USDA Forest Service, Intermountain Research Station, Ogden, UT.

Demarchi, D., Bonner, L., Simpson, K., Andrusia, K.L. & Lashmar, M. (1996) *Standards for Wildlife Habitat Capability/Suitability Ratings in British Columbia*. Wildlife Interpretations Subcommittee. Resources Inventory Committee. Ministry of Environment, Lands and Parks, Victoria.

Department of Environment Affairs (1992) *Checklist of Environmental Characteristics*. Guideline Document 5. Department of Environment Affairs, Pretoria.

Department of the Environment (1989) *Environmental Assessment: a Guide to the Procedures*. HMSO, London.

Department of the Environment (1990) *Integrated Pollution Control — A Practical Guide*. HMSO, London.

Department of the Environment (1991) *Interpretation of major accidents to the environment for purposes of the CIMAH regulations*. A guidance note by the

Department of the Environment. Department of the Environment, London.

Department of the Environment (1995) *A Guide to Risk Assessment and Risk Management for Environmental Protection*. HMSO, London.

Department of the Environment (1996) *Indicators of Sustainable Development for the United Kingdom*. HMSO, London.

Detwyler, T.R. (1971) *Man's Impact on the Environment*. McGraw-Hill, London.

Diamond, J.M. (1975) The island dilemma: lessons of modern biogeographic studies for the design of natural preserves. *Biological Conservation* 7, 129–146.

Dobberteen, R.A. & Nickerson, N.H. (1991) Use of created cattail (*Typha*) wetlands in mitigation strategies. *Environmental Management* 15 (6), 797–808.

Donn, S. & Wade, M. (1994) *UK Directory of Ecological Information*. Packard Publishing, Chichester.

Duel, H., Specken, B.P.M., Denneman, W.D. & Kwakernaak, C. (1995) The habitat evaluation procedure as a tool for ecological rehabilitation of wetlands in the Netherlands. *Water Science and Technology* 31 (8), 387–391.

Durell, S.E.A. le V. Dit, Goss-Custard, J.D. & Clarke, R.T. (1997) Differential response of migratory subpopulations to winter habitat loss. *Journal of Applied Ecology* 34, 1155–1164.

DWW (1995) *Wildlife Crossings for Roads and Waterways*. Ministry of Transport, Public Works and Water Management, Road and Hydraulic Engineering Division, Delft.

Eberhardt, L.L. (1976) Quantitative ecology and impact assessment. *Journal of Environmental Management* 4, 27–70.

Eckholm, E.P. (1982) *Down to Earth: Environment and Human Needs*. Pluto Press, London.

Ecosystem Working Group (1998) *Standard for Terrestrial Ecosystem Mapping in British Columbia*. Ecosystems Working Group of the Terrestrial Ecosystem Task Force, British Columbia Resources Inventory Committee, Victoria.

Edmonds, C.P., Keating, D.M. & Stanwick, S. (1997) Wetland mitigation. *The Appraisal Journal*, January 1997, 72–76.

EEC (1982) Council Directive 82/501/EEC on the Major-Accident Hazards of Certain Industrial Activities. (OJ No. L230, 5.8.82.)

Ehrlich, P. & Ehrlich, A. (1992) The value of biodiversity. *Ambio* 21 (3), 219–226.

EIA Centre (1995) *EIA Newsletter* 10 (eds N. Lee, J. Hughes, C. Wood & C.E. Jones). EIA Centre, Manchester.

Elliott, J.M. (1977) *Some Methods for the Statistical Analysis of Samples of Benthic Invertebrates*, 2nd edn. Freshwater Biological Association, Ambleside.

Emlen, J.M. (1989) Terrestrial population models for ecological risk assessment—a state of the art review. *Environmental Toxicology and Chemistry* 8 (9), 831–842.

English Nature (1992) *Strategic Planning and Sustainable Development*. English Nature, Peterborough.

English Nature (1993) *Roads and Nature Conservation. Guidance on Impacts, Mitigation and Enhancement*. English Nature, Peterborough.

English Nature (1994a) *Nature Conservation in Environmental Impact Assessment*. English Nature, Peterborough.

English Nature (1994b) *Are habitat corridors conduits for animals and plants in a fragmented landscape? A review of the scientific evidence*. English Nature Research Report 94. English Nature, Peterborough.

English Nature (1997) *Wildlife and Freshwater: An Agenda for Sustainable Development*. English Nature, Peterborough.

Environment Agency (1996) *The Environment Agency and Sustainable Development*. Explanatory document accompanying the statutory guidance.

Department of Environment, London.

Environment Canada (1994) *A Framework for Ecological Risk Assessment at Contaminated Sites in Canada: Review and Recommendations.* Scientific Series no. 199. Ecosystem Conservation Directorate Evaluation and Interpretation Branch, Environment Canada, Ottawa.

Environment Canada, Canadian Parks Service (1993) *Kouchibouguac National Park Management Plan.* Resource and Conservation, Kouchibouguac National Park, Halifax.

Epp, H.T. (1995) Application of science to environmental impact assessment in boreal forest management: the Saskatchewan example. *Water, Air and Soil Pollution* 82, 179–188.

European Community (1994) Proposal for a Council Directive amending Directive 85/337/EEC on the assessment of the effects of certain public and private projects on the environment. *Official Journal of the European Communities No. C 130*, 8–12.

Eversham, B.C., Harding, P.T., Loder, N., Arnold, H.R. & Fenton, R.W. (1992) Research applications using data from species surveys in Britain. In: *Faunal Inventories of Sites for Cartography and Nature Conservation* (eds J.L. van Goethem & P. Grootaert), pp. 29–40. Bulletin de l'Institute Royal des Sciences Naturelles de Belgique.

Foppen, R. & Reijnen, R. (1994) The effects of car traffic on breeding bird populations in woodland II. Breeding dispersal of male willow warblers *Phylloscopus trochilus* in relation to the proximity of a highway. *Journal of Applied Ecology* 31, 95–101.

Forbes, J. & Heath, J. (1990) *The Ecological Impact of Road Schemes.* Nature Conservancy Council and the Department of Transport, Peterborough.

Frodin, D.G. (1990) *Guide to the Standard Floras of the World. An annotated, geographically arranged systematic bibliography of the principal floras, enumerations, checklists and chorological atlases of different areas.* Cambridge University Press, Cambridge.

Fry, C. (1996) Helping nature. *GIS Europe*, September 1996, 28–30.

Fuller, R.J. (1980) A method for assessing the ornithological interest of sites for conservation. *Biological Conservation* 17, 229–239.

Fuller, R.J., Stuttard, P. & Ray, C.M. (1989) The distribution of breeding songbirds within mixed coppiced woodland in Kent, England in relation to vegetation age and structure. *Annales Zoologici Fennici* 26, 265–275.

Fuller, R.M., Groom, G.B. & Jones, A.R. (1994) The Land Cover Map of Great Britain: an automated classification of Landsat Thematic Mapper data. *Photogrammetric Engineering and Remote Sensing* 60, 553–562.

Gaston, K.J. (1996a) (ed.) *Biodiversity: A Biology of Numbers and Difference.* Blackwell Science, Oxford.

Gaston, K.J. (1996b) Species richness: measure and measurement. In: *Biodiversity: A Biology of Numbers and Difference* (ed. K.J. Gaston), pp. 77–114. Blackwell Science, Oxford.

Geppert, R.R., Lorenz, C.W. & Larson, A.G. (1984) *Cumulative effects of forest practice on the environment: a state of knowledge.* Report prepared for the Washington Forest Practice Board, Olympia.

Gibeau, M.L. (1993) *Grizzly habitat effectiveness model for Banff, Yoho and Kootnay National Parks.* Parks Canada, Banff.

Gibson, R.B. (1993) Environmental assessment design: lessons from the Canadian experience. *The Environmental Professional* 15 (1), 12–24.

Gillespie, J. & Shepherd, P. (1995) *Establishing criteria for identifying critical natural capital in the terrestrial environment.* English Nature Research Report No.141. English Nature, Peterborough.

Gilpin, M.E. & Soule, M.E. (1986) Minimum viable populations: processes of species extinction. In: *Conservation Biology: The Science of Scarcity and Diversity* (ed. M.E. Soule), pp. 19–34. Sinauer Associates, Sunderland, MA.

Goss-Custard, J.D., Durell, S.E.A. le V. Dit, Caldow, R.W.G. & Ens, B.J. (1995) Population consequences of winter habitat loss in a migratory shorebird: II model predictions. *Journal of Applied Ecology* 32, 337–351.

Gotmark, F., Ahlund, M. & Eriksson, M. (1986) Are indices reliable for assessing conservation value of natural areas. An avian case study. *Biological Conservation* 38, 55–73.

Green, M. (1995) Assessing oil and gas developments. *Ecos* 16 (3/4), 80.

Hagemeijer, W.J.M. & Blair, M.J. (1997) *The European Bird Census Council's Atlas of European Breeding Birds.* T. and A.D. Poyser, London.

Haimes, Y.Y., Barry, T. & Lambert, J.H. (eds) (1994) *When and How Can You Specify a Probability Distribution When You Don't Know Much?* Proceedings of a workshop organized by the US Environment Protection Agency and the University of Virginia, April 18–20, 1993.

Hamilton-Wright, D. (1983) Species–energy theory: an extension of species–area theory. *Oikos* 41, 496–506.

Hanley, N. & Spash, C.L. (1993) *Cost–Benefit Analysis and the Environment.* Edward Elgar, Aldershot.

Harris, J.A. (1983) *Birds and Coniferous Plantations.* The Royal Forestry Society of England, Wales and Northern Ireland, Tring.

Harris, S. (1989) Taking stock of the brock. *BBC Wildlife* 7, 460–464.

Harris, S. & Jefferies, D.J. (1991) Working within the law: guidelines for veterinary surgeons and wildlife rehabilitators on the rehabilitation of wild mammals. *British Veterinary Journal* 147, 12–16.

Harrison, S., Murphy, D.D. & Ehrlich, P.R. (1988) Distribution of the Bay checkerspot butterfly, *Euphrydryas editha bayensis*: evidence for a metapopulation model. *American Naturalist* 132, 360–382.

Hawksworth, D.L. & Rose, F. (1976) *Lichens as Pollution Monitors.* Edward Arnold, London.

Henriques, R.G. (1996) *The Portuguese National Network of Geographical Information (SNIG Network).* Proceedings of the Joint European Conference (JEC) 27–29 March 1996, Barcelona.

Herson, A.I. (1986) in McCold, L.N. (1991) Reducing global, regional and cumulative impacts with the national environmental policy act. *The Environmental Professional* 13, 107–113.

Herson, A.I. & Bogdan, K.M. (1991) Cumulative impact analysis under NEPA: recent legal developments. *The Environmental Professional* 13, 11–106.

Heyer, W.R., Donnelly, M.A., McDiarmid, R.W., Hayek, L.C. & Foster, M.S. (eds) (1994) *Measuring and Monitoring Biological Diversity: Standard Methods for Amphibians.* Smithsonian Institution Press, Washington, DC.

Hill, D. & Hockin, D. (1992) Can roads be bird friendly? *Landscape Design,* February, 38–41.

Hill, D., Hockin, D., Price, G., Tucker, R., Morris, R. & Treweek, J. (1997) Bird disturbance: improving the quality and utility of disturbance research. *Journal of Applied Ecology* 34 (2), 275–289.

Hill, M.O., Preston, C.D. & Smith, A.G. (eds) (1994) *Atlas of Bryophytes of Britain and Ireland,* Vol. 3. *Mosses (Diplolepidae).* Harley, Colchester.

Hinsley, S.A., Bellamy, P.E. & Newton, I. (1995) Bird species turnover and stochastic extinction in woodland fragments. *Ecography* **18**, 41–50.

HMSO (1996) *Biodiversity: the UK Steering Group Report*. HMSO, London.

Hockin, D., Ounsted, M., Gorman, M., Hill, D., Keller, V. & Barker, M. (1992) Examination of the effects of disturbance on birds with reference to the role of environmental impact assessments. *Journal of Environmental Management* **36**, 253–286.

Hodson, N.L. & Snow, D.W. (1965) The Road Deaths Enquiry, 1960–1961. *Bird Study* **12**, 90–99.

Hollick, M. (1981) The role of quantitative decision-making methods in environmental impact assessment. *Journal of Environmental Management* **12**, 65–78.

Holling, C.S. (1973) Resilience and stability of ecological systems. *Annual Review of Ecology and Systematics* **4**, 1–23.

Holling, C.S. (ed.) (1978) *Adaptive Environmental Assessment and Management*. John Wiley and Sons, London.

Holling, C.S. (1986) The resilience of terrestrial ecosystems: local surprise and global change. In: *Sustainable Development of the Biosphere* (eds W.C. Clark & R.E. Munn), pp. 292–317. Cambridge University Press, Cambridge.

Holling, C.S. (1992) Cross-scale morphology, geometry and dynamics of ecosystems. *Ecological Monographs* **62** (4), 447–502.

Holling, C.S. & Clark, W.C. (1975) Notes towards a science of ecological management. In: *Unifying Concepts in Ecology*, pp. 247–251. Report of the plenary sessions, First International Congress of Ecology, PUDOC, Wageningen.

Hood, C.C., Jones, D.K.C., Pidgeon, N.F., Turner, B.A. & Gibson, R. (1992) Risk management. In: *Risk Analysis, Perception, Management*, pp. 135–201. Report of a Royal Society Study Group. The Royal Society, London.

Hooper, M.D. (1970) Dating hedges. *Area* **4**, 63–65.

Horak, G.C., Vlachos, E.C. & Cline, E.W. (1983) *Methodological guidance for assessing cumulative impacts on fish and wildlife*. Report for Eastern Energy and Land Use Team, US Fish and Wildlife Service, Washington, DC.

Hughes, D. (1992) *Environmental Law*. Butterworths, London.

Humphrey, C.L., Faith, D.P. & Dostine, P.L. (1995) Baseline requirements for assessment of mining impact using biological monitoring. *Australian Journal of Ecology* **20**, 150–166.

Hunsaker, C.T. & Carpenter, D.E. (eds) (1990) *Ecological indicators for the Environmental Assessment and Monitoring Program*. EPA/600/3-91/023. US Environmental Protection Agency. Environmental Research Laboratory, Corvallis, OR.

Hunter, M.D. & Price, P.W. (1992) Playing chutes and ladders: heterogeneity and the relative roles of bottom-up and top-down forces in natural communities. *Ecology* **73** (3), 724–732.

Hurlbert, S.H. (1984) Pseudoreplication and the design of ecological field experiments. *Ecological Monographs* **54** (2), 187–211.

Institute of Environmental Assessment (1995) *Guidelines for Baseline Ecological Impact Assessment*. E. and F.N. Spon, London.

IUCN (1987) *The IUCN Position Statement on Translocation of Living Organisms*. IUCN, Gland.

IUCN (1995) *Guidelines for Reintroductions*. IUCN/Species Survival Commission Reintroduction Specialist Group. IUCN, Gland.

James, F.C., Hess, C.A. & Kufrin, D. (1997) Species-centered environmental analysis: indirect effects of fire history on red-cockaded woodpeckers. *Ecological Applications* **7** (1), 118–129.

Janzen, D.H. & Hallwachs, W. (1994) *All Taxa Biodiversity Inventory (ATBI) of terrestrial systems.* Draft report of an NSF Workshop, 16–18 April 1993, Philadelphia, PA.

Jenny, H. (1941) *Factors of Soil Formation.* McGraw-Hill, New York.

Joao, E. & Fonseca, A. (1996) *Current Use of Geographical Information Systems for Environmental Assessment: A Discussion Document.* LSE Research Papers in Environmental and Spatial Analysis No. 36. London School of Economics, London.

Johnson, A.R. (1988) Diagnostic variables as predictors of ecological risk. *Environmental Management* **12**, 515–523.

Johnston, C.A. (1998) *Geographic Information Systems in Ecology.* Blackwell Science, Oxford.

Johnston, C.A., Detenbeck, N.E., Bonde, J.P. & Neimi, G.J. (1988) Geographical information systems for cumulative impact assessment. *Photogrammetric Engineering and Remote Sensing* **54** (11), 1609–1615.

Joint Nature Conservation Committee (1993) *Handbook for Phase 1 Habitat Survey: A Technique for Environmental Audit.* Joint Nature Conservation Committee, Peterborough.

Jones, C.A. & Pease, K.A. (1996) *Restoration-Based Measures of Compensation in Natural Resource Liability Statutes.* Seventh Annual Conference of the European Association of Environmental and Resource Economists (EAERE), Lisbon, June 27–29.

Jones, P.H. (1981) Snow and ice control and the transport environment. *Environmental Conservation* **8**, 33–38.

Kalff, S.A. (1995) *A proposed framework to assess cumulative environmental effects in Canadian national parks.* Parks Canada Technical Report in Ecosystem Science, No. 1. Parks Canada, Halifax.

Kansas, J.L., Raine, R.M. & Gibeau, M.L. (1989) *Ecological studies of the black bear in Banff National Park, Alberta.* Final Report. Prepared by Beak Associates Consulting Ltd for Canadian Parks Service, Western Region, Calgary.

Karr, J.R., Fausch, K.D., Angermeier, P.L., Yant, P.R. & Schlosser, I.J. (1986) *Assessing Biological Integrity in Running Waters: A Method and its Rationale.* Special Publication No 5. Illinois Natural History Survey, Champaign, IL.

Kersting, K. (1988) Normalised ecosystem strain in micro-ecosystems using different sets of state variables. *Verh. Int. Verein. Limnol.* **23**, 1641–1646.

Kirby, P. (1992) *Habitat Management for Invertebrates: A Practical Handbook.* RSPB, Sandy.

Kleinschmidt, V. (1994) Rahmenkonzept fur windraftanlagen und -parks im Binnenland. *Natur und Landschaft* **69**, 9–18.

Komex International Ltd (1995) *Atlas of the Central Rockies ecosystem. Towards an ecologically sustainable landscape.* A Status Report to the Central Rockies Ecosystem Interagency Liaison Group (CREILG). Calgary.

Korte, G.B. (1994) *The GIS Book*, 3rd edn. OnWord Press, Santa Fe, FL.

La Point, T.W. (1995) Signs and measurements of ecotoxicity in the aquatic environment. In: *Handbook of Ecotoxicology* (ed. D.J. Hoffman), pp. 13–24. CRC Press, Boca Raton, FL.

Landis, W.G., Matthews, G.B., Matthews, R.A. & Sergeant, A. (1994) Application of multivariate techniques to end-point determination, selection and evaluation in ecological risk assessment. *Environmental Toxicology and Chemistry* **13** (12), 1917–1927.

Langton, T.E.S. (ed.) (1989) *Amphibians and Roads.* ACO Polymer Products,

Shefford.

Langton, T.E.S. & Beckett, C.L. (1995) Home range size of Scottish amphibians and reptiles. *Scottish Natural Heritage Review* 53.

Lawton, J.H., Prendergast, J.R. & Eversham, B.C. (1994) The numbers and spatial distributions of species: an analysis of British data. In: *Systematics and Conservation Evaluation* (eds P. Forey, C.J. Humphries & R.I. Vane-Wright), pp. 177–195. Oxford University Press, Oxford.

Lee, N. & Colley, R. (1990) *Reviewing the Quality of Environmental Statements.* Occasional Paper 24. EIA Centre, University of Manchester.

Lee, N. & Walsh, F. (1992) Srategic environmental assessment: an overview. *Project Appraisal* 7, 126–136.

Lee, N. & Wood, C.M. (1978) EIA—a European perspective. *Built Environment* 4, 101–110.

Lefkovitch, L.P. & Fahrig, L. (1985) Spatial characteristics of habitat patches and population survival. *Ecological Modelling* 30, 297–308.

Leibowitz, N.C., Ernst, T.L., Urquhart, N.S., Stehman, S. & Roose, D. (1993) *Evaluation of EMAP-Wetlands Sampling Design Using National Wetlands Inventory Data.* EPA/620/R-93/773. US Environmental Protection Agency, Environmental Research Laboratory, Corvallis, OR.

Leopold, L.B., Clarke, F.E., Hanshaw, B.B. & Balsey, J.R. (1971) A procedure for evaluating environmental impact. *US Geological Survey Circular* 645. US Geological Survey, Washington, DC.

Levy, S. (1997) Ultimate sacrifice. *New Scientist* 6, September 1997, 39–41.

Likens, G.E. (1985) An experimental approach for the study of ecosystems. *Journal of Ecology* 73, 381–396.

Liu, J.H. & Hills, P. (1997) Environmental planning, biodiversity and the development process: the case of Hong Kong's Chinese white dolphins. *Journal of Environmental Management* 50, 351–367.

Loder, N. (1992) *The habitats of the moths and butterflies of Great Britain.* Unpublished report to the University of York, as dissertation for BSc in ecology.

Lovejoy, T.E., Bierregaard, R.O., Rylands, A.B., Malcolm, J.R., Quintela, C.E. & Hayes, M.B. (1986) Edge and other effects of isolation on Amazon forest fragments. In: *Conservation Biology, the Science of Scarcity and Diversity* (ed. M.E. Soule), pp. 257–285. Sinauer Associates, MA.

Lubchenko, J., Olson, A.M., Brubaker, L.B., Carpenter, S.R., Holland, M.M., Hubbell, S.P., Levin, S.A., MacMahon, J.A., Matson, P.A., Melillo, J.M., Mooney, H.A., Peterson, C.H., Pulliam, H.P., Real, L.A., Regal, P.J. & Risser, P.G. (1991) The sustainable biosphere initiative: an ecological research agenda. *Ecology* 72, 371–412.

Ludwig, D. (1996) The end of the beginning. *Ecological Applications* 6 (1), 16–17.

Mack, E.L., Firbank, L.G., Bellamy, P.E., Hinsley, S.A. & Veitch, N. (1997) The comparison of remotely sensed and ground-based habitat area data using species–area models. *Journal of Applied Ecology* 34 (5), 1222–1229.

Mader, H.-J. (1984) Animal habitat isolation by roads and agricultural fields. *Biological Conservation* 29, 81–96.

Major, J. (1951) A functional, factorial approach to plant ecology. *Ecology* 32, 392–412.

Mallet, J. (1996) The genetics of biological diversity: from varieties to species. In *Biodiversity: A Biology of Numbers and Difference* (ed. K.J. Gaston), pp. 13–47. Blackwell Science, Oxford.

Mannan, R.W., Morrison, M.L. & Meslow, E.C. (1984) Comment: the use of guilds in forest bird management. *Wildlife Society Bulletin* 12, 426–430.

Mapstone, B.D. (1995) Scalable decision rules for environmental impact studies. *Ecological Applications* **5** (2), 401–410.

Marcot, B.G. & Holthausen, R. (1987) Analysing population viability of the Spotted Owl in the Pacific Northwest. *Transactions of the North American Wildlife Natural Resources Conference* **52**, 333–347.

Margules, C.R. & Usher, M.B. (1981) Criteria used in assessing wildlife potential. *Biological Conservation* **21**, 79–109.

Margules, C.R., Nicholls, A.O. & Pressey, R.L. (1988) Selecting networks of reserves to maximise biological diversity. *Biological Conservation* **43**, 663–676.

Markandya, A. & Pearce, D.W. (1991) Development, the environment and the social rate of discount. *World Bank Research Observer* **6** (2), 137–152.

Marsh, L.L., Porter, D.R. & Salvesen, D.A. (eds) (1996) *Mitigation Banking: Theory and Practice.* Island Press, Washington, DC.

Mason, J. & Sadoff, C.W. (1994) Why protect protected areas? *Dissemination Notes on the Environment*, No 8. World Bank, Washington, DC.

May, R.M. (1990) How many species? *Philosophical Transactions of the Royal Society London B* **330**, 293–304.

McAuley, I. (1991) Environmental impact analysis: a cost-effective GIS application? *Mapping Awareness* **5** (4), 36–40.

McCold, L.N. (1991) Reducing global, regional and cumulative impacts with the national environmental policy act. *The Environmental Professional* **13**, 107–113.

Merenlander, A.M., Woodruff, D.S., Ryder, O.A., Kock, R. & Vahala, J. (1989) Allozyme variation and differentiation in African and Indian Rhinoceroses. *Journal of Heredity* **80**, 377–382.

Merriam, G. (1994) *Is the Metaphor of Ecosystem Health Useful in Landscape Ecology?* Paper presented at the First International Symposium on Ecosystem Health and Medicine, 19–23 June, Ottawa.

Merriam, G., Wegner, J. & Pope, S. (1993) *Generic Ecological Framework for National Park Planning and Management.* Landscape Ecology Research Laboratory, Ottawa.

Messer, J.J., Linthurst, R.A. & Overton, W.S. (1991) An EPA program for monitoring ecological status and trends. *Environmental Monitoring and Assessment* **17**, 67–78.

Middleton, D.A.J. & Nisbet, R.M. (1997) Population persistence time: estimates, models and mechanisms. *Ecological Applications* **7** (1), 107–117.

More, T.A., Averill, J.R. & Stevens, T.H. (1996) Values and economics in environmental management: a perspective and critique. *Journal of Environmental Management* **48**, 397–409.

Morton, J.K. (1982) Preservation of endangered species by transplantation. *Canadian Botanical Association Bulletin* **15**, 32.

Munro, A. & Hanley, N. (1991) *Shadow Projects and the Stock of Natural Capital: A Cautionary Note.* Stirling discussion papers in economics, finance and investment. Department of Economics, University of Sterling.

Murphy, D.D., Freas, K.E. & Weiss, S.B. (1990) An environment-metapopulation approach to population viability analysis for a threatened invertebrate. *Conservation Biology* **4**, 41–51.

Naeem, S., Thompson, L.J., Lawler, S.P., Lawton, J.H. & Woodfin, R.M. (1994) Declining biodiversity can alter the performance of ecosystems. *Nature* **368**, 734–737.

Nanson, B., Smith, N. & Davey, A. (1995) *What is the British National Geospatial Database?* Proceedings of the AGI '95 Conference, Birmingham, pp. 1.41–1.45. The Association for Geographic Information, London.

Natural Environment Research Council (1992) *Evolution and biodiversity. The new taxonomy*. Report of the Committee set up by the NERC and chaired by Prof. J.R. Krebs, FRS. NERC, Swindon.

Nature Conservancy Council (1990) *Review of NCC Policy on Species Translocations in Great Britain*. NCC, Peterborough.

Nicholson, A.M. (1980) *Ecology of the sand lizard* Lacerta agilis *L. in southern England and comparisons with the common lizard* Lacerta vivipara *Jacquin*. PhD thesis, University of Southampton.

Nilsson, C. & Grelsson, G. (1995) The fragility of ecosystems — a review. *Journal of Applied Ecology* **32** (4), 677–692.

NOAA (1996) *Habitat Equivalency Analysis: An Overview*. Policy and Technical Paper Series, number 95–1. Damage Assessment and Restoration Program. National Oceanic and Atmospheric Administration Department of Commerce.

Norton, B.G. (1991) Ecological health and sustainable resource management. In: *Ecological Economics: the Science and Management of Sustainability* (ed. R Costanza), pp. 102–117. Columbia University Press, New York.

Novajosky, W. (1993) Present and future capabilities of SPOT satellite systems. In: *Proceedings of the 12th Pecora Memorial Conference: Land Information from Space Based Systems*, Sioux Falls, 24–26 August, pp. 319–321. American Society for Photogrammetry and Remote Sensing, Bethesda, MD.

Novitzki, R.P. (1995) EMAP-wetlands: a sampling design with global application. *Vegetatio* **118**, 171–184.

O'Byrne, P., Nelson, J. & Seneca, J. (1985) Housing values, census estimates, disequilibrium and the environmental cost of airport noise. *Journal of Environmental Economics and Management* **12**, 169–178.

Odum, E.P. (1977) The emergence of ecology as a new integrative discipline. *Science* **195**, 1289–1293.

Odum, W.E. (1982) Environmental degradation and the tyranny of small decisions. *BioScience* **32** (9), 728–729.

Okey, B.W. (1996) Systems approaches and properties and agroecosystem health. *Journal of Environmental Management* **48**, 187–199.

Ortolano, L. & Shepherd, A. (1995) Environmental impact assessment. In: *Environmental and Social Impact Assessment* (eds F. Vanclay & D.A. Bronstein), pp. 3–31. John Wiley and Sons, Chichester.

Osieck, E.R. (1986) Bedreigde en karakteristieke vogels in Nederland. *Ned. Ver. tot bescherming van vogels*, Zeist.

Paine, R.T. (1974) Intertidal community structure: experimental studies on the relationship between a dominant competitor and its principal competitor. *Oecologia* **15**, 93–120.

Parker, D.M. (1995) *Habitat Creation: A Critical Guide*. English Nature, Peterborough.

Peach, W. & Baillie, S.R. (1989) Population changes on constant effort sites, 1987–1988. *BTO News* **161**, 12–13.

Pearce, D., Markandya, A. & Barbier, E.B. (1989) *Blueprint for a Green Economy*. Earth Scan Publications, London.

Pearce, F. (1997) Northern Exposure. *New Scientist* **31**, May 1997, 24–27.

Penny, M. (1988) *Rhinos: Endangered Species*. Facts on File, New York.

Perring, F.H. & Farrell, L. (1977) *British Red Data Books*, 1, *Vascular Plants*, 1st edn. Royal Society for Nature Conservation, Nettleham.

Perring, F.H. & Farrell, L. (1983) *British Red Data Books*, 1, *Vascular Plants*, 2nd edn. Royal Society for Nature Conservation, Nettleham.

Peterken, G.F. (1974) A method for assessing woodland flora for conservation using indicator species. *Biological Conservation* 6, 239–245.

Peters, R.H. (1991) *A Critique for Ecology*. Cambridge University Press, Cambridge.

Peterson, E.B., Chan, Y.H., Peterson, N.M., Constable, G.A., Caton, R.B., Davies, C.S., Wallace, R.R. & Yarranton, G.A. (1987) *Cumulative Effects Assessment in Canada: An Agenda for Action and Research*. Canadian Environmental Impact Assessment Research Council, Ottawa.

Pickett, S.T.A. & Thompson, J.N. (1978) Patch dynamics and the design of nature reserves. *Biological Conservation* 13, 27–37.

Pielou, E.C. (1981) The usefulness of ecological models: a stock-taking. *The Quarterly Review of Biology* 56, 17–31.

Plummer, S.E., Danson, F.M. & Wilson, A.K. (1995) Advances in remote sensing technology. In: *Advances in Environmental Remote Sensing* (eds F.M. Danson & S.E. Plummer), pp. 1–9. John Wiley and Sons, Chichester.

Pollard, E., Hooper, M.D. & Moore, N.W. (1974) *Hedges*. Collins, London.

Porneluzi, P., Bednarz, J.C., Goodrich, L.G., Zawada, N. & Hoover, J. (1993) Reproductive performance of territorial overnbirds occupying forest fragments and a continuous forest in Pennsylvania. *Conservation Biology* 7, 618–622.

Potter, C., Barr, C. & Lobley, M. (1996) Environmental change in Britain's countryside: an analysis of recent patterns and socio-economic processes based on the Countryside Survey 1990. *Journal of Environmental Management* 48, 169–186.

Power, M.E. (1992) Top-down and bottom-up forces in food webs: do plants have primacy? *Ecology* 73 (3), 733–746.

Power, M.E. & Mills, L.S. (1995) The Keystone cops meet in Hilo. *Trends in Ecology and Evolution* 10 (5), 182–184.

Prance, G.T. (1991) Rates of loss of biological diversity: a global view. In: *Scientific Management of Temperate Communities for Conservation* (eds I.F. Spellerberg, F.B. Goldsmith & M.G. Morris), pp. 27–44. Blackwell Scientific Publications, Oxford.

Pressey, R.L. & Nicholls, A.O. (1989) Efficiency in conservation evaluation: scoring versus iterative approaches. *Biological Conservation* 50, 199–218.

Preston, C.D. & Croft, J.M. (1997) *Aquatic Plants in Britain and Ireland*. Harley Books, Colchester.

Preston, C.D. & Hill, M.O. (1997) The geographical relationships of British and Irish vascular plants. *Botanical Journal of the Linnean Society* 124, 1–120.

Preston, E.M. & Bedford, B.L. (1988) Evaluating cumulative effects in wetland functions: a conceptual view and generic framework. *Environmental Management* 12 (5), 565–583.

Pritchard, D. (1993) Towards sustainability in the planning process: the role of EIA. *Ecos* 14 (3/4), 10–15.

Pritchard, D.E., Housden, S.D., Mudge, G.P., Galbraith, C.A. & Pienkowski, M.W. (eds) (1992) *IBAs in the UK including the Channel Islands and the Isle of Man*. RSPB, Sandy.

Prosser, M.V. & Wallace, H.L. (1995) *West Sedgemoor vegetation and hydrological monitoring 1993–1994*. Report. RSPB, Sandy.

Pulliam, H.R. (1988) Sources, sinks and population regulation. *American Naturalist* 132, 652–661.

Pulliam, H.R. & Danielson, B.J. (1991) Sources, sinks and habitat selection: a landscape perspective on population dynamics. *American Naturalist* 137, S50–S66.

Purves, H.D., White, C.A. & Paquet, P.C. (1992) *Wolf and grizzly bear habitat use and displacement by human use in Banff, Yoho, and Kootenay National Parks: a preliminary analysis.* Canadian Parks Service Report, Banff National Park, Banff.

Pywell, R.F., Webb, N.R. & Putwain, P.D. (1995) A comparison of techniques for restoring heathland on abandoned farmland. *Journal of Applied Ecology* 32, 397–409.

Raine, R.M. & Riddell, R.N. (1991) *Grizzly bear research in Yoho and Kootenay National Parks, 1988–1990.* Final Report. Prepared by Beak Associates Consulting Ltd for the Canadian Parks Service, Western Region, Calgary.

Ralls, K., Harvey, P.H. & Lyles, A.M. (1986) Inbreeding in natural populations of birds and mammals. In: *Conservation Biology: The Science of Scarcity and Diversity* (ed. M.E. Soule), pp. 35–56. Sinauer Associates, Sunderland, MA.

Rapport, D.J. (1989) What constitutes ecosystem health? *Perspectives in Biology and Medicine* 33 (1), 120–132.

Ratcliffe, D.A. (1977) *A Nature Conservation Review.* Cambridge University Press, Cambridge.

Ravenscroft, N.O.M. & Young, M.R. (1996) Habitat specificity, restricted range and metapopulation persistence of the slender scotch burnet moth *Zygaena loti* in western Scotland. *Journal of Applied Ecology* 33, 993–1000.

RCEP (1994) *18th Report: Transport and the Environment.* HMSO, London.

Reading, C.J. (1997) A proposed standard method for surveying reptiles on dry lowland heath. *Journal of Applied Ecology* 34 (4), 1057–1070.

Reijnen, M.S.J.M. & Foppen, R.P.B. (1991a) *Effect van wegen met autoverkeer op de dichtheid van broedvogels; hoofdrapport.* IBN-rapport 91/1, DLO-Instituut voor Bos-en Natuuronderzoek, Leersum.

Reijnen, M.S.J.M. & Foppen, R.P.B. (1991b) *Effect van wegen met autokerveer op de dichtheid van broedvogels; opzet en methoden.* IBN-rapport 91/2, DLO-Instituut voor Bos-en Natuuronderzoek, Leersum.

Reijnen, R. & Foppen, R. (1994) The effects of car traffic on breeding bird populations in woodland I. Evidence of a reduced habitat quality for willow warblers *Phylloscopus trochilus* breeding close to a highway. *Journal of Applied Ecology* 31, 85–94.

Reijnen, R. & Foppen, R. (1995) The effects of car traffic on breeding bird populations in woodland. III: Reduction in density in relation to the proximity of main roads. *Journal of Applied Ecology* 32, 187–202.

Reijnen, R., Foppen, R., ter Braak, C. & Thissen, J. (1995a) The effects of car traffic on breeding bird populations in woodland III. Reduction of density in relation to the proximity of main roads. *Journal of Applied Ecology* 32, 187–202.

Reijnen, R., Veenbas, G. & Foppen, R.P.B. (1995b) *Predicting the Effects of Motorway Traffic on Breeding Bird Populations.* Road and Hydraulic Engineering Division, DLO-Institute for Forestry and Nature Research, Delft.

Rhodes, O.E. Jr & Chesser, R.K. (1994) Genetic concepts for habitat conservation: the transfer and maintenance of genetic variation. *Landscape and Urban Planning* 28, 55–62.

Richardson, C.J. (1994) Ecological values and human values in wetlands: a framework for assessing forestry impacts. *Wetlands* 14 (1), 1–9.

Richardson, J.H., Shore, R.F., Treweek, J.R. & Larkin, S.B.C. (1997) Are major roads a barrier to small animals? *Journal of Zoology, London* 243, 840–846.

Robinson, N. (1992) International trends in environmental impact assessment. *Boston College Environmental Affairs Law Review* 19 (3), 591–621.

Rodwell, J.S. (ed.) (1991a) *British Plant Communities,* Vol. 1. *Woodland and Scrub.* Cambridge University Press, Cambridge.

Rodwell, J.S. (ed.) (1991b) *British Plant Communities*, Vol. 2. *Mires and Heaths*. Cambridge University Press, Cambridge.

Rodwell, J.S. (ed.) (1992) *British Plant Communities*, Vol. 3. *Grassland and Montane Communities*. Cambridge University Press, Cambridge.

Rodwell, J.S. (ed.) (1994) *British Plant Communities*, Vol. 4. *Aquatic Communities, Swamp and Tall-herb Fens*. Cambridge University Press, Cambridge.

Rogers, J.W. (1996a) Wetland mitigation banking and watershed planning. In: *Mitigation Banking: Theory and Practice* (eds L.L. Marsh, D.R. Porter & D.A. Salvesen), pp. 159–183. Island Press, Washington, DC.

Rogers, M. (1996b) Poverty and degradation. In: *The Economics of Environmental Degradation. Tragedy for the Commons?* (ed. T.M. Swanson), pp. 109–127. Edward Elgar, Cheltenham.

Root, R. (1967) The niche exploitation pattern of the blue-grey gnatcatcher. *Ecological Monographs* **37**, 317–350.

Rowe, W.D. (1994), Understanding uncertainty. *Risk Analysis* **14** (5), 743–750.

Roy, D.B. & Eversham, B.C. (1994) *Biological Records Centre single species and species richness data for commissioned research*. Report, Institute of Terrestrial Ecology, Huntingdon.

Sadler, B. (1993) NSDS and environmental impact assessment: post Rio perspectives. *Environmental Assessment* **1**, 29–30.

Sadler, B. (1996) *Environmental impact assessment in a changing world: evaluating practice to improve performance*. Final Report of the International Study of the Effectiveness of Environmental Impact Assessment. Canadian Environmental Assessment Agency and the International Association for Impact Assessment, Ottawa.

Salveson, D. (1990) *Wetlands: Mitigating and Regulating Development Impacts*. Urban Land Institute, Washington, DC.

Salwasser, H. (1988) Managing ecosystems for viable populations of vertebrates: a focus on biodiversity. In: *Ecosystem Management for Parks and Wilderness*, pp. 87–104. University of Washington Press, Seattle.

Sanderson, R.A., Rushton, S.P., Pickering, A.T. & Byrne, J.P. (1995) A preliminary method of predicting plant species distributions using the British National Vegetation Classification. *Journal of Environmental Management* **43**, 265–288.

Sausman, K.A. (1984) Survival of captive-born *Ovis canadensis* in North American zoos. *Zoo Biology* **3**, 111–121.

Savitsky, B.G., Lacher, T.E., Burnett, G.W., Fallas, J. & Vaughn, C. (1995) Applying proven GIS techniques in a new setting: GAP analysis in Costa Rica. In: *Environmental Impact Assessment and Development* (eds R. Goodland & V. Edmundsen). World Bank, Washington, DC.

Savoie, A. & Woodley, S. (1993) The mandate of the national parks and the stresses on their ecosystems. In: *The ecological integrity of La Maurice National Park: a week of reflection*. Workshop Summary, Parks Canada, Halifax.

Schaeffer, D.J. (1991) A toxicological perspective on ecosystem characteristics to track sustainable development. *Ecotoxicology and Environmental Safety* **22** (2), 225–239.

Schafer, J.A. & Penland, S.T. (1985) Effectiveness of Swareflex reflectors in reducing deer–vehicle accidents. *Journal of Wildlife Management* **49** (3), 774–776.

Schall, J.J. & Pianka, E.R. (1978) Geographic trends in numbers of species. *Science* **201**, 679–686.

Schroeder, R.L. & Haire, S.L. (1993) *Guidelines for the development of community-level habitat evaluation models*. Biological Report 8. US Fish and

Wildlife Service, Washington, DC.

Schroeter, S.C., Dixon, J.D., Kastendiek, J., Smith, R.O. & Bence, J.R. (1993) Detecting the ecological effects of environmental impacts: a case study of kelp forest invertebrates. *Ecological Applications* **3**, 330–349.

Scott, J.M., Davis, F., Csuti, B., Noss, R., Butterfield, B., Groves, C., Anderson, H., Caicco, S., D'Erchia, F., Edwards, T.C. Jr , Ullman, J. & Wright, R.G. (1993) *Gap Analysis, a Geographic Approach to the Protection of Biological Diversity*. Wildlife Monographs 123. The Wildlife Society, Bethesda, MD.

Scottish Office (1994) *Setting Forth: Environmental Appraisal of Alternative Strategies*. Scottish Office, Edinburgh.

Sebastini, M., Sambrano, A., Villamizar, A. & Villalba, C. (1989) Cumulative impact and sequential geographical analysis as tools for land use planning. A case study: Laguna La Reina, Miranda State, Venezuela. *Journal of Environmental Management* **29**, 237–248.

Semenza, J.C, Tolbert, P.E., Rubin, C.H., Guillette, L.J. & Jackson, R.J. (1997) Reproductive toxins and alligator abnormalities at Lake Apopka, Florida. *Environmental Health Perspectives* **105** (10), 1030–1032.

Severinghaus, W.D. (1981) Guild theory development as a mechanism for assessing environmental impact. *Environmental Management* **5**, 187–190.

Shaffer, M.L. (1981) Minimum population sizes for species conservation. *BioScience* **31**, 131–134.

Shaffer, M.L. (1990) Population viability analysis. *Conservation Biology* **4** (1), 39.

Sharrock, J.T.R. (1976) *The Atlas of Breeding Birds in Britain and Ireland*. British Trust for Ornithology/Irish Wildbird Conservancy. T. and A.D. Poyser, Berkhamsted.

Sheail, J. (1983) The availability and use of archival sources of information. In: *Ecological Mapping from Ground, Air and Space* (ed. R.M. Fuller), pp. 22–28. Institute of Terrestrial Ecology Symposium No. 10. NERC, HMSO, London.

Sheppard, D. (1995) Guidance notes for invertebrate translocations and introductions. In: *Species Conservation Handbook*. English Nature, Peterborough.

Shirt, D.B. (ed.) (1987) *British Red Data Books, 2, Insects*. Nature Conservation Council, Peterborough.

Shopley, J., Sowman, M. & Fuggle, R. (1990) Extending the capability of the component interaction matrix as a technique for addressing secondary impacts in environmental assessment. *Journal of Environmental Management* **31**, 197–213.

Short, H.L. & Burnham, K.P. (1982) *Technique for structuring wildlife guilds to evaluate impacts on wildlife communities*. Special scientific report on wildlife 244. US Fish and Wildlife Service, Fort Collins, CO.

Shrader-Frechette, K.S. & McCoy, E.D. (1992) Statistics, costs and rationality in ecological inference. *Trends in Ecology and Evolution* **7** (3), 96–99.

Shrader-Frechette, K.S. & McCoy, E.D. (1994) *Ethics, Integrity and Health*. Paper presented at the First International Symposium on Ecosystem Health and Medicine, 19–23 June, Ottawa.

Simenstad, C.A. & Thom, R.M. (1996) Functional equivalency trajectories of the restored Gog-Le-Hi-Te Estuarine Wetland. *Ecological Applications* **6** (1), 38–56.

Skye, E. (1979) Lichens as biological indicators of air pollution. *Annual Reviews of Phytopathology* **17**, 325–341.

Small, M. (1997) China's mountain monkeys. *New Scientist*, 31 May 1997.

Smith, L.G. (1993) *Impact Assessment and Sustainable Resource Development*. Longman, Harlow.

Smith, M.H. & Rhodes, O.E. Jr (1992) Genetics and biodiversity in wildlife management. *Transactions of the North American Wildlife and Natural Resources Conference* **57**, 243–251.

Smith, P.G.R. & Theberge, J.B. (1986) A review of criteria for evaluating natural areas. *Environmental Management* **10**, 715–734.

Smith, P.G.R. & Theberge, J.B. (1987) Evaluating natural areas using multiple criteria: theory and practice. *Environmental Management* **11**, 445–460.

Snoeijs, P.J.M. (1991) Monitoring pollution effects by diatom community composition. A comparison of sampling methods. *Archiv für Hydrobiologie* **121**, 497–510.

Snoeijs, P.J.M. & Prentice, I.C. (1989) Effects of cooling water discharge on the structure and dynamics of epilithic algal communities in the northern Baltic. *Hydrobiologia* **184**, 99–123.

Solbrig, O.T. (1991) Biodiversity. Scientific issues and collaborative research proposals. *Man and Biosphere Digest* 9. UNESCO, Paris.

Sondheim, M.W. (1978) A comprehensive methodology for assessing environmental impact. *Journal of Environmental Management* **6** (1), 27–42.

Sonntag, N.C., Everitt, R.R., Rattie, L.P., Colnett, D.L., Wolf, C.P., Truett, J.C., Dorcey, A.H.J. & Holling, C.S. (1987) *Cumulative effects assessment: a context for further research and development.* Report to the Canadian Environmental Assessment Review Commission, Ministry of Supply and Services, Ottawa.

Sparks, T.H., Mountford, J.O., Manchester, S.J., Rothery, P. & Treweek, J.R. (1997). Sample size for estimating species lists in vegetation surveys. *The Statistician* **46**, 253–260.

Spellerberg, I.F. (1991) *Monitoring Ecological Change.* Cambridge University Press, Cambridge.

Spellerberg, I.F. (1992) *Evaluation and Assessment for Conservation: ecological guidelines for determining priorities for nature conservation.* Chapman and Hall, London.

Spellerberg, I. & Minshull, A. (1992) An investigation into the nature and use of ecology in environmental impact assessments. *British Ecological Society Bulletin* **13**, 38–45.

Spencer, C.N., McClelland, B.R. & Stanford, J.A. (1991) Shrimp stocking, salmon collapse, and eagle displacement: cascading interactions in the food web of a large aquatic system. *BioScience* **41** (1), 14–21.

Stacey, P.B. & Taper, M. (1992) Environmental variation and the persistence of small populations. *Ecological Applications* **2** (1), 18–29.

Stewart, A., Pearman, C.D. & Preston, C.D. (1994) *Scarce Plants in Britain.* Joint Nature Conservation Committee, Peterborough.

Stewart-Oaten, A. (1993) Evidence and statistical summaries in environmental assessment. *Trends in Ecology and Evolution* **8** (5), 156–158.

Stewart-Oaten, A., Murdoch, W. & Parker, K. (1986) Environmental impact assessment: 'pseudoreplication' in time? *Ecology* **67**, 929–940.

Stewart-Oaten, A., Bence, J.R. & Osenberg, C.W. (1992) Assessing the effects of unreplicated perturbations: no simple solutions. *Ecology* **73**, 1396–1404.

Stiehl, R.B. (ed.) (1994) *Habitat Evaluation Procedures Workbook.* National Biological Survey, Fort Collins, CO.

Stohlgren, T.J., Coughenour, M.B., Chong, G.W., Binkley, D., Kalkhan, M.A., Schell, L.D., Buckley, D.J. & Berry, J.K. (1997) Landscape analysis of plant diversity. *Landscape Ecology* **12**, 155–170.

Stork, N.E. (1988) Insect diversity: facts, fiction and speculation. *Biological Journal of the Linnean Society.* **35**, 321–337.

Stork, N.E. (1994) Inventories of biodiversity: more than a question of numbers.

In: *Systematics and Conservation Evaluation*, the Systematics Association Special Volume No. 50 (eds P.L. Forey, C.J. Humphries & R.I. Vane-Wright), pp. 81–100. Oxford University Press, Oxford.

Stork, N. & Davies, J. (1996) Biodiversity inventories. In: *Biodiversity Assessment. A Guide to Good Practice. Field Manual 1. Data and Specimen Collection of Plants, Fungi and Microorganisms.* HMSO, London.

Stork, N.E. & Samways, M. (1995) *Inventorying and Monitoring of Biodiversity.* Section 5, UNEP Global Biodiversity Assessment. Cambridge University Press, Cambridge.

Stribosch, H. (1978) *Long term study into the population of* Lacerta agilis *on heathland.* Internal research report of the Animal Ecology Department, University of Nijmegen, the Netherlands.

Suter, G.W. II. (1993a) *Ecological Risk Assessment.* Lewis Publishers, Boca Raton, FL.

Suter, G.W. II. (1993b) A critique of ecosystem health concepts and indexes. *Environmental Toxicology and Chemistry* 12, 1533–1539.

Tans, W. (1974) Priority ranking of biotic natural areas. *Michigan Botanist* 13, 31–39.

Taylor, P.D., Fahrig, L., Henein, K. & Merriam, G. (in press) Connectivity is a vital element of landscape structure. *Oikos.*

ter Braak, C.J.F. (1987a) The analysis of vegetation–environment relationships by canonical correspondence analysis. *Vegetatio* 69, 69–77.

ter Braak, C.J.F. (1987b) Ordination. In: *Data Analysis in Community and Landscape Ecology* (ed. by R.H.G. Jongman, C.J.F. ter Braak & O.F.R. van Tongeren), pp. 91–173. Pudoc, Wageningen.

TEST (1991) *Wrong Side of the Tracks? Impacts of Road and Rail Transport on the Environment: A Basis for Discussion.* TEST.

Therivel, R. & Thompson, S. (1996) *Strategic Environmental Assessment and Nature Conservation.* English Nature, Peterborough.

Therivel, R., Wilson, E., Thompson, S., Heaney, D. & Pritchard, D. (1992) *Strategic Environmental Assessment.* Earthscan, London.

Thomas, J.A. (1984) Conservation of butterflies in temperate countries. In: *The Biology of Butterflies* (eds R.I. Vane-Wright & P.R. Ackery), pp. 333–353. Symposium of the Royal Entomological Society II. Academic Press, London.

Thompson, S. (1995) *The status of ecology in the British EIA process.* PhD thesis, Oxford Brookes University.

Thompson, S., Treweek, J.R. & Thurling, D.J. (1995) The potential application of strategic environmental assessment (SEA) to the farming of atlantic salmon (*Salmo salar* L.) in mainland Scotland. *Journal of Environmental Management* 45, 219–229.

Thompson, S., Treweek, J.R. & Thurling, D.J. (1997) The ecological component of environmental impact assessment: a critical review of British environmental statements. *Journal of Environmental Planning and Management* 40 (2), 157–171.

Treweek, J.R. (1995) Ecological impact assessment. In: *Environmental and Social Impact Assessment* (eds F. Vanclay & D.A. Bronstein), pp. 171–193. John Wiley and Sons, Chichester.

Treweek, J. (1996) Ecology and environmental impact assessment. *Journal of Applied Ecology* 33, 191–199.

Treweek, J. & Thompson, S. (1997) A review of ecological mitigation measures in UK environmental statements with respect to sustainable development. *International Journal of Sustainable Development and World Ecology* 4, 40–50.

Treweek, J. & Veitch, N. (1996) The potential application of GIS and remotely sensed data to the ecological assessment of proposed new road schemes. *Global Ecology and Biogeography Letters* **5**, 249–257.

Treweek, J., Thompson, S., Veitch, N. & Japp, C. (1993) Ecological assessment of proposed road developments: a review of environmental statements. *Journal of Environmental Planning and Management* **36**, 295–307.

Treweek, J., Hankard, P., Roy, D., Arnold, H. & Thompson, S. (1998) Scope for strategic ecological assessment of trunk road development in England with respect to potential impacts on lowland heathland, the Dartford warbler (*Sylvia undata*) and the sand lizard (*Lacerta agilis*). *Journal of Environmental Management* **53** (2), 147–163.

Tubbs, C.R. & Blackwood, J.W. (1971) Ecological evaluation of land for planning purposes. *Biological Conservation* **3**, 169–172.

Tucker, G.M. & Heath, M.F. (1994) *Birds in Europe, their Conservation Status.* Birdlife International, Cambridge.

Tucker, K., Rushton, S.P., Sanderson, R.A., Martin, E.B. & Blaiklock, J. (1997) Modelling bird distributions—a combined GIS and Bayesian rule-based approach. *Landscape Ecology* **12** (2), 77–93.

Turner, R.K. (ed.) (1993) *Sustainable Environmental Economics and Management. Principles and Practice.* Belhaven Press, London.

Underwood, A.J. (1993) The mechanics of spatially replicated sampling programmes to detect environmental impacts in a variable world. *Australian Journal of Ecology* **18**, 99–116.

UNECE (1992) *Application of Environmental Impact Assessment Principles to Policies, Plans and Programmes.* Environmental Series No. 5, ECE/ENVWA/27. United Nations Economic Commission for Europe, New York.

UNECE (1996) *Current Policies, Strategies and Aspects of Environmental Impact Assessment in a Transboundary Context.* Environmental Series No. 6. United Nations Economic Commission for Europe, New York.

UNEP (1992) Convention on Biological Diversity, June 1992. United Nations Environment Programme, Nairobi.

UNEP (1993) *Guidelines for Country Studies on Biological Diversity.* United Nations Environment Programme, Nairobi.

UNEP-IETC (1996) *Environmental Risk Assessment.* Winter Newsletter. United Nations Environment Programme–International Environmental Technology Centre, Nairobi.

Unsworth, R.E. & Bishop, R.C. (1994) Assessing natural resource damages using environmental annuities. *Ecological Economics* **11** (1), 35–41.

US Army Corps of Engineers (1983) *Generic Environmental Mitigation Guidelines Manual.* San Francisco District, San Francisco, CA.

US CEQ (1978) *National Environmental Policy Act—Final Regulations.* Federal Register 43(230): 55978–56007. US Council on Environmental Quality, Washington, DC.

US CEQ (1992) *Incorporating Biodiversity Considerations under the National Environmental Policy Act.* US Council on Environmental Quality, Washington, DC.

US CEQ (1993) *Environmental Quality.* 23rd Annual Report. US Council on Environmental Quality, Washington, DC.

US Department of the Interior (1980) *Habitat Evaluation Procedures.* Division of Ecological Services, Fish and Wildlife Service, Washington, DC.

US Environmental Protection Agency (1992) *Framework for Ecological Risk Assessment.* EPA/630/R-92/001. Risk Assessment Forum. US Environmental

Protection Agency, Washington, DC.

US Environmental Protection Agency (1993) *Wildlife Exposure Factors Handbook*, Vol. I. US Environmental Protection Agency, Washington, DC.

US Environmental Protection Agency (1994a) *The Indicator Development Strategy for the Environmental Monitoring and Assessment Program.* EPA/620/R-94/022 (ed. C. Barber). US Environmental Protection Agency, Washington, DC.

US Environmental Protection Agency (1994b) *Landscape Monitoring and Assessment Research Plan.* EPA 620/R-94–009. Office of Research and Development, Washington, DC.

US Environmental Protection Agency (1996) *Proposed Guidelines for Ecological Risk Assessment.* EPA/630/R-95/002B. Risk Assessment Forum. US Environmental Protection Agency, Washington, DC.

US Executive Office of the President (1994) *Coordinating Geographic Data Acquisition and Access: The National Spatial Data Infrastructure* (Executive Order 12906). Executive Office of the President, Washington, DC.

US Fish and Wildlife Service (1980) *Habitat Evaluation Procedures (HEP).* ESM 102. Division of Ecological Services, Department of the Interior, Washington, DC.

US Fish and Wildlife Service (1981) *Standards for the Development of Habitat Suitability Index Models for Use in the Habitat Evaluation Procedures (HEP).* Division of Ecological Services, Department of the Interior, Washington, DC.

US Fish and Wildlife Service (1985) *Red-cockaded Woodpecker Recovery Plan.* US Fish and Wildlife Service, Atlanta, GA.

US Soil Conservation Service. (1977) *Guide for Environmental Impact Assessment.* US Soil Conservation Service, Washington, DC.

Usher, M.B. (1986) *Wildlife Conservation Evaluation.* Chapman and Hall, London.

Usher, M.B. (1992) Management and diversity of arthropods in *Calluna* heathland. *Biodiversity and Conservation* 1, 63–79.

Usher, R.G. & Kansas, J.L. (1994) *Ecologically-Integrated Landscape Mapping for Purposes of Wildlife Habitat Assessment: Husky Oil Moose Mountain and Quirk Creek Study Areas.* Gaia Consultants, Calgary.

Van den Brink (1997) *The Science of the Total Environment* **198**, 43.

Van der Fluit, N., Cuperus, R. & Canters, K.J. (1990) *Mitigating and Compensating Measures to the Main Road Network, for Promoting Ecological Features.* Centrum voor Milieukunde, Rijksuniversiteit, Leiden.

van der Ploeg, S. & Vlijm, L. (1978) Ecological evaluation, nature conservation and land use planning with particular reference to methods used in the Netherlands. *Biological Conservation* **14**, 197–221.

van der Zande, A.N., ter Keurs, W.J. & van der Weijden, W.J. (1980) The impact of roads on the densities of four species in an open field habitat—evidence of a long distance effect. *Biological Conservation* **18**, 299–321.

Van Leeuwen, B.H. (1982) Protection of migrating common toad (*Bufo bufo*) against car traffic in the Netherlands. *Environmental Conservation* **9** (1), 34.

Vanclay, F. & Bronstein, D.A. (eds) (1995) *Environmental and Social Impact Assessment.* John Wiley and Sons, Chichester.

Veitch, N., Webb, N.R. & Wyatt, B.K. (1995) The application of geographic information systems and remotely sensed data to the conservation of heathland fragments. *Biological Conservation* **72**, 91–97.

Verheem, R. (1992) Environmental Impact Assessment at the strategic level in the Netherlands. *Project Appraisal* **7** (3), 150–156.

Vlachos, E. (1982) Cumulative impact analysis. *Impact Assessment Bulletin* **1** (4), 61–64.

Wadsworth, R. & Treweek, J. (1999) *Geographical Information Systems for*

Ecology: an introduction. Addison Wesley Longman, Harlow.

Walker, D.A., Webber, P.J., Walker, M.S., Lederer, N.D., Meehan, R.H. & Nordstrand, E.A. (1986) Use of geobotanical maps and automated mapping techniques to examine cumulative impacts in the Prudhoe Bay oilfield, Alaska. *Environmental Conservation* 13, 149–160.

Walker, D.A., Webber, P.J., Binnian, E.F., Everett, K.R., Lederer, N.D., Nordstrand, E.A. & Walker, M.D. (1987) Cumulative effects of oil fields on northern Alaskan landscapes. *Science* 238, 757–761.

Walters, C.J. (1993) Dynamic models and large scale field experiments in environmental impact assessment and management. *Australian Journal of Ecology* 18, 53–61.

Ward, L.K. & Stevenson, M.J. (1994) *M3 Bar End to Compton: botanical monitoring: Arethusa Clump chalk grassland restoration 1994.* ITE Report to Mott MacDonald Civil Engineering Ltd.

Warnken, J. & Buckley, R. (1998) Scientific quality of tourism environmental impact assessment. *Journal of Applied Ecology* 35 (1), 1–9.

Wathern, P. & Russell, S. (1993) The five year review. *Environmental Impact Assessment* 1 (2), 51–53.

Watt, T.A. (1997) *Introductory Statistics for Biology Students*, 2nd edn. Chapman and Hall, London.

WCED (1987) *Our Common Future.* World Commission on Environment and Development, Oxford University Press, Oxford.

Webb, J.W. & Sigal, L.L. (1992) Strategic environmental assessment in the United States. *Project Appraisal* 7, 137–142.

Wells, M. (1992) Biodiversity conservation, affluence and poverty: mismatched costs and benefits and efforts to remedy them. *Ambio* 21 (3), 237–243.

Western, D. (1987) Africa's elephants and rhinos: flagships in crisis. *Trends in Ecology and Evolution* 2, 343–345.

Westman, W.E. (1985) *Ecology, Impact Assessment and Environmental Planning.* John Wiley and Sons, New York.

Whittaker, R.H. (1974) 'Stability' in plant communities. In: *Structure, Functioning and Management of Ecosystems*, p. 68. Proceedings of the First International Congress on Ecology. PUDOC, Wageningen.

Wiens, J.A., Crawford, C.S. & Gosz, J.R. (1985) Boundary dynamics: a conceptual framework for studying landscape ecosystems. *Oikos* 45, 421–427.

Wildlife Interpretations Subcommittee (1998) *British Columbia Wildlife Habitat Rating Standards.* Review Draft. Wildlife Interpretations Subcommittee, British Columbia Resources Inventory Committee, Victoria.

Wilkinson, G.G. (1996) A review of current issues in the integration of GIS and remote sensing data. *International Journal of Geographical Information Systems* 10 (1), 85–101.

Williams, B.K. (1996) Assessment of accuracy in the mapping of vertebrate biodiversity. *Journal of Environmental Management* 47, 269–282.

Williamson, K. (1968) Buntings on a barley farm. The bird community of farmland. *Bird Study* 15, 34–37.

Willis, K. & Garrod, G. (1991) An individual travel cost method of evaluating forest recreation. *Journal of Agricultural Economics* 42 (1), 33–42.

Willmot, A. (1980) The woody species of hedges with special reference to age in Church Broughton Parish, Derbyshire. *Journal of Ecology* 68, 269–285.

Wilson, E.O. (1990) Threats to biodiversity. In: *Managing Planet Earth. Readings from Scientific American*, pp. 49–60. W.H. Freeman, New York.

Wood, C. (1992) Strategic environmental assessment in Australia and New

Zealand. *Project Appraisal* 7 (3), 143–149.

Wood, C. (1995) *Environmental Impact Assessment. A Comparative Review.* Longman, Harlow.

Wood, C.M. & Djeddour, M. (1992) Strategic environmental assessment: EA of policies, plans and programmes. *Impact Assessment Bulletin* 10, 3–22.

Wright, D. (1977) A site evaluation scheme for use in the assessment of potential nature reserves. *Biological Conservation* 11, 293–305.

Wright, J.F., Moss, D., Armitage, P.D. & Furse, M.T. (1984) A preliminary classification of running water sites in Great Britain based on macroinvertebrate species and prediction of community type using environmental data. *Freshwater Biology* 14, 221–256.

Wright, S. (1931) Evolution in Mendelian populations. *Genetics* 16, 97–159.

Wright, S. (1938) Size of population and breeding structure in relation to evolution. *Science* 87, 430–431.

Wynne, G. (ed.) (1993) *Biodiversity Challenge: An Agenda for Conservation Action in the UK.* RSPB for the Biodiversity Challenge Group, Sandy.

Yapp, W. (1972) Ecological evaluation of a linear landscape. *Biological Conservation* 5, 45–47.

Zedler, J.B. (1996) Ecological issues in wetland mitigation — an introduction to the forum. *Ecological Applications* 6 (1), 33–37.

Zheng, D., Wallin, D.O. & Hao, Z. (1997) Rates and patterns of landscape change between 1972 and 1988 in the Changbai mountain area of China and North Korea. *Landscape Ecology* 12 (4), 241–254.

Zimmerman, B.L. & Rodrigues, M.T. (1990) Frogs, snakes and lizards of the INPA-WWF reserves near Manaus, Brazil. In: *Four Neotropical Forests* (ed. A.H. Gentry), pp. 426–454. Yale University Press, New Haven, CT.

Index

Page numbers in *italics* refer to figures,
those in **bold** refer to tables and boxes